Bernard Kipsang Rop
Micah Cheruiyot Keter
Naomi Chepkemoi Sang

Environmental Geology

Bernard Kipsang Rop
Micah Cheruiyot Keter
Naomi Chepkemoi Sang

Environmental Geology

Insights from Applied Geology on Environmental Impact

LAP LAMBERT Academic Publishing

Imprint
Any brand names and product names mentioned in this book are subject to trademark, brand or patent protection and are trademarks or registered trademarks of their respective holders. The use of brand names, product names, common names, trade names, product descriptions etc. even without a particular marking in this work is in no way to be construed to mean that such names may be regarded as unrestricted in respect of trademark and brand protection legislation and could thus be used by anyone.

Cover image: www.ingimage.com

Publisher:
LAP LAMBERT Academic Publishing
is a trademark of
Dodo Books Indian Ocean Ltd. and OmniScriptum S.R.L publishing group

120 High Road, East Finchley, London, N2 9ED, United Kingdom
Str. Armeneasca 28/1, office 1, Chisinau MD-2012, Republic of Moldova, Europe
Managing Directors: Ieva Konstantinova, Victoria Ursu
info@omniscriptum.com

Printed at: see last page
ISBN: 978-620-7-63955-7

PREFACE

The Earth's geological systems underpin all environmental processes and human activities. However, as global populations expand and industrial activities proliferate, the fragile equilibrium between geological stability and environmental sustainability is increasingly jeopardised. *Environmental Geology: Insights from Applied Geology on Environmental Impact* addresses this imbalance, offering a synthesis of scientific rigor, practical methodologies, and ethical considerations to guide responsible interactions with our planet. This book serves as a comprehensive resource for students, educators, and practitioners, fostering a deeper understanding of the intricate relationships between geology and environmental health.

Context and Motivation

The 21st century has ushered in unprecedented environmental challenges—climate change, resource depletion, biodiversity loss, and intensified geohazards—all intricately linked to geological processes. Traditional approaches to geology often fall short in addressing these complexities. This book bridges that gap, connecting academic geology with real-world environmental management and responding to the demand for interdisciplinary solutions that integrate geological science with policy, engineering, and community engagement.

A comprehensive understanding of geological processes is essential for developing effective strategies for environmental stewardship. For example, grasping the geological underpinnings of aquifer systems is crucial for sustainable water resource management, especially in arid regions. By equipping readers with theoretical insights and practical applications, we hope to inspire a new generation of geoscientists and environmental professionals committed to sustainable practices.

Structure and Scope

This book is structured to guide readers from foundational principles to advanced applications, organized into four thematic sections:

Section I: Foundations (Chapters 1–4)

These chapters establish the theoretical bedrock of environmental geology, introducing the interplay between geological processes and environmental systems, including soil mechanics and hydrogeology. We also explore the historical evolution of geological studies in response to environmental challenges, illustrating how past events inform current practices.

Section II: Tools and Techniques (Chapters 5–7)

This section delves into methodologies for environmental impact assessments (EIAs), risk modeling, and climate resilience strategies. It offers frameworks for conducting EIAs aligned with international standards and highlights technological advancements, such as remote sensing and GIS, that enhance environmental assessments. We emphasize the importance of stakeholder engagement in fostering sustainable decision-making.

Section III: Applications and Case Studies (Chapters 8–12)

The heart of the book focuses on applied geology, exploring geohazard mitigation, sustainable resource management, and the geopolitics of mineral extraction, including Kenya's role in the rare earth element supply chain. Case studies illustrate the real-world relevance of geological principles and the socio-economic implications of geological practices, fostering an inclusive dialogue around environmental geology.

Section IV: Innovation and Future Directions (Chapter 13)

This chapter underscores the critical role of subsurface characterization in mitigating environmental risks, synthesizing emerging technologies like AI-driven surveys and advocating for standardized protocols. It encourages readers to consider innovative approaches and interdisciplinary collaboration to address future challenges in environmental geology.

Unique Contributions

This book distinguishes itself through three key contributions:

1. **Interdisciplinary Integration:** By combining geology with hydrology, engineering, and policy, it offers a holistic framework for tackling environmental challenges.

2. **Global-Local Balance:** Emphasizing region-specific case studies, such as Kenya's Mrima Hill REE deposits, it illustrates the nuances of resource management in developing economies.

3. **Ethical Imperatives:** Each chapter addresses the ethical dimensions of environmental decision-making, promoting a responsible approach to resource management that prioritizes social equity and environmental justice.

A Call to Action

Geology is not merely the study of rocks; it is the study of time, change, and interconnectedness. This book serves as a call to action for geologists, engineers, and policymakers to wield their knowledge as a force for sustainability. We invite you to explore these pages as active participants in shaping a sustainable future. Together, let us embark on this journey toward a more equitable world, recognizing that our collective actions today will determine the health of our planet for generations to come.

Geol. Prof. Bernard Kipsang Rop, PhD

MGSK, Reg. Geol. & Lead Expert EIA

ACKNOWLEDGEMENTS

The completion of this book would not have been possible without the invaluable support and contributions of numerous individuals and organizations. We extend our heartfelt gratitude to our colleagues and peers in the fields of geology and environmental science, whose insights and expertise have significantly enriched the content of this work. Their dedication to advancing the understanding of geological processes and their implications for environmental management has been a constant source of inspiration, motivating us to delve deeper into the complexities of our subject matter. The collaborative discussions and exchanges of ideas have not only enhanced our knowledge but have also fostered a sense of community among those committed to addressing the pressing environmental challenges of our time.

We are particularly grateful to the institutions that provided access to essential resources, data, and case studies. Their commitment to advancing knowledge in environmental geology has been instrumental in shaping the discussions presented in this book. Notably, the Kenya Ministries of Forests and Environment, as well as Mining, Maritime, and Blue Economy Affairs, have played crucial roles in facilitating research and providing valuable insights that have informed our work. Their support has not only enhanced the academic rigor of our research but has also ensured that our findings are relevant to current policy discussions and practical applications in the field. The collaboration with these institutions has underscored the importance of integrating scientific research with policy-making, ultimately contributing to more effective environmental governance.

We would also like to acknowledge the contributions of our colleagues at Stanford University's School of Earth, Energy & Environmental Sciences, as well as those at JKUAT's Institute of Geomatics, GIS, and Remote Sensing, and the Mining, Materials, and Petroleum Engineering Department. Their cutting-edge insights into emerging technologies, such as AI-driven geospatial analysis, have greatly enhanced our understanding of the field. The collaborative spirit and innovative thinking fostered within these institutions have been vital in addressing the complex challenges we face in environmental geology, allowing us to incorporate the latest advancements into our work and ensuring that our discussions remain at the forefront of the discipline. The interdisciplinary approach championed by these institutions has encouraged us to think critically about how various fields intersect and how we can leverage these connections to develop more comprehensive solutions.

We are especially thankful for the graduate students whose fresh perspectives and enthusiasm have significantly enriched this project. Their ability to frame complex geological concepts for broader audiences has underscored the importance of engaging the next generation of geoscientists and environmental stewards. Their contributions have not only added depth to our discussions but have also inspired us to consider how we can better communicate our findings to diverse audiences, ensuring that our work resonates beyond academic circles. The energy and creativity that these students bring to the table remind us of the vital role that youth play in shaping the future of environmental science and advocacy.

Equally vital were the contributions of field researchers and local communities in Kenya's artisanal small-scale mining regions. Their lived experiences and hospitality provided a human dimension to our discussions on sustainable resource extraction, reminding us of the real-world implications of our work. Engaging with these communities has deepened our understanding of the socio-economic and environmental challenges they face, and their resilience has inspired us to advocate for more equitable and sustainable practices in resource management. Their stories and insights have been invaluable in grounding our research in the realities of those most affected by the issues we explore. By listening to their voices, we have gained a richer perspective on the complexities of resource management and the importance of incorporating local knowledge into our analyses.

On a personal note, we owe immeasurable gratitude to our families and friends for their unwavering patience and encouragement throughout the writing process. Their support has been a cornerstone of our ability to complete this project. To Micah Cheruiyot Keter and Naomi Chepkemoi Sang, whose curiosity about the natural world in water resources management and sustainability—particularly their day-to-day Hydrogeological Survey Reports—has inspired us daily: thank you for your boundless enthusiasm, which has significantly contributed to the case studies in this book. Your questions and insights have continually motivated us to explore deeper and think critically about the issues at hand, reminding us of the importance of curiosity and inquiry in the pursuit of knowledge. The encouragement from our loved ones has provided us with the strength to persevere through challenges and has reinforced our commitment to making a meaningful contribution to the field.

Finally, we extend our appreciation to you, the reader. Whether you are a student, practitioner, or engaged citizen, your interest in environmental geology is a vital step toward fostering a sustainable future. We hope this book inspires you to ask critical questions, challenge conventional practices, and advocate for science-driven policies. Your engagement with the material is essential for driving the change needed to address the pressing environmental challenges we face today. We believe that informed and active participation from individuals like you is crucial for creating a more sustainable and equitable world.

We would also like to express our gratitude to the authors referenced throughout this book, whose foundational work has informed our research and discussions. Their contributions to the field of geology and environmental science have been invaluable, and we are thankful for their dedication to advancing knowledge in these critical areas.

Geol. Prof. Bernard K. Rop   Mr. Micah C. Keter & Naomi C. Sang

TABLE OF CONTENTS

CHAPTER 1. INTRODUCTION TO ENVIRONMENTAL GEOLOGY.... 11

CHAPTER 2. FUNDAMENTALS OF APPLIED GEOLOGY 22

CHAPTER 3. GEOLOGICAL HAZARDS AND RISKS 81

CHAPTER 4. SOIL AND WATER RESOURCES 110

CHAPTER 5. ENVIRONMENTAL IMPACT ASSESSMENT (EIA) 168

CHAPTER 6. GEOLOGY AND CLIMATE CHANGE 226

CHAPTER 7. GEOLOGY AND NATURAL RESOURCES 258

CHAPTER 8. GEOHAZARDS AND RISK MANAGEMENT 274

CHAPTER 9. GEOLOGY AND ENVIRONMENTAL POLICY 288

CHAPTER 10. THE FUTURE OF GEOLOGY AND SUSTAINABILITY 301

CHAPTER 11. ROCK BLASTING CONSTRAINTS 336

CHAPTER 12. STRATEGIC METALS, MINERALS, AND RARE EARTH ELEMENTS (REES) IN THE MINING SECTOR IN KENYA 367

CHAPTER 13. FUNDAMENTALS OF GROUND INVESTIGATION ... 391

REFERENCES 479

ABOUT THE BOOK 491

ENGAGING WITH ENVIRONMENTAL GEOLOGY: CRITICAL QUESTIONS AND ANSWERS 497

GEOLOGICAL PERSPECTIVES ON ENVIRONMENTAL CHALLENGES: AN IN-DEPTH QUESTIONS AND ANSWERS 507

CHAPTER 1

INTRODUCTION TO ENVIRONMENTAL GEOLOGY

1.1 Definition and Scope of Environmental Geology

Definition: Environmental geology is a specialized branch of geology that examines the interactions between geological processes and the environment. It focuses on understanding how geological materials, processes, and hazards influence human activities and the natural world. This field integrates geological science with environmental concerns, providing a framework for addressing the complex challenges posed by human interaction with the Earth's systems. By studying these interactions, environmental geologists can offer valuable insights that inform sustainable practices and policies aimed at protecting both the environment and public health.

Scope: The scope of environmental geology is broad and multifaceted, encompassing a wide range of issues that are critical to sustainable development and environmental management.

1. **Land Use Planning**: Environmental geologists assess geological conditions to inform decisions about land development, ensuring that projects are sited in locations that minimize risks associated with geological hazards, such as landslides, flooding, and earthquakes. By integrating geological knowledge into land use planning, communities can enhance their resilience to environmental challenges and promote sustainable growth.

2. **Resource Management**: This includes the identification, evaluation, and sustainable management of natural resources such as minerals, water, and energy. Environmental geologists play a crucial role in assessing the availability and quality of these resources, providing essential information for their responsible extraction and utilization. For example, in groundwater management, geologists evaluate aquifer characteristics and recharge rates to ensure that water resources are used sustainably and protected from contamination.

3. **Pollution Control**: Geologists study the movement and behavior of contaminants in soil, water, and air, providing insights into how

pollutants spread and their potential impacts on human health and ecosystems. By understanding the geological factors that influence contamination, environmental geologists can develop effective remediation strategies and pollution prevention measures.

4. **Natural Hazard Assessment**: Environmental geologists evaluate geological hazards such as earthquakes, volcanic eruptions, landslides, and floods to understand their potential impacts on communities and infrastructure. By conducting hazard assessments, geologists can inform risk management strategies and disaster preparedness efforts, helping to protect lives and property.

Conclusion: In summary, environmental geology is a vital field that examines the interactions between geological processes and the environment. Its broad scope encompasses land use planning, resource management, pollution control, and natural hazard assessment. By understanding these interactions, environmental geologists provide valuable insights that inform sustainable practices and policies, ultimately contributing to the protection of the environment and the well-being of society.

1.2 Importance of Geology in Understanding Environmental Issues

Role of Geology: Geology plays a crucial role in understanding environmental issues, serving as the foundation for comprehending the Earth's processes and the interactions between geological systems and the environment. Geological knowledge is essential for a variety of reasons, including the understanding of soil formation, groundwater movement, and the distribution of natural resources.

1. **Soil Formation and Composition**: The geological characteristics of an area, including the types of rocks and sediments present, influence the properties of the soil that develops over time. This understanding is vital for agricultural practices, land development, and environmental conservation. Knowledge of soil types can inform farmers about the best crops to plant and the appropriate management practices to enhance soil health.

2. **Groundwater Movement**: Geologists study the characteristics of aquifers, including their porosity and permeability, to understand how

water moves through subsurface materials. This information is crucial for managing water resources, particularly in regions facing water scarcity or contamination issues. Understanding the geological framework of aquifers enables the development of sustainable water management practices.

3. **Natural Hazard Risks**: Geological assessments are vital for evaluating the risks associated with natural hazards, such as earthquakes, landslides, and floods. By studying the geological features and processes that contribute to these hazards, geologists can assess potential risks to communities and infrastructure, informing land-use planning and disaster preparedness efforts.

4. **Interdisciplinary Collaboration**: The interdisciplinary nature of environmental geology enhances our ability to address complex environmental challenges. Geology intersects with various fields, including ecology, hydrology, and environmental science, allowing for collaborative approaches to problem-solving.

Conclusion: In summary, geology is fundamental to understanding environmental issues, providing essential insights into processes such as soil formation, groundwater movement, and natural hazard assessment. The knowledge gained from geological studies informs decision-making in land use, resource management, and environmental protection, ultimately contributing to sustainable development and the well-being of both human and ecological communities.

1.3 Overview of Applied Geology

Definition: Applied geology refers to the practical application of geological principles and techniques to solve real-world problems. This dynamic field encompasses a variety of applications that are essential for addressing contemporary challenges in environmental management, resource utilization, and public safety.

1. **Environmental Assessments**: Applied geologists conduct thorough evaluations of geological conditions to determine the potential impacts of proposed projects on the environment. This includes assessing soil stability, groundwater quality, and the presence of geological hazards.

Comprehensive environmental assessments ensure that development projects are carried out responsibly and sustainably.

2. **Resource Exploration**: Resource exploration is another critical aspect of applied geology. Geologists play a vital role in identifying and managing natural resources, including minerals, water, and energy. Through geological surveys and assessments, applied geologists evaluate the availability and quality of these resources, providing essential information for sustainable extraction and utilization. For example, in groundwater resources, applied geologists assess aquifer characteristics and recharge rates to ensure that water is used sustainably and that potential contamination risks are mitigated.

3. **Hazard Mitigation**: Hazard mitigation is a further key area of focus within applied geology. Geologists assess geological hazards such as earthquakes, landslides, and volcanic eruptions to develop strategies that minimize their impacts on communities and infrastructure. By understanding the geological factors that contribute to these hazards, applied geologists can inform risk management strategies and enhance community preparedness. For example, they may recommend land-use planning measures that avoid high-risk areas or design engineering solutions that improve the resilience of structures to geological events. This proactive approach to hazard mitigation is essential for safeguarding public safety and reducing the economic costs associated with natural disasters.

4. **Site Investigations**: Applied geologists conduct site investigations to gather data on subsurface conditions before construction or development projects. This involves drilling, sampling, and testing soil and rock materials to assess their properties and suitability for specific uses. The information obtained from site investigations is crucial for designing foundations, determining construction methods, and ensuring compliance with environmental regulations.

5. **Environmental Remediation**: Applied geology also plays a significant role in environmental remediation efforts. Geologists assess contaminated sites to determine the extent of pollution and develop strategies for cleanup. This may involve identifying the sources of contamination, evaluating the movement of pollutants through soil and

groundwater, and designing remediation techniques to restore the affected environment.

Conclusion: By bridging the gap between theoretical geology and practical application, applied geology plays a vital role in environmental management and decision-making. It integrates scientific knowledge with real-world considerations, enabling stakeholders to make informed choices that balance economic development with environmental protection. The interdisciplinary nature of applied geology fosters collaboration among geologists, engineers, environmental scientists, and policymakers, ensuring that diverse perspectives are considered in addressing complex challenges.

In summary, applied geology is a crucial field that addresses a wide range of real-world problems through the application of geological principles and techniques. Its contributions to environmental assessments, resource exploration, hazard mitigation, site investigations, and environmental remediation are essential for promoting sustainable development and enhancing community resilience. As the world faces increasing environmental challenges, the role of applied geology will continue to grow in importance, providing the knowledge and tools necessary to navigate the complexities of our changing planet.

1.4 Historical Context and Development of Environmental Geology

The field of environmental geology has evolved significantly over the past century. Its roots can be traced back to the early 20th century when geologists began to recognize the importance of geological factors in environmental issues. Key milestones in the development of environmental geology include the establishment of the first environmental impact assessments in the 1960s and the growing awareness of environmental degradation and sustainability in the latter half of the century. Influential figures, such as John Wesley Powell and Ian McHarg, contributed to the integration of geology with environmental planning and management, laying the groundwork for modern environmental geology.

1.5 Structure of the Book

This book is organized into thirteen chapters, each addressing a specific aspect of environmental geology. The structure is meticulously designed to provide a comprehensive understanding of the field, guiding readers from foundational concepts to contemporary issues and future directions. By following this structured approach, the book aims to equip readers with the knowledge and skills necessary to navigate the complexities of environmental geology and its applications in real-world scenarios.

Chapter Overview

1. **Introduction to Environmental Geology**: This chapter sets the stage by providing an overview of the field and highlighting the importance of geological understanding in addressing environmental issues. It establishes the context for subsequent discussions, emphasizing how geological principles are integral to solving pressing environmental challenges.

2. **Fundamentals of Applied Geology**: This chapter delves into key concepts in applied geology, discussing the role of geology in environmental management. It covers essential topics such as geological mapping, stratigraphy, and the significance of understanding geological formations in relation to environmental issues.

3. **Geological Hazards and Risks**: Readers will explore various types of geological hazards, including earthquakes, landslides, and volcanic eruptions. This chapter provides insights into assessing the risks associated with these hazards, equipping readers with tools to evaluate potential impacts on communities and infrastructure.

4. **Soil and Water Resources**: This chapter emphasizes the significance of soil and water in environmental geology, discussing their critical roles in sustaining ecosystems and human activities. It outlines strategies for their management and conservation, highlighting the need for sustainable practices.

5. **Environmental Impact Assessment (EIA)**: This chapter introduces the principles and methodologies of EIA, outlining the steps involved in conducting thorough assessments of proposed projects. It emphasizes

the importance of EIAs in identifying potential environmental impacts and recommending mitigation measures.

6. **Geology and Climate Change**: This chapter examines the impact of climate change on geological processes and highlights the contributions of geology to climate change mitigation. It explores how geological factors influence climate patterns and how understanding these processes can inform strategies for adaptation and resilience.

7. **Geology and Natural Resources**: An overview of natural resources and their geological context, focusing on sustainable management practices. This chapter discusses the extraction and utilization of resources such as minerals, water, and energy, emphasizing responsible management.

8. **Geohazards and Risk Management**: Strategies for managing geohazards are discussed, along with case studies illustrating effective risk management approaches. This chapter emphasizes proactive planning and preparedness in mitigating the impacts of geological hazards.

9. **Geology and Environmental Policy**: This chapter explores the intersection of geology and environmental policy, examining regulatory frameworks and their implications. It discusses how geological knowledge informs policy decisions and the importance of integrating scientific understanding into environmental regulations.

10. **The Future of Geology and Sustainability**: Addressing emerging trends in geology and the role of geological practices in promoting sustainable development. This chapter explores innovative approaches and technologies shaping the future of the field.

11. **Rock Blasting Constraints**: Focuses on the techniques and challenges associated with rock blasting, including environmental considerations. This chapter provides insights into practical aspects of rock blasting and emphasizes careful planning to minimize environmental impacts.

12. **Strategic Metals, Minerals, and Rare Earth Elements (REEs) in the Mining Sector in Kenya**: This chapter provides an overview of strategic resources in Kenya and discusses the environmental and economic implications of mining REEs. It highlights the significance of

these resources in the global market and the challenges associated with their extraction.

13. **Fundamentals of Ground Investigation**: Emphasizes the critical role of ground investigation in environmental geology. It outlines key objectives and methodologies essential for conducting effective assessments, particularly in understanding how geological factors influence environmental conditions.

Conclusion

The structured approach of this book enables readers to establish a solid foundation in environmental geology while delving into specific topics in depth. Each chapter builds upon the previous ones, creating a cohesive narrative that underscores the interconnectedness of geological processes and environmental issues. By exploring both theoretical concepts and practical applications, the book equips readers with the knowledge and skills necessary to address contemporary challenges in environmental geology.

Moreover, the book emphasizes the importance of interdisciplinary collaboration in tackling environmental issues. It encourages readers to consider the roles of various stakeholders, including geologists, environmental scientists, policymakers, and community members, in developing effective solutions. This collaborative approach is essential, as environmental challenges often transcend disciplinary boundaries and require a multifaceted response. By fostering a holistic understanding of environmental geology, the book aims to inspire future generations of geologists and environmental professionals to engage in responsible practices that promote sustainability and protect the planet for future generations.

In summary, this comprehensive text serves as a valuable resource for students, professionals, and anyone interested in the field of environmental geology. By covering a wide range of topics, from foundational principles to contemporary challenges, the book provides a thorough exploration of how geological knowledge can be applied to address pressing environmental issues. Through its detailed discussions and practical insights, it aims to empower readers to contribute to sustainable development and effective environmental management in their respective fields.

1.6 Objectives of the Book

The primary objective of this book is to provide comprehensive insights into the impact of applied geology on environmental issues. By exploring the principles and practices of environmental geology, the book aims to achieve several key objectives that are essential for fostering a deeper understanding of the intricate relationships between geological processes and environmental conditions.

Educate: One of the foremost goals of this book is to enhance the understanding of how geological processes influence environmental conditions and human activities. This involves elucidating fundamental concepts in geology, such as the formation of soils, the dynamics of groundwater systems, and the behavior of geological hazards. By presenting these concepts in an accessible manner, the book seeks to equip readers with the foundational knowledge necessary to appreciate the complexities of environmental geology. This educational objective is particularly important for students and early-career professionals who are entering the field, as it lays the groundwork for more advanced studies and practical applications.

Inform: The book aims to provide practical knowledge and real-world case studies that illustrate the application of geology in addressing environmental challenges. By integrating theoretical concepts with practical examples, readers will gain insights into how geological assessments inform decision-making in various contexts, such as urban planning, resource management, and environmental remediation. The inclusion of case studies serves to contextualize the principles discussed, demonstrating their relevance and applicability in real-world scenarios. This objective is particularly valuable for professionals and policymakers who require evidence-based information to guide their actions and policies in the field of environmental management.

Promote Sustainability: Another critical objective of the book is to encourage the adoption of sustainable practices in resource management, land use, and environmental protection. As environmental challenges become increasingly complex, the need for sustainable solutions that balance human needs with ecological integrity is paramount. The book emphasizes the role of applied geology in promoting sustainable practices, such as responsible resource extraction, effective waste management, and the preservation of natural habitats. By highlighting successful examples of sustainable practices and the geological principles that underpin them, the book aims to inspire readers to consider sustainability as a fundamental aspect of their work and decision-making processes.

The target audience for this book includes students, professionals, and policymakers who are interested in the intersection of geology and environmental science. Students will benefit from the educational content that prepares them for future careers in environmental geology, while professionals will find practical insights that can be applied in their work. Policymakers will gain a deeper understanding of the geological factors that influence environmental policies, enabling them to make informed decisions that promote sustainable development.

In summary, the objectives of this book are designed to foster a holistic understanding of applied geology's role in addressing environmental issues. By educating, informing, and promoting sustainability, the book aims to contribute to the development of a knowledgeable and proactive community of individuals who are equipped to tackle the pressing environmental challenges of our time.

Conclusion

In conclusion, this book's structured approach provides a comprehensive understanding of environmental geology while emphasizing the critical role of geological knowledge in addressing pressing challenges. By integrating theoretical insights with practical applications, it empowers readers to recognize the significance of geology in everyday life and its implications for environmental health and sustainability.

The focus on interdisciplinary collaboration reflects the reality that environmental issues often transcend single disciplines. By integrating perspectives from geologists, environmental scientists, policymakers, and community members, the book fosters a nuanced understanding of the complexities involved in environmental decision-making, essential for developing effective and equitable solutions.

Additionally, the book promotes sustainability in an era of rapid environmental change and resource depletion. By showcasing successful sustainable practices and the geological principles that support them, it inspires readers to prioritize ecological integrity alongside human development, encouraging consideration of the long-term impacts of their actions.

Covering a diverse range of topics—from foundational geological concepts to contemporary environmental challenges—this book equips readers to engage with the complexities of environmental geology. The inclusion of real-world case studies contextualizes theoretical concepts, demonstrating their practical relevance and highlighting the importance of evidence-based decision-making in environmental management.

In summary, this book is a vital resource for anyone interested in the intersection of geology and environmental science. Its structured approach, emphasis on collaboration, and commitment to sustainability make it essential for students, professionals, and policymakers. By equipping readers with the knowledge and skills to address contemporary environmental challenges, the book aims to inspire a new generation of geoscientists and environmental professionals dedicated to promoting sustainable practices and protecting our planet.

As we navigate an increasingly complex world, the insights presented in this book will guide responsible decision-making and foster a sustainable future. We invite readers to engage with the material, reflect on their practices, and contribute to the ongoing dialogue surrounding environmental geology and its critical role in shaping a sustainable world. Together, we can work towards solutions that address immediate environmental challenges and ensure the health of our planet for generations to come.

CHAPTER 2

FUNDAMENTALS OF APPLIED GEOLOGY

2.0 Introduction

The exploration and utilization of natural resources demand a skilled workforce equipped with a strong foundation in engineering and geoscientific knowledge, complemented by technical expertise. Engineers and geoscientists play a crucial role at the forefront of these endeavors, benefiting from their relevant knowledge and proficiency in various geological processes and resource management.

This comprehensive overview of applied geology underscores the importance of essential skills required for the identification and description of prevalent rock-forming minerals. These minerals are fundamental to understanding the intricate processes of rock formation and the geological history of an area. By acquiring and honing these vital skills, professionals in the field can effectively contribute to the successful exploration and characterization of geological resources, such as minerals, hydrocarbons, and groundwater.

Applied geology also encompasses the study of earth processes, commonly referred to as geohazards. These include phenomena such as earthquakes, landslides, volcanic eruptions, and flooding, which can negatively impact infrastructure, communities, and the environment. Geologists must study these processes to understand their mechanisms, assess risks, and recommend mitigation measures necessary for minimizing damage in geological terrains that are vulnerable to such hazards. This proactive approach is essential for safeguarding lives and property, as well as ensuring the sustainability of development projects.

Furthermore, applied geology plays a critical role in infrastructure development. Knowledge of geological conditions is vital for the design and construction of buildings, roads, bridges, and other structures. Understanding soil and rock properties, groundwater behavior, and potential geohazards allows engineers to make informed decisions that enhance the safety and longevity of infrastructure.

This chapter aims to bolster essential knowledge and practical skills for both academics and professionals involved in infrastructural development. By emphasizing the integration of geological principles with engineering practices, it highlights the importance of interdisciplinary collaboration in addressing the challenges associated with resource exploration and management.

The successful exploration and sustainable exploitation of natural resources necessitate a skilled workforce well-versed in engineering and geoscientific knowledge, complemented by technical proficiency. Engineers and geoscientists play a vital role at the forefront of these endeavors, equipped with relevant expertise and capabilities that enable them to navigate the complexities of resource management and environmental stewardship.

This comprehensive overview of applied geology emphasizes the significance of acquiring essential skills for the typical identification and detailed description of common rock-forming minerals. These minerals are integral components of the intricate processes involved in rock formation and provide critical insights into the geological history and resource potential of an area. By possessing and cultivating these critical skills, professionals in the field can actively contribute to effective exploration and characterization of geological resources, promoting responsible and informed resource management (Rop et al., 2022).

Applied geology also encompasses the study of earth processes, generally termed geohazards, which include phenomena such as earthquakes, tsunamis, floods, landslides, and volcanic activity. These geohazards can have an enormous negative impact on civilization, threatening lives, infrastructure, and economic stability. Applied geologists study these processes to understand their mechanisms and impacts, and they recommend mitigation measures necessary for minimizing damage in geological terrains that are vulnerable to such hazards. This proactive approach is essential for enhancing community resilience and ensuring the safety of infrastructure development.

In this chapter, the exploration rationale in applied geology is addressed, with recent illustrations of rare earth elements (REEs) included. These elements are vital as critical raw materials for advancing technologies, including clean-energy applications, high-tech military components, and electronics. The growing demand for REEs underscores the importance of sustainable exploration practices that balance resource extraction with environmental protection and community well-being.

By integrating geological knowledge with engineering practices, this overview aims to highlight the importance of interdisciplinary collaboration in addressing the challenges associated with resource exploration and management. Ultimately, the insights presented in this chapter will contribute to the ongoing discourse on sustainable development and the responsible stewardship of natural resources.

The research incorporates illustrations to underscore the importance of rare earth elements (REEs) in modern technologies and to emphasize their relevance in the context of applied geology. Furthermore, the chapter provides a detailed description and identification properties of selected minerals. These descriptions serve as a basis for comparison with the results obtained from field studies.

By presenting this comparison, readers can assess the accuracy of the provided mineral descriptions and gain confidence in the research findings. The integration of recent illustrations and detailed mineral analyses enhances the overall understanding of the significance of rare earth elements and their potential impact on evolving technologies (Rop et al., 2022).

2.1 Definition and Scope of Applied Geology

Applied geology is a specialized branch of geology that emphasizes the practical application of geological knowledge to address real-world challenges. This field encompasses a diverse array of activities, including site investigations, environmental assessments, resource exploration, and hazard mitigation. The scope of applied geology is extensive, as it addresses critical issues that impact society, the environment, and the economy. By leveraging geological expertise, applied geologists contribute to informed decision-making processes that promote sustainable development and resource management.

One of the key areas of focus within applied geology is land use planning. Geologists evaluate geological conditions to inform land development and zoning decisions, ensuring that construction and infrastructure projects are sited in locations that minimize risks associated with geological hazards. This involves assessing factors such as soil stability, groundwater availability, and the potential for natural hazards like landslides or flooding. By integrating geological insights into land use planning, communities can enhance their resilience to environmental challenges and promote sustainable growth.

Environmental impact assessments (EIAs) are another critical component of applied geology. These assessments evaluate the potential environmental effects of proposed projects, such as construction, mining, or infrastructure development. Applied geologists play a vital role in conducting EIAs by identifying potential impacts on geological resources, ecosystems, and human health. They also recommend mitigation measures to minimize adverse effects, ensuring that projects comply with environmental regulations and contribute to sustainable development.

Resource management is a fundamental aspect of applied geology, involving the identification and management of natural resources, including minerals, water, and energy. Applied geologists assess the availability and quality of these resources, providing essential information for sustainable extraction and utilization. This includes conducting geological surveys, evaluating resource potential, and developing management strategies that balance economic needs with environmental protection. By effectively managing natural resources, applied geologists contribute to the long-term sustainability of communities and industries.

Natural hazard assessment is another critical area of applied geology, focusing on the evaluation of geological hazards and the development of strategies to minimize their impacts. Applied geologists assess risks associated with hazards such as earthquakes, landslides, and flooding, providing valuable insights for disaster preparedness and response. By understanding the geological factors that contribute to these hazards, geologists can inform risk management strategies, enhance infrastructure resilience, and protect communities from potential disasters.

Key Principles of Applied Geology

Several key principles guide the practice of applied geology, ensuring that geological knowledge is effectively applied to real-world problems. One fundamental principle is geological mapping, which involves the creation of geological maps that illustrate the distribution and characteristics of geological materials. These maps are essential tools for land use planning, resource exploration, and hazard assessment, providing critical information that informs decision-making processes.

Stratigraphy, the study of the layering of rocks and sediments, is another key principle in applied geology. Understanding stratigraphy is crucial for interpreting geological history and assessing subsurface conditions. Stratigraphic analysis helps geologists identify potential resources, such as groundwater or fossil fuels, and evaluate geological hazards, such as the stability of slopes or the potential for subsidence. By analyzing stratigraphic relationships, applied geologists can make informed recommendations for resource management and hazard mitigation.

Structural geology, which focuses on the study of rock deformation and structural features such as faults and folds, is also essential in applied geology. Understanding the stability of geological formations is critical for risk assessments related to construction projects and natural hazard mitigation. Structural geology informs engineers and planners about the potential for ground movement, enabling them to design structures that can withstand geological forces and minimize risks to public safety.

Finally, hydrogeology, a subfield of applied geology, focuses on the distribution and movement of groundwater. Hydrogeological studies are vital for managing water resources, assessing contamination risks, and designing effective remediation strategies. By understanding groundwater flow patterns and the interactions between surface water and groundwater, applied geologists can develop sustainable water management practices that protect this essential resource.

In summary, applied geology is a dynamic field that addresses a wide range of real-world challenges through the practical application of geological knowledge. By focusing on key principles such as geological mapping, stratigraphy, structural geology, and hydrogeology, applied geologists contribute to informed decision-making processes that promote sustainable development, resource management, and hazard mitigation. Their expertise is essential for building resilient communities and ensuring the responsible use of natural resources.

Applied Geology in Infrastructural Development

Applied geology is a critical discipline in the context of infrastructural development, as it provides essential insights into the geological factors that influence the design, construction, and sustainability of infrastructure projects. By integrating various geological specialties, applied geologists can assess site conditions, evaluate resource availability, and mitigate potential geological hazards, ensuring that infrastructure is built safely and effectively. Below are some of the specialized branches of geology that contribute to applied geology in infrastructural development:

- **Geophysics**: This branch explores the Earth's physical properties and processes, utilizing techniques such as seismic surveys, magnetic and gravitational measurements, and electrical resistivity testing. Geophysical methods provide essential data for understanding subsurface conditions, helping engineers and planners identify suitable locations for construction and assess the integrity of existing structures. For example, geophysical surveys can detect voids, fractures, or other anomalies in the subsurface that may pose risks to infrastructure.

- **Geochemistry**: Geochemistry investigates the chemical composition of rocks, minerals, and soils, which is crucial for assessing resource potential and environmental impacts. In infrastructural development, geochemical analyses can inform decisions regarding material selection, contamination assessment, and environmental remediation. Understanding the chemical properties of construction materials ensures that they are suitable for their intended use and helps identify any potential environmental risks associated with their extraction or use.

- **Petrology**: This branch examines the origin, composition, and classification of rocks, providing valuable information for identifying suitable materials for construction and other applications. Petrologists analyze rock samples to determine their physical and chemical properties, which can influence their performance in construction projects. For instance, understanding the durability and weathering characteristics of different rock types is essential for selecting materials for roads, bridges, and buildings.

- **Hydrogeology**: Hydrogeology focuses on the study of groundwater, which is essential for water resource management and understanding aquifer systems. Infrastructural development often requires a thorough understanding of groundwater conditions, particularly when designing foundations, drainage systems, and water supply networks. Hydrogeologists assess aquifer characteristics, recharge rates, and potential contamination sources to ensure that water resources are managed sustainably and that infrastructure is designed to minimize impacts on groundwater.

- **Paleontology**: While primarily concerned with the history of life on Earth through fossil analysis, paleontology can also contribute to infrastructural development by providing insights into past environments and climate changes. Understanding the geological history of a site can inform decisions about construction methods and materials, particularly in areas where sedimentary deposits may indicate past environmental conditions that could affect current engineering practices.
- **Engineering Geology**: This specialized branch assesses geological factors relevant to construction and infrastructure projects, ensuring that designs are safe and effective. Engineering geologists evaluate site conditions, including soil stability, rock strength, and the presence of geological hazards such as landslides or earthquakes. Their assessments inform the design of foundations, retaining walls, and other structural elements, helping to mitigate risks and enhance the safety and longevity of infrastructure.
- **Global Tectonics**: This field investigates the movements and interactions of Earth's tectonic plates, which are fundamental to understanding seismic risks and land stability. Knowledge of tectonic processes is crucial for infrastructure development in seismically active regions, where engineers must design structures that can withstand earthquakes and other tectonic-related hazards. Understanding tectonic settings also informs land-use planning and risk management strategies.

Autonomous Robotic Technologies in Mining

The integration of autonomous robotic technologies in mining represents a significant advancement in addressing the challenges faced by the industry, particularly in terms of safety, efficiency, sustainability, and environmental impacts. As the mining sector plays a crucial role in supplying raw materials that drive global economic growth, the adoption of these technologies is essential for modernizing operations and mitigating risks associated with traditional mining methods.

Challenges in Traditional Mining Methods

Traditional mining methods, especially in underground environments, are often labor-intensive and expose workers to various hazards, including:

- Cave-Ins: The risk of ground collapse poses a significant threat to miners, particularly in poorly supported tunnels.
- Toxic Gases: The presence of harmful gases, such as methane and carbon monoxide, can lead to dangerous working conditions.
- Poor Visibility: Limited visibility in underground settings can hinder navigation and increase the likelihood of accidents.
- Unstable Geological Structures: Variability in geological conditions can lead to unexpected challenges during mining operations.

These risks not only compromise worker safety but also limit operational efficiency and increase production costs. As a result, there is a pressing need for innovative solutions that can enhance safety and streamline mining processes.

Key Developments in AI Robotics in Mining

1. **Complex Underground Environments**: The unstructured nature of underground mining environments presents unique challenges for deploying conventional robotic systems. Traditional robots are often designed for controlled settings, making it difficult for them to navigate narrow tunnels, detect hazards, and adapt to unpredictable conditions. This limitation has spurred the development of specialized autonomous systems tailored for the complexities of underground mining.

2. **Historical Context of Mining Methods**: Early mining relied heavily on human labor and rudimentary tools. The introduction of mechanized equipment, such as drilling machines and conveyors in the 20th century, improved efficiency but still required significant human oversight. The evolution of mining technology has paved the way for more advanced automation solutions.

3. **Automation in Surface Mining**: Large mining companies, including Rio Tinto, BHP Billiton, and Vale, began deploying autonomous haul trucks and drilling rigs in the 2000s. These innovations have led to substantial improvements in productivity and safety. For instance, a case study by Corke et al. (2019) highlights that autonomous haul trucks at Rio Tinto's Pilbara Mine increased productivity by 15-20%. This demonstrates the potential of autonomous technologies to enhance operational efficiency in surface mining.

4. **Broader Adoption of Autonomous Systems**: The mining industry has seen a broader adoption of autonomous technologies, including GPS-based navigation, machine learning algorithms for predictive maintenance, and remote monitoring systems. These advancements have primarily been observed in surface mining operations, where the risks associated with human labor have been significantly reduced. The implementation of autonomous systems has contributed to lower human fatalities and operational inefficiencies, marking a transformative shift in mining practices.

5. **Future Prospects**: As the mining industry continues to evolve, the potential for autonomous robotic technologies to address the challenges of underground mining remains significant. Ongoing research and development efforts are focused on enhancing the capabilities of autonomous systems, enabling them to navigate complex environments, detect hazards in real-time, and operate safely alongside human workers.

Environmental Impacts of Mining

While autonomous robotic technologies offer numerous benefits, it is essential to consider their environmental impacts as well. Mining operations can have significant effects on the environment, including:

- **Habitat Destruction**: Mining activities often lead to the clearing of large areas of land, resulting in habitat loss for flora and fauna. Autonomous systems can help minimize land disturbance by optimizing extraction processes and reducing the footprint of mining operations.

- **Soil and Water Contamination**: The use of chemicals in mineral processing can lead to soil and water contamination if not managed properly. Autonomous technologies can enhance monitoring and management practices, allowing for real-time detection of leaks or spills and minimizing environmental risks.

- **Waste Generation**: Mining generates substantial amounts of waste, including tailings and overburden. Autonomous systems can improve waste management practices by optimizing the transportation and disposal of waste materials, reducing the environmental impact associated with waste storage.
- **Carbon Emissions**: Traditional mining operations often rely on fossil fuels for machinery and transportation, contributing to greenhouse gas emissions. The integration of autonomous electric vehicles and renewable energy sources can help reduce the carbon footprint of mining operations.
- **Ecosystem Disruption**: Mining can disrupt local ecosystems, affecting biodiversity and water quality. Autonomous technologies can facilitate more precise mining practices, reducing the impact on surrounding ecosystems and promoting sustainable resource extraction.

Conclusion on Applied Geology in Infrastructural Development

In conclusion, applied geology plays a vital role in infrastructural development by providing essential insights from various geological specialties. By understanding the geological context of a project, applied geologists can assess site conditions, evaluate resource availability, and mitigate potential hazards, ensuring that infrastructure is designed and constructed safely and sustainably. The integration of geological knowledge into engineering practices not only enhances the safety and effectiveness of infrastructure projects but also contributes to the long-term sustainability of communities and the environment. As urbanization and infrastructure demands continue to grow, the importance of applied geology in guiding responsible development will only increase.

Simultaneously, the integration of autonomous robotic technologies in mining represents a critical step toward improving safety, efficiency, sustainability, and environmental stewardship within the industry. Traditional mining methods present numerous challenges, but the adoption of advanced automation solutions has the potential to revolutionize operations, particularly in hazardous underground environments. As the mining sector continues to embrace these technologies, it is essential to ensure that small and medium-sized enterprises, especially in developing regions, have access to affordable and effective solutions. By fostering innovation and collaboration, the mining industry can enhance its resilience and contribute to sustainable economic growth.

Both themes emphasize the importance of integrating specialized knowledge and innovative technologies to enhance safety, efficiency, and sustainability in their respective fields—applied geology in infrastructure development and autonomous robotics in mining. They highlight the need for responsible practices that consider environmental impacts and the accessibility of solutions for smaller enterprises, particularly in developing regions. As both sectors face increasing demands and challenges, the role of advanced methodologies and collaborative efforts will be crucial for achieving long-term sustainability and resilience. Together, the integration of applied geology and autonomous technologies can pave the way for a more sustainable future, addressing the pressing needs of infrastructure and resource management in an ever-evolving world.

The Role of Applied Geology in Understanding Earth's Processes

The integration of diverse geological disciplines significantly enhances our understanding of the planet's complex and dynamic nature (Rop and Namwiba, 2018). Scientific studies comprehensively investigate natural materials and geological processes, often with a focus on applying fundamental principles to improve the quality of life for humanity. These principles are typically applied in engineering-related projects, highlighting the critical role of applied geology in contributing to the sustainability of modern civilization. By integrating geological processes and natural resources into infrastructural development, applied geology plays a vital role in addressing contemporary challenges (Rop and Namwiba, 2018; 2019).

Understanding Geological Processes

For the effective application of selected aspects of applied geology, it is essential to thoroughly understand various geological processes. Applied geology provides valuable insights into phenomena such as shown in Figure 1:

- **Lithospheric Movements**: The study of movements within the Earth's lithosphere, which can lead to significant geological events such as earthquakes and volcanic activity.
- **Rifting and Subduction**: These processes are crucial in understanding how tectonic plates interact, leading to the formation of new geological features and the recycling of materials within the Earth.
- **Mantle-Crust Interaction**: The ongoing interaction between the mantle and crust is fundamental to understanding the dynamics of plate tectonics and the formation of different rock types.

By studying these phenomena, researchers gain a deeper understanding of the interconnected plate tectonic processes (Patwardhan, 1999) that contribute to the formation of igneous, sedimentary, and metamorphic rocks. This knowledge is essential for predicting geological hazards and managing natural resources effectively.

Insights into Mineralization

Additionally, the insights gained from applied geology shed light on episodes of mineralization, elucidating the genesis and distribution of valuable mineral deposits. Understanding the geological processes that lead to mineral formation is crucial for resource exploration and sustainable extraction practices. The close examination of these geological processes through applied geology unveils the intricate workings of our planet and its geological history, enriching our understanding of Earth's dynamic and diverse geological features (Figure 1).

Conclusion

In summary, applied geology serves as a bridge between fundamental geological research and practical applications that enhance the quality of life. By integrating knowledge of geological processes into engineering and infrastructural development, applied geology contributes to the sustainability of modern civilization. As we continue to explore and understand the complexities of our planet, the insights provided by applied geology will be invaluable in addressing the challenges of resource management, environmental protection, and disaster mitigation.

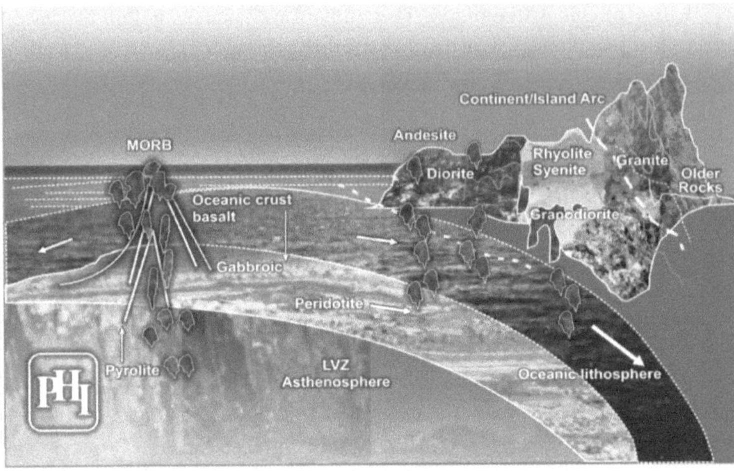

Figure 1: The Dynamic Earth System (Patwardhan, 1999)

In conclusion, the integration of applied geology into infrastructural development is a multifaceted approach that significantly enhances our understanding of geological processes while ensuring that resource management and construction practices are both sustainable and informed by scientific principles. This holistic approach is essential for addressing the myriad challenges faced by modern civilization, including urbanization, climate change, and resource depletion, all while safeguarding the environment and promoting the well-being of communities.

Enhancing Understanding of Geological Processes

Applied geology provides critical insights into the Earth's processes, such as tectonic movements, erosion, sedimentation, and mineralization. By understanding these processes, engineers and planners can make informed decisions that take into account the geological context of a project. This knowledge is vital for predicting potential hazards, such as landslides, earthquakes, and flooding, which can have devastating impacts on infrastructure and communities. For instance, incorporating geological assessments into site selection and design can help mitigate risks associated with unstable ground conditions, ensuring that structures are built to withstand natural forces.

Sustainable Resource Management

The integration of geological insights into infrastructure projects also plays a crucial role in sustainable resource management. By understanding the distribution and availability of natural resources, such as water, minerals, and energy sources, applied geologists can guide the responsible extraction and use of these resources. This is particularly important in an era where resource scarcity is becoming increasingly prevalent. Sustainable practices, such as minimizing waste, recycling materials, and using renewable resources, can be informed by geological studies that identify the most efficient and environmentally friendly methods of resource utilization.

Promoting Resilient Infrastructure

Creating resilient infrastructures that can withstand natural hazards is a key objective of modern engineering practices. By leveraging geological insights, engineers can design structures that are not only safe but also adaptable to changing environmental conditions. For example, understanding local geology can inform the design of flood defenses, earthquake-resistant buildings, and landslide mitigation measures. This proactive approach not only protects human life and property but also reduces the economic costs associated with disaster recovery and infrastructure repair.

Support Sustainable Development Goals

The integration of applied geology into infrastructural development aligns with global sustainable development goals (SDGs), particularly those related to sustainable cities and communities, responsible consumption and production, and climate action. By prioritizing geological considerations in planning and construction, we can create urban environments that are not only functional but also environmentally sustainable. This includes promoting green building practices, enhancing urban green spaces, and ensuring that infrastructure projects contribute positively to the local ecosystem.

Community Well-Being and Environmental Stewardship

Ultimately, the integration of applied geology into infrastructure development fosters community well-being and environmental stewardship. By engaging with local communities and incorporating their knowledge and needs into geological assessments, we can ensure that infrastructure projects are socially responsible and culturally sensitive. This collaborative approach not only builds trust between stakeholders but also empowers communities to take an active role in the management of their natural resources.

Conclusion

In summary, the integration of applied geology into infrastructural development is essential for creating a sustainable future. By enhancing our understanding of geological processes, promoting sustainable resource management, and designing resilient infrastructures, we can address the challenges of modern civilization while safeguarding the environment and supporting community well-being. As we continue to face pressing global issues, the collaboration between geology and engineering will be crucial in developing innovative solutions that prioritize both human safety and the health of our planet. Through this integrated approach, we can work towards achieving sustainable development goals and ensuring a better quality of life for current and future generations.

Applied Geology in Infrastructural Development

Applied geology is a critical field that focuses on understanding geological processes and their implications for human activities, particularly in the context of infrastructural development. Applied geologists study geological processes and recommend mitigation measures necessary for minimizing damage in geological terrains that are vulnerable to geohazards. For instance, by analyzing flood patterns of rivers, geologists can delineate residential neighborhoods that are safe, thereby avoiding damage from future flood episodes. This proactive approach not only protects lives and property but also informs urban planning and development strategies.

Mitigating Geological Hazards

Knowledge in earthquake science is crucial for minimizing damage caused by significant seismic events. In this context, identifying localities for the generation of earth tremors (epicenters) allows engineers to factor in seismic considerations when designing structures in earthquake-prone regions. This includes implementing building codes that require earthquake-resistant designs, thereby enhancing the resilience of infrastructure against potential seismic threats.

In general, applied geology examines how geological issues affect human lives, including the impacts of geological hazards and associated environmental problems, as well as methods for accessing and managing natural resources for infrastructural development (Rop and Namwiba, 2018; 2019). The importance of applied geology in infrastructural development is further emphasized by Patwardhan (1999), who highlights the role of plate tectonics in various geological processes, including the formation of igneous, sedimentary, and metamorphic rocks, as well as the genesis of mineral deposits. This understanding is essential for resource management and infrastructure planning, as it informs the sustainable extraction and utilization of geological resources.

Addressing Geo-Environmental Hazards

Applied geology also addresses critical geo-environmental hazards such as volcanic eruptions, earthquakes, floods, tsunamis, droughts, and desertification. By studying these hazards, applied geologists can develop strategies to mitigate their impacts, ensuring that communities are better prepared and more resilient in the face of natural disasters. For example, hazard mapping and risk assessment can guide land-use planning and emergency response strategies, ultimately reducing vulnerability and enhancing community resilience.

Renewable Energy and Geological Context

In today's context, applied geology has become increasingly important in managing the growing adoption of renewable energy sources, including solar, hydro, wind, and nuclear energy. Understanding the geological context of these energy sources is vital for their sustainable implementation and for minimizing environmental impacts (Patwardhan, 1999). For example, site selection for wind farms requires knowledge of local geology and topography to optimize energy capture while minimizing ecological disruption. Similarly, hydroelectric projects must consider geological stability to prevent landslides and other hazards, ensuring that infrastructure is both safe and effective.

Conclusion

In summary, applied geology plays a pivotal role in infrastructural development by providing essential insights into geological processes, hazards, and resource management. By integrating geological knowledge into engineering practices, we can enhance the safety, sustainability, and resilience of infrastructure, ultimately contributing to the well-being of communities and the environment. The ongoing collaboration between geologists, engineers, and policymakers is crucial for addressing the challenges posed by natural hazards and for promoting sustainable development in an ever-changing world. As we continue to face the impacts of climate change and urbanization, the role of applied geology will be increasingly vital in guiding responsible development practices that prioritize both human safety and environmental stewardship. Through this integrated approach, we can work towards a more sustainable future, where infrastructure supports the needs of society while respecting the planet's natural systems.

Concerns Requiring Applied Geology for Infrastructures

(a) Building and Civil Engineering Structures

Engineering geological expertise is essential in the design and construction of residential, commercial, and industrial buildings, as well as other civil engineering projects. The foundation of any structure is critical to its stability and longevity, making it paramount to ensure that all structures are sited on suitable ground. This requires a thorough understanding of the geological conditions at the site, which can significantly influence the design and construction processes.

Conducting adequate site investigations is a fundamental step in interpreting ground conditions and identifying potential weak zones. These investigations typically involve a combination of geological mapping, soil sampling, and geophysical surveys to assess the subsurface characteristics. By analyzing the geological profile of a site, engineers can identify issues such as soil type, moisture content, and the presence of bedrock or other geological features that may affect construction. This information is crucial for making informed decisions about the design and engineering of structures.

In instances where loose or unstable ground is encountered, appropriate construction measures must be implemented to ensure the stability and safety of the structures. Applied geology provides valuable insights into the nature of the ground and informs the selection of suitable construction techniques. For example, if a site is found to have loose or unconsolidated soils, engineers may need to employ soil stabilization techniques to enhance the load-bearing capacity of the ground. These techniques can include mechanical densification, where soil is compacted to increase its density and strength, as well as chemical stabilization methods that involve the use of additives to improve soil properties.

Other methods of soil stabilization may involve the use of geosynthetics, which are synthetic materials used to reinforce soil and improve its mechanical properties. Techniques such as soil nailing, where steel bars are inserted into the ground to provide additional support, or the use of retaining walls to prevent soil movement, may also be employed. The choice of stabilization method depends on the specific geological conditions and the requirements of the project.

Applied geology plays a vital role in providing guidance throughout the planning and execution of building and civil engineering projects. By integrating geological knowledge into the design process, engineers can develop structures that are not only safe and stable but also resilient to potential geological hazards. This proactive approach helps to minimize risks associated with ground instability, flooding, and other environmental factors that could compromise the integrity of the structure.

In summary, the expertise of applied geology is indispensable in the field of building and civil engineering. Through comprehensive site investigations and the application of appropriate construction measures, engineering geologists ensure that structures are built on suitable ground, enhancing their safety and longevity. By employing soil stabilization techniques and other geological insights, engineers can effectively address the challenges posed by varying ground conditions, ultimately contributing to the successful execution of construction projects. As urban development continues to expand, the importance of integrating geological considerations into engineering practices will remain a critical aspect of sustainable infrastructure development.

(b) Mineral Resources

Mineral resources, encompassing both ore deposits and industrial minerals, are vital natural raw materials that play a crucial role in various industries and are in high demand globally. These resources are essential for the production of a wide range of products, from construction materials to energy sources. The extraction and utilization of these minerals require a comprehensive understanding of geological processes, as well as the application of modern exploration techniques to ensure sustainable and responsible management.

Ore deposits, such as those rich in iron, are fundamental to the manufacturing of steel, which is a critical material used in construction, infrastructure, and various industrial applications. Iron ores, particularly magnetite and hematite, are the primary sources of iron for steel production. The demand for steel continues to rise as urbanization and industrialization expand, making the identification and extraction of high-quality iron ore deposits increasingly important. Applied geologists play a key role in this process by conducting geological surveys and assessments to locate and evaluate these ore deposits, ensuring that mining operations are both economically viable and environmentally responsible.

In addition to ore deposits, industrial minerals are essential for a variety of applications. These minerals, which include materials such as clay, limestone, and gypsum, are used in construction, manufacturing, and agriculture. For example, limestone is a key ingredient in cement production, while clay is used in ceramics and pottery. The extraction of these minerals must be carefully managed to minimize environmental impacts and ensure that resources are used efficiently. Sustainable practices, such as land reclamation and responsible mining techniques, are critical to maintaining the balance between resource extraction and environmental protection.

Crude oil, a significant hydrocarbon resource, occurs as a natural fluid trapped in geological formations. It is one of the leading sources of energy worldwide, powering transportation, heating, and electricity generation. The exploration for crude oil involves the application of advanced geological and geophysical techniques to identify potential reservoirs in remote and often challenging geological terrains. This process requires a thorough understanding of subsurface geology, as well as the use of technologies such as seismic imaging and drilling to locate and assess oil deposits.

The by-products of crude oil extraction are numerous and have a wide range of applications, including bituminous products used in construction, such as asphalt for road paving. However, the extraction and processing of crude oil also present significant environmental challenges. The uncontrolled emission of oil by-products into the atmosphere can lead to air pollution, contributing to climate change and posing health risks to communities. Therefore, it is imperative that the oil industry implements stringent management practices and regulatory measures to mitigate these environmental hazards. This includes adopting cleaner technologies, reducing emissions, and ensuring that waste products are handled responsibly.

In conclusion, mineral resources, including ore deposits and industrial minerals, are essential for modern society and economic development. The responsible exploration and extraction of these resources require a collaborative effort among geologists, engineers, and environmental scientists to ensure that they are utilized sustainably. By employing modern exploration techniques and adhering to environmental regulations, we can harness the benefits of these natural resources while minimizing their impact on the environment. This balanced approach is crucial for securing a sustainable future in which mineral resources continue to support economic growth and development.

(c) Geohazards Mitigation

Geohazards mitigation is a critical aspect of applied geology that focuses on identifying and managing potential natural and man-made hazards that can pose significant risks to human life, property, and the environment. An applied geologist plays a vital role in this process by utilizing their expertise to assess geological conditions and recommend strategies that enhance safety and resilience. This proactive approach is essential in a world where urbanization, climate change, and other factors are increasing the frequency and intensity of geohazards.

One of the primary responsibilities of an applied geologist is to identify potential hazards associated with geological processes, such as landslides, earthquakes, flooding, and subsidence. For example, slope degradation is a common concern in areas with steep terrain or where human activities, such as construction and mining, have altered the natural landscape. By incorporating an understanding of slope stability into the design stage of construction projects, applied geologists can help engineers and planners implement effective slope cuts and support measures. These measures may include retaining walls, drainage systems, and vegetation stabilization techniques, all aimed at ensuring the long-term stability of slopes and minimizing the risk of landslides.

In addition to slope stability, applied geologists assess various geological conditions that can contribute to flooding, such as soil permeability, drainage patterns, and the presence of natural barriers. By analyzing these factors, they can recommend appropriate engineering solutions, such as the construction of levees, flood control channels, and retention basins. These interventions are designed to manage water flow and reduce the likelihood of flooding in vulnerable areas, thereby protecting communities and infrastructure from potential damage.

Furthermore, applied geologists are instrumental in evaluating the risks associated with man-made hazards, such as those arising from industrial activities, mining operations, and urban development. For instance, the extraction of natural resources can lead to ground instability and contamination of surrounding areas. By conducting thorough geological assessments, applied geologists can identify potential risks and recommend mitigation strategies, such as proper waste management practices, monitoring systems, and remediation techniques. This holistic approach not only safeguards the environment but also ensures compliance with regulatory standards and promotes sustainable development.

The integration of geohazards mitigation into urban planning and infrastructure development is essential for building resilient communities. Applied geologists work collaboratively with engineers, urban planners, and policymakers to develop comprehensive risk management plans that address both current and future hazards. This interdisciplinary approach allows for the incorporation of geological insights into land-use planning, zoning regulations, and emergency preparedness strategies. By fostering a culture of risk awareness and proactive planning, communities can better withstand the impacts of geohazards and recover more quickly in the aftermath of an event.

In conclusion, geohazards mitigation is a fundamental responsibility of applied geologists, who play a crucial role in identifying and managing both natural and man-made hazards. Through their expertise in assessing geological conditions and recommending engineering solutions, they contribute to the safety and resilience of infrastructure and communities. By prioritizing geohazards mitigation in planning and development processes, we can create safer environments that are better equipped to handle the challenges posed by geological hazards.

Rare Earth Elements

Rare Earth Elements (REEs) are critical components in modern technologies, playing a vital role in various applications ranging from electronics and renewable energy to defense technologies. These elements, which include the 15 lanthanides along with scandium and yttrium, are essential for the production of high-performance magnets, catalysts, phosphors, and batteries. However, the supply chains for REEs are increasingly vulnerable to disruption, primarily due to geopolitical factors, environmental concerns, and the concentration of production in a few countries.

Vulnerability of Supply Chains

The potential risk of supply shortages underscores the need for increased efforts in searching for new reserves and developing alternative sources of REEs. Currently, the world's known resources (endowment) of REEs are likely to increase over time, with significant deposits primarily located in the following countries, in descending order: China, Russia, the United States, India, and Australia. China dominates the global supply of REEs, accounting for a substantial portion of production and processing. This concentration raises concerns about supply security, particularly as demand for REEs continues to grow.

Growing Demand and Sustainable Management

As demand for REEs continues to rise, particularly in sectors such as electronics, renewable energy, and defense technologies, the exploration and sustainable management of these resources become increasingly important. The transition to green technologies, such as electric vehicles and wind turbines, relies heavily on REEs, making their availability crucial for achieving sustainability goals. Therefore, it is essential to develop strategies for responsible mining, recycling, and efficient use of these materials to mitigate environmental impacts and ensure long-term supply.

The Role of Applied Geology

In conclusion, applied geology plays a crucial role in addressing various infrastructural concerns, from ensuring the stability of engineering structures to managing mineral resources and mitigating geohazards. The ongoing exploration for rare earth elements further highlights the importance of geological expertise in securing essential materials for future technological advancements. Geologists are instrumental in identifying and evaluating potential REE deposits, assessing their economic viability, and developing sustainable extraction methods.

By integrating geological knowledge into infrastructure planning and resource management, we can promote sustainable development and enhance the resilience of our communities. This includes not only the responsible extraction of REEs but also the implementation of recycling programs and the development of alternative materials to reduce dependency on rare earth elements. As we navigate the challenges of resource management in a rapidly changing world, the expertise of applied geologists will be essential in ensuring that we meet the demands of modern technology while safeguarding the environment for future generations.

Geological Formation and Occurrence: Rare Earth Elements (REEs) are formed through various geological processes, including magmatic, hydrothermal, and sedimentary activities. Magmatic processes occur when molten rock cools and crystallizes, leading to the formation of REE-rich minerals. Hydrothermal processes involve the circulation of hot, mineral-laden fluids that can deposit REEs in veins or fractures. Sedimentary processes can concentrate REEs in certain environments, such as riverbeds. Notable types of REE deposits include carbonatites, which are igneous rocks rich in carbonate minerals, and ion-adsorption clays, primarily found in weathered granitic terrains, both of which are significant sources of REEs.

Environmental Impact of REE Mining: The extraction of REEs poses several environmental challenges, including habitat destruction, water pollution, and the generation of radioactive waste. Mining activities can disrupt local ecosystems and lead to soil erosion and deforestation. Additionally, the chemicals used in the extraction process can contaminate water sources, affecting both wildlife and human populations. To mitigate these impacts, it is crucial to implement environmentally responsible mining practices, such as using less harmful extraction methods, rehabilitating mined areas, and adhering to strict environmental regulations.

Recycling and Circular Economy: Recycling plays a vital role in the sustainable management of REEs, particularly as demand for these elements continues to rise. Initiatives aimed at recovering REEs from electronic waste, batteries, and other products can significantly reduce the need for new mining operations. Technologies such as hydrometallurgical processes and advanced separation techniques are being developed to enhance the efficiency of REE recovery. By promoting a circular economy, where materials are reused and recycled, we can minimize waste and lessen the environmental footprint associated with REE extraction.

Technological Innovations: Advancements in technology are crucial for reducing reliance on REEs and improving their efficiency in applications. Research into alternative materials that can replace REEs in certain applications is ongoing, with promising results in fields such as magnet manufacturing and battery technology. Additionally, innovations in manufacturing processes can enhance the efficiency of REE use, reducing the overall quantity required for specific applications. These technological developments are essential for ensuring the sustainability of industries that depend on REEs.

Geopolitical Considerations: The geopolitical landscape surrounding REE supply chains is complex, with significant implications for global trade and national security. Countries that dominate REE production, such as China, hold considerable leverage in the market, leading to trade tensions and export restrictions. Securing a stable supply of REEs is strategically important for nations reliant on these materials for advanced technologies and defense applications. As a result, countries are increasingly seeking to diversify their sources of REEs and invest in domestic production capabilities.

Future Trends and Research Directions: Looking ahead, the demand for REEs is expected to grow, driven by advancements in technology and the transition to renewable energy. Ongoing research in geology and materials science aims to identify new sources of REEs, particularly in under-explored regions. Additionally, the development of synthetic alternatives to REEs could reshape the market landscape. As industries evolve and the need for sustainable practices intensifies, the exploration of innovative solutions will be critical in addressing the challenges associated with REE supply and management.

The Role of Applied Geology in Mineral and Rock Identification

Applied geology plays a fundamental role in the identification and detailed description of minerals and rocks, which is crucial for understanding Earth's materials and processes. The process begins with field observations, where students and professionals examine geological formations and collect hand specimens. This initial exploration is often followed by laboratory analysis, allowing for a comprehensive understanding of the physical and chemical properties of the materials. This hands-on approach not only reinforces theoretical knowledge but also enhances practical skills essential for geologists.

To aid in this learning process, illustrative figures of field occurrences are provided, enabling students to visualize real-life scenarios that are prevalent in various geological settings. These visual aids help bridge the gap between theory and practice, allowing students to grasp the principles of applied geology more effectively and understand how these principles manifest in the field.

A laboratory manual dedicated to mineral and rock identification serves as an essential resource for students, providing them with the foundational skills necessary for engaging in geological studies effectively. The manual is designed to equip students with specific competencies critical for their academic and professional development in geology. By engaging with the manual, students enhance their technical skills and develop a deeper appreciation for the complexities of mineralogy and petrology.

One of the primary competencies outlined in the manual is the demonstration of an understanding of standard laboratory practices. Students learn to conduct identification and description procedures on hand specimens of typical minerals and rocks, adhering to established laboratory protocols. This foundational knowledge is crucial, as it ensures that students are familiar with safety measures, equipment, and methodologies used in geological laboratories. Mastering these practices prepares students to conduct their own research confidently and accurately, contributing to the scientific community.

Another key competency involves the ability to create labeled sketches of typical mineral and rock samples. Students are encouraged to produce detailed sketches that depict the physical characteristics of the specimens they study, including color, luster, texture, and crystal form. This skill not only enhances their observational abilities but also improves their capacity to communicate geological information effectively. Labeled sketches serve as valuable tools for documentation and analysis, allowing students to convey complex information in a clear and concise manner. This practice fosters a deeper understanding of the distinguishing features of minerals and rocks, which is essential for accurate identification.

In addition to identification skills, the manual emphasizes the importance of assessing the quality of minerals and rocks based on their physical characteristics. Students learn to evaluate specimens under standard laboratory conditions, considering factors such as hardness, cleavage, and specific gravity. This assessment is crucial for determining the suitability of materials for various applications, including construction, manufacturing, and environmental management. By developing the ability to assess quality, students gain insights into the practical implications of their findings, preparing them for careers in geology, engineering, and related fields.

Finally, the manual encourages students to use laboratory results as a springboard for further studies. After examining minerals or rock samples, students are prompted to recommend additional investigations or field confirmations. This aspect of the manual promotes a continuous learning and exploration mindset, encouraging students to think critically about their findings and consider the broader implications of their research. By fostering curiosity and a desire for deeper understanding, the manual prepares students to engage with the dynamic and evolving nature of geological sciences.

In summary, the laboratory manual for mineral and rock identification is a comprehensive guide that equips students with essential skills and knowledge. Through a focus on standard laboratory practices, effective communication through sketches, quality assessment, and the encouragement of further studies, the manual prepares students to become competent and confident geologists. These competencies not only enhance their academic experience but also lay the groundwork for successful careers in the geosciences, enabling them to contribute meaningfully to the field and address real-world geological challenges.

The Role of Applied Geology in Mineral and Rock Identification

Introduction

Applied geology plays a fundamental role in the identification and detailed description of minerals and rocks, which is crucial for understanding Earth's materials and processes. The process begins with field observations, where students and professionals examine geological formations and collect hand specimens. This initial exploration is often followed by laboratory analysis, allowing for a comprehensive understanding of the physical and chemical properties of the materials. This hands-on approach not only reinforces theoretical knowledge but also enhances practical skills essential for geologists.

Field Observations and Laboratory Analysis

To aid in this learning process, illustrative figures of field occurrences are provided, enabling students to visualize real-life scenarios that are prevalent in various geological settings. These visual aids help bridge the gap between theory and practice, allowing students to grasp the principles of applied geology more effectively and understand how these principles manifest in the field.

Laboratory Manual for Mineral and Rock Identification

A laboratory manual dedicated to mineral and rock identification serves as an essential resource for students, providing them with the foundational skills necessary for engaging in geological studies effectively. The manual is designed to equip students with specific competencies critical for their academic and professional development in geology. By engaging with the manual, students enhance their technical skills and develop a deeper appreciation for the complexities of mineralogy and petrology.

Competencies in Laboratory Practices

One of the primary competencies outlined in the manual is the demonstration of an understanding of standard laboratory practices. Students learn to conduct identification and description procedures on hand specimens of typical minerals and rocks, adhering to established laboratory protocols. This foundational knowledge is crucial, as it ensures that students are familiar with safety measures, equipment, and methodologies used in geological laboratories. Mastering these practices prepares students to conduct their own research confidently and accurately, contributing to the scientific community.

Sketching and Communication Skills

Another key competency involves the ability to create labeled sketches of typical mineral and rock samples. Students are encouraged to produce detailed sketches that depict the physical characteristics of the specimens they study, including color, luster, texture, and crystal form. This skill not only enhances their observational abilities but also improves their capacity to communicate geological information effectively. Labeled sketches serve as valuable tools for documentation and analysis, allowing students to convey complex information in a clear and concise manner. This practice fosters a deeper understanding of the distinguishing features of minerals and rocks, which is essential for accurate identification.

Quality Assessment of Minerals and Rocks

In addition to identification skills, the manual emphasizes the importance of assessing the quality of minerals and rocks based on their physical characteristics. Students learn to evaluate specimens under standard laboratory conditions, considering factors such as hardness, cleavage, and specific gravity. This assessment is crucial for determining the suitability of materials for various applications, including construction, manufacturing, and environmental management. By developing the ability to assess quality, students gain insights into the practical implications of their findings, preparing them for careers in geology, engineering, and related fields.

Encouraging Further Studies

Finally, the manual encourages students to use laboratory results as a springboard for further studies. After examining minerals or rock samples, students are prompted to recommend additional investigations or field confirmations. This aspect of the manual promotes a continuous learning and exploration mindset, encouraging students to think critically about their findings and consider the broader implications of their research. By fostering curiosity and a desire for deeper understanding, the manual prepares students to engage with the dynamic and evolving nature of geological sciences.

Conclusion

In summary, the laboratory manual for mineral and rock identification is a comprehensive guide that equips students with essential skills and knowledge. Through a focus on standard laboratory practices, effective communication through sketches, quality assessment, and the encouragement of further studies, the manual prepares students to become competent and confident geologists. These competencies not only enhance their academic experience but also lay the groundwork for successful careers in the geosciences, enabling them to contribute meaningfully to the field and address real-world geological challenges.

Economic Considerations in REE Development

The development of Rare Earth Element (REE) mines involves substantial economic considerations, primarily due to the high costs associated with extraction and the establishment of necessary facilities (Figure 2). The expenses related to REE extraction are significant, reflecting the complexity and technical challenges inherent in mining and processing these critical materials. Establishing extraction facilities requires advanced technologies and infrastructure, which can lead to considerable initial capital investments. These facilities must be equipped to handle the unique characteristics of REEs, including their often low concentrations in ore and the need for specialized processing techniques to separate and purify the elements. Additionally, the environmental regulations and safety standards that govern mining operations can further increase operational costs, as companies must invest in sustainable practices and technologies to minimize their ecological footprint.

Moreover, fluctuations in global demand for REEs can impact the economic viability of mining projects. As industries such as electronics, renewable energy, and defense technologies increasingly rely on REEs, the demand can drive prices up, making new projects more attractive. However, market volatility can also pose risks, as sudden drops in demand or changes in trade policies can affect profitability.

In summary, while the potential for profit in REE mining is significant, the high costs of extraction, facility establishment, and compliance with environmental regulations must be carefully managed. Companies must conduct thorough economic analyses to assess the feasibility of REE projects, considering both the initial investments and the long-term market dynamics that influence the profitability of these essential materials.

Figure 2: Screens for Cell Phones, Laptops, and Television Sets (A), Light Bulbs (B), and Wind-Powered Turbines (C) (Adopted after USGS, 2014)

The demand for Rare Earth Elements (REEs) in modern technologies, such as electronics and renewable energy systems, highlights the necessity for efficient and sustainable mining practices. As global reliance on these materials continues to grow, the need for skilled professionals in applied geology becomes increasingly critical. These experts play a vital role in ensuring that extraction processes are not only economically viable but also environmentally responsible, balancing the demands of industry with the need to protect natural ecosystems.

In conclusion, applied geology is essential for more than just the identification and description of minerals and rocks; it is also crucial for addressing the economic and environmental challenges associated with resource extraction.

By equipping students with the necessary skills and knowledge, we can prepare the next generation of geologists to contribute effectively to the sustainable management of natural resources. This preparation is particularly important for the vital Rare Earth Elements that underpin modern technological advancements, ensuring that their extraction and use align with principles of sustainability and environmental stewardship. Through this focus on education and responsible practices, we can support the ongoing development of technologies that benefit society while safeguarding the planet's resources for future generations.

Applied Geology plays a fundamental role in the identification and detailed description of minerals and rocks.

Applied geology plays a fundamental role in the identification and detailed description of minerals and rocks. The process begins with observing field scenarios and examining hand specimens, ideally in laboratory settings following field excursions. To enhance students' understanding of the practical aspects of applied geology, selected illustrations of field occurrences are provided. These visual aids enable students to visualize real-life scenarios that are prevalent in various geological locations, thereby deepening their comprehension of applied geology principles and their practical applications in mineral and rock identification (Rop, Wangari, and Namwiba, 2022).

A laboratory manual designed for the identification and description of minerals and rocks aims to equip students with specific competencies essential for their geological studies (Figure 3). These competencies include:

1. **Understanding Standard Laboratory Practices**: Students will demonstrate an understanding of standard laboratory procedures for identifying and describing hand specimens of typical minerals and rocks. This foundational knowledge is crucial for ensuring safety and accuracy in laboratory work.

2. **Illustrating Labeled Sketches**: Students will create labeled sketches of typical mineral and rock samples, depicting their physical characteristics. This skill is vital for identifying and distinguishing between different minerals and rocks, enhancing observational and communication abilities.

3. **Assessing Quality**: Students will learn to assess the quality of minerals and rocks based on their physical characteristics, as determined under standard laboratory conditions. This assessment is important for evaluating the suitability of materials for various applications.

4. **Recommending Further Studies**: Finally, students will use laboratory results to recommend further studies for minerals or rock samples already examined. This encourages critical thinking and a proactive approach to geological research, prompting students to consider additional investigations or field confirmations.

By developing these competencies, students are better prepared to engage in geological studies and contribute to the field of applied geology effectively.

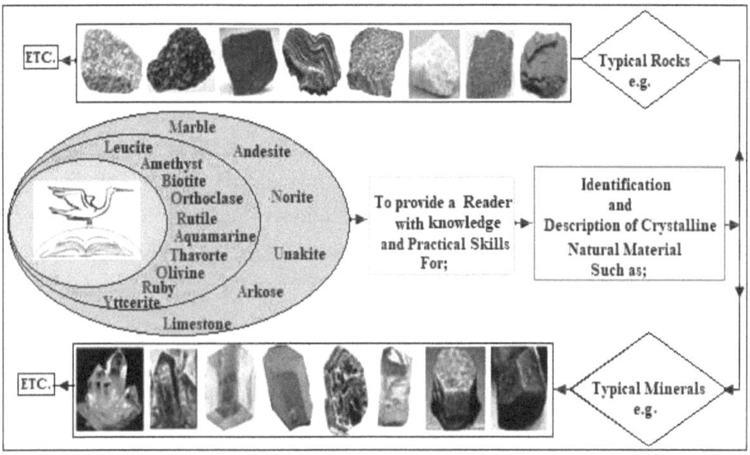

Figure 3: Showing typical laboratory illustrative communications of minerals and rocks identification, descriptions and uses (Rop, et al. 2023).

Environmental Impacts on the Role of Applied Geology

Introduction

Applied geology plays a crucial role in understanding and mitigating the environmental impacts associated with various geological processes and human activities. The intersection of applied geology with environmental concerns is evident in several key areas, each highlighting the importance of geological expertise in addressing contemporary challenges.

Resource Extraction and Management

One of the most significant areas where applied geology intersects with environmental issues is in resource extraction and management. Mining activities, including the extraction of metals and rare earth elements, can lead to substantial environmental degradation. Applied geologists assess geological conditions to minimize impacts such as habitat destruction, soil erosion, and water contamination. By recommending sustainable mining practices and rehabilitation strategies, they help restore ecosystems post-extraction, ensuring that the long-term health of the environment is prioritized alongside economic interests.

Integration of Robotics in Mining

The integration of robotics in mining operations is an emerging trend that can further enhance these efforts. Robotic technologies can be employed for remote sensing, automated drilling, and monitoring environmental conditions, reducing the need for human presence in potentially hazardous areas. This not only improves safety but also allows for more precise and efficient extraction methods, minimizing the environmental footprint of mining activities.

Hydrocarbon Exploration

Similarly, hydrocarbon exploration poses risks such as spills, groundwater contamination, and habitat disruption. Applied geologists evaluate geological formations to identify potential risks associated with oil and gas extraction, developing strategies to mitigate environmental impacts and protect surrounding ecosystems.

Geohazards Assessment and Mitigation

Geohazards assessment and mitigation is another critical area where applied geology contributes to environmental protection. Applied geologists study slope stability and soil composition to predict and mitigate landslide risks, providing recommendations for land use planning and engineering solutions that stabilize slopes and prevent erosion. This work is essential for protecting both human infrastructure and natural habitats from the destructive effects of landslides. Additionally, understanding the geological factors that contribute to seismic activity is vital for developing building codes and land-use policies in earthquake-prone areas. By assessing fault lines and ground conditions, applied geologists inform construction practices that enhance resilience to earthquakes, ultimately safeguarding communities and reducing potential damage.

Water Resource Management

Water resource management is another domain where applied geology plays a vital role. Groundwater studies are essential for assessing aquifer recharge and contamination risks. Geologists analyze the geological formations that influence groundwater flow and quality, helping to manage water resources sustainably and protect them from pollution. This is particularly important in regions where groundwater serves as a primary source of drinking water. Furthermore, understanding the geology of river basins and watersheds is crucial for managing surface water resources. Applied geologists analyze sediment transport, flood risks, and water quality to inform effective management strategies that ensure the health of aquatic ecosystems and the availability of clean water for communities.

Climate Change Adaptation

As climate change continues to impact natural systems, applied geology is increasingly important for climate change adaptation. Coastal geology, for instance, involves studying coastal processes and sediment dynamics to assess the impacts of sea-level rise and coastal erosion. Insights from applied geologists inform coastal management practices and the design of infrastructure that can withstand changing environmental conditions. Additionally, as climate change affects weather patterns and natural hazards, applied geologists provide critical information for land use planning. By assessing geological risks and recommending sustainable practices, they help communities adapt to future environmental changes and enhance their resilience.

Pollution and Remediation Efforts

Pollution and remediation efforts also benefit significantly from the expertise of applied geologists. They play a key role in identifying and assessing contaminated sites, such as former industrial areas or landfills. Through soil and groundwater investigations, applied geologists determine the extent of contamination and develop remediation strategies to restore affected environments. This work is essential for protecting public health and ensuring that contaminated sites are safely rehabilitated. Moreover, understanding geological formations is crucial for the safe disposal of hazardous waste. Applied geologists evaluate sites for landfills and waste storage to minimize environmental risks and ensure compliance with regulations, thereby safeguarding ecosystems and communities from potential harm.

Conclusion

In conclusion, the role of applied geology is integral to addressing environmental impacts associated with human activities and natural processes. By providing essential insights into geological conditions and processes, applied geologists contribute to sustainable resource management, hazard mitigation, and environmental protection. Their work is vital for ensuring that development projects are conducted responsibly, balancing economic needs with the preservation of ecosystems and public health. As environmental challenges continue to evolve, the expertise of applied geologists will be increasingly important in fostering resilience and sustainability in our communities. The incorporation of advanced technologies, including robotics, will further enhance their ability to navigate the complexities of environmental stewardship and ensure a sustainable future for generations to come.

2.3 Methodologies in Applied Geology

Introduction

Applied geologists employ a variety of methodologies to gather data and analyze geological conditions, ensuring that their findings are accurate and relevant for practical applications. Field investigations are a critical component of applied geology, providing firsthand observations and data collection essential for understanding geological conditions.

Field Investigations

Key techniques involved in field investigations include site visits, soil sampling, rock core drilling, and geophysical surveys.

- **Site Visits:** During site visits, geologists observe geological features, assess environmental conditions, and identify potential hazards, allowing them to gather qualitative data that informs their analyses.
- **Soil Sampling:** Soil sampling involves collecting samples from various depths and locations to analyze their composition, structure, and properties, which is crucial for understanding soil behavior and contamination levels.
- **Rock Core Drilling:** Rock core drilling extracts cylindrical samples of rock from the subsurface, providing valuable information about geological strata, including rock types and structural features.
- **Geophysical Surveys:** Geophysical surveys, such as seismic and resistivity methods, investigate subsurface conditions without direct excavation, helping to identify geological structures and potential mineral deposits.

Integration of Robotics

In recent years, the integration of robotics into field investigations has emerged as a transformative approach. Robotic systems can be deployed for remote sensing, automated sampling, and monitoring geological conditions in hazardous or hard-to-reach areas. For instance, drones equipped with sensors can conduct aerial surveys, capturing high-resolution images and data that enhance traditional field methods. This technology not only improves safety by reducing the need for human presence in dangerous environments but also increases the efficiency and accuracy of data collection.

Laboratory Analysis

Once samples are collected, they are analyzed in laboratories to determine their physical and chemical properties. Common laboratory tests include:

- **Soil Texture Analysis:** This determines the proportions of sand, silt, and clay in soil samples, influencing water retention and nutrient availability.
- **Mineral Identification:** Techniques such as X-ray diffraction (XRD) and scanning electron microscopy (SEM) help geologists understand the mineral composition of rock and soil samples, which is essential for assessing resource potential and environmental impacts.
- **Contaminant Concentration Assessments:** These measure the levels of pollutants, such as heavy metals or hydrocarbons, in soil and water samples, providing critical information for evaluating environmental health and developing remediation strategies.

Geographic Information Systems (GIS)

Geographic Information Systems (GIS) technology is another powerful tool used in applied geology, allowing geologists to analyze spatial data and create detailed maps. GIS enables the modeling of geological processes, such as erosion and groundwater flow, helping predict future changes and inform land-use planning. It is also used to assess the potential environmental impacts of proposed projects by overlaying geological data with environmental factors, allowing geologists to identify areas at risk and recommend mitigation measures. Additionally, GIS provides a platform for visualizing complex geological data, aiding stakeholders in making informed decisions based on geological assessments.

Remote Sensing

Remote sensing is a valuable technique that involves the use of satellite imagery and aerial photography to gather information about geological features and land use. This method allows geologists to conduct large-scale assessments of geological features and environmental changes over extensive areas, particularly in regions that are difficult to access. By analyzing historical satellite images, geologists can monitor changes in land use and geological features over time, which is essential for understanding the impacts of human activities and natural processes on the environment. Remote sensing data can also be used to create detailed geological maps, highlighting features such as fault lines and mineral deposits, which are crucial for resource management and hazard assessment.

Conclusion

In conclusion, the methodologies employed in applied geology are diverse and multifaceted, combining field investigations, laboratory analyses, GIS technology, remote sensing techniques, and the innovative use of robotics. Together, these approaches enable applied geologists to gather comprehensive data, analyze geological conditions, and provide valuable insights for resource management, environmental protection, and infrastructure development. By integrating these methodologies, applied geologists can address complex geological challenges and contribute to sustainable practices in various sectors. As the field continues to evolve, the adoption of innovative methodologies, including robotics, will be essential for enhancing the effectiveness and relevance of applied geology in addressing contemporary environmental issues.

2.4 Case Studies in Applied Geology

To illustrate the principles and methodologies of applied geology, this section presents several case studies that highlight real-world applications of geological techniques and insights.

Case Study 1: Site Investigation for Urban Development – Hydrogeological and/or Geophysical Survey for Giftmart Supermarket in Kericho, Kenya

Location: Kapkatet Area, Kapkatet Location, Bureti Division, Bureti Sub-County, Kericho County, Kenya

Overview

This case study examines a site investigation conducted for a proposed urban development project in the Kapkatet Area of Kericho County, which included plans for the Giftmart Supermarket. The primary goal of the investigation was to identify a suitable site for drilling a productive borehole that would serve as the sole water source for domestic purposes within the plot. The estimated water demand for the supermarket is approximately 25 cubic meters per day (m^3/day). Given the rapid urbanization in the region, it was essential to conduct a thorough assessment to inform the design and planning processes.

Methodologies

The investigation employed a combination of hydrogeological and geophysical survey techniques to gather data on the subsurface conditions. Hydrogeological assessments focused on understanding groundwater availability, quality, and flow patterns, while geophysical surveys utilized methods such as electrical resistivity and seismic refraction to delineate geological layers and identify potential aquifers. The survey began with a comprehensive review of existing geological maps and previous studies in the area. Fieldwork included drilling boreholes and conducting in-situ tests to collect soil and rock samples. The geophysical surveys were conducted to complement the borehole data, providing a broader understanding of the subsurface geology without extensive excavation.

Findings

The project area is situated in a zone characterized by medium to low groundwater potential. The surrounding boreholes have varying yields ranging from 2.2 to 4.7 cubic meters per hour (m^3/hr), with total depths ranging from 131 meters to 238 meters. Based on the findings, a borehole is recommended to be drilled at Point 4 to a maximum depth of 230 meters to ensure that the deeper aquifer is fully penetrated. The yield of a borehole drilled at the recommended location is expected to fall within the observed range, but careful construction and development will be essential to maximize productivity, efficiency, and longevity. The climate in Kericho is classified as warm and temperate, with colder seasons experiencing higher rainfall. The average temperature is around 14°C, and the average annual rainfall is approximately 1951 mm. The topography of the area is undulating, as it is located on the slopes of the Mau Escarpment, a significant geological feature formed by the East African Rift System. Geologically, the project area is underlain by superficial deposits of volcanic soils, which are found above layers of vitric pumice tuffs, ignimbrites, and welded tuffs. Groundwater is typically found at the contact zones between different volcanic lava flows, at old land surfaces, and within weathered and fractured zones of the volcanic rocks.

Implications for Urban Development

The findings from the site investigation provided critical insights for the urban development project. The identification of groundwater resources allowed for the planning of a sustainable water supply system for the supermarket.

Additionally, the assessment of geological hazards, such as potential landslides or subsidence, informed the design of the building's foundation and overall site layout. The study also highlighted the importance of integrating geological assessments into urban planning processes. As urbanization continues to accelerate in regions like Kericho County, understanding the geological context is essential for mitigating risks and ensuring the long-term sustainability of development projects. Furthermore, the investigation considered potential environmental impacts, such as the effects of groundwater extraction on local ecosystems and the need for responsible water management practices to prevent depletion of aquifers.

Environmental Impacts

The investigation also took into account potential environmental impacts associated with the urban development project. Key considerations included the effects of groundwater extraction on local ecosystems, particularly in relation to the Mau Forest and surrounding areas. Sustainable water management practices were recommended to prevent over-extraction of groundwater, which could lead to depletion of aquifers and negatively affect nearby vegetation and wildlife habitats. Additionally, the project aimed to minimize soil erosion and sedimentation in local waterways through effective landscaping and drainage designs. By implementing these measures, the development sought to protect the integrity of the local environment while accommodating urban growth. The assessment of potential geological hazards, such as landslides and flooding, also contributed to environmental protection by ensuring that the site was developed in a manner that mitigated risks to both human safety and ecological health.

Conclusion

In conclusion, the investigation for the Giftmart Supermarket exemplifies the critical role of applied geology in urban development. The methodologies employed—geological mapping, soil testing, and hazard assessment—provided essential data that informed the design and planning of the project. The integration of geological insights not only enhances the safety and sustainability of developments but also contributes to the responsible management of natural resources in rapidly urbanizing areas. Furthermore, the consideration of environmental impacts highlights the importance of sustainable practices in urban planning. This case study serves as a model for future urban development projects, demonstrating how geological investigations can effectively address the challenges posed by rapid urbanization while ensuring the safety and sustainability of new constructions. By prioritizing both geological and environmental considerations, urban planners can create resilient and sustainable communities that thrive in harmony with their natural surroundings.

Case Study 2: Groundwater Contamination Remediation

Location: Industrial Area, Nairobi, Kenya

Overview: This case study analyzes a groundwater contamination incident that occurred near an industrial facility in Nairobi, Kenya. The contamination was primarily attributed to the leakage of hazardous chemicals, including heavy metals and volatile organic compounds (VOCs), from the industrial operations. The incident raised significant public health and environmental concerns, prompting the need for a comprehensive remediation plan to address the contamination and restore groundwater quality.

Methodologies: The remediation process began with a thorough contamination assessment conducted by applied geologists. This involved the installation of monitoring wells strategically placed around the industrial facility to collect groundwater samples. These samples were analyzed in laboratories to determine the concentration of various contaminants, including heavy metals such as lead and cadmium, as well as VOCs like benzene and toluene. The assessment provided critical data on the spatial distribution and concentration levels of the contaminants, which was essential for planning the remediation efforts.

To understand the dynamics of groundwater flow and the potential spread of contaminants, geologists developed hydrogeological models. These models incorporated data on geological formations, hydraulic conductivity, and groundwater recharge rates in the area. By simulating different scenarios, the models helped identify the source of contamination and the areas most at risk of further exposure. This step was crucial for prioritizing remediation efforts and ensuring that resources were allocated effectively.

Based on the findings from the contamination assessment and hydrogeological modeling, a comprehensive remediation plan was designed. The plan included several techniques tailored to the specific contaminants and site conditions. Key remediation strategies proposed included pump-and-treat systems, which involved extracting contaminated groundwater, treating it to remove pollutants, and then reinjecting the treated water back into the aquifer. This approach aimed to lower contaminant concentrations in the groundwater over time. Additionally, bioremediation was employed, utilizing microorganisms to degrade organic contaminants in the groundwater. By introducing specific bacteria that can metabolize VOCs, the remediation process aimed to enhance the natural attenuation of contaminants. For VOCs present in the unsaturated zone, soil vapor extraction was proposed, which involved extracting vapor from the soil to reduce the concentration of contaminants in both the soil and groundwater.

Outcomes: The remediation efforts were successful in significantly reducing contaminant levels in the groundwater. Continuous monitoring was established to assess the effectiveness of the remediation strategies and to ensure that contaminant levels remained below regulatory thresholds. The restoration of the aquifer allowed for the safe use of groundwater for local communities, which had previously been at risk due to contamination. Additionally, the case study highlighted the importance of ongoing monitoring and community engagement in preventing future contamination incidents and ensuring the long-term sustainability of groundwater resources.

In conclusion, this case study exemplifies the critical role of applied geology in addressing groundwater contamination issues. Through systematic assessment, modeling, and the implementation of targeted remediation strategies, the project not only restored groundwater quality but also contributed to the protection of public health and the environment in Nairobi, Kenya.

Case Study 3: Natural Hazard Mitigation

Location: Landslide-Prone Region, Muranga County, Kenya

Overview: This case study focuses on a region in Muranga County, Kenya, that is prone to landslides. In response to the increasing frequency and severity of landslides in the area, applied geological assessments were conducted to inform land-use planning and develop effective hazard mitigation strategies. The goal was to enhance community safety and resilience by understanding the geological factors contributing to landslide risks and implementing measures to mitigate these hazards.

Methodologies: The first step in the assessment involved a comprehensive evaluation of the area's landslide risk. Geologists conducted a landslide risk assessment by analyzing slope stability, soil composition, and historical landslide occurrences to identify patterns and potential triggers for landslides. Field surveys were conducted to gather data on slope angles, vegetation cover, and soil moisture levels. Additionally, remote sensing techniques, such as satellite imagery and aerial photography, were employed to identify vulnerable slopes and monitor changes in land use over time. This multifaceted approach provided a detailed understanding of the factors contributing to landslide susceptibility in the region.

Following the risk assessment, detailed geotechnical investigations were performed to evaluate the physical properties of the soil and rock in the area. Soil samples were collected from various depths and locations, and laboratory testing was conducted to assess parameters such as shear strength, moisture content, and compaction. This information was critical for understanding the geotechnical characteristics of the slopes and identifying the factors that contribute to landslide risk, such as soil cohesion and internal friction angles.

Based on the findings from the risk assessment and geotechnical investigations, geologists developed a range of mitigation strategies aimed at reducing landslide risks. Recommended strategies included slope stabilization techniques, which involved engineering solutions such as retaining walls, terracing, and soil nailing to enhance slope stability and prevent soil movement. Additionally, effective drainage improvements were proposed to manage surface water runoff and reduce soil saturation, a key factor in triggering landslides. This included the installation of culverts, ditches, and drainage channels to divert water away from vulnerable slopes. To limit development in high-risk areas, geologists also recommended implementing land-use regulations that restrict construction on steep slopes and promote sustainable land management practices.

Outcomes: The implementation of the recommended mitigation strategies led to a significant reduction in the risk of landslides in the region. The stabilization of vulnerable slopes and improvements in drainage systems contributed to enhanced safety for local communities. Additionally, community awareness programs were established to educate residents about landslide risks and preparedness measures. These programs provided valuable information on recognizing warning signs of landslides, emergency response protocols, and the importance of responsible land use.

Conclusion: This case study illustrates the diverse applications of applied geology in addressing real-world challenges. Through site investigations, contamination assessments, and hazard mitigation strategies, applied geologists play a vital role in ensuring the safety, sustainability, and resilience of communities. By employing a range of methodologies, they provide critical insights that inform decision-making and promote responsible land use and resource management. The proactive measures taken in Muranga County serve as a model for other regions facing similar geological hazards, highlighting the importance of integrating geological assessments into land-use planning and disaster risk reduction efforts.

2.5 Challenges and Future Directions in Applied Geology

Applied geology is a dynamic field that plays a crucial role in addressing complex geological and environmental issues. As the discipline continues to evolve, it faces several significant challenges that must be navigated to ensure that applied geologists can effectively contribute to sustainable development. Understanding these challenges is essential for shaping the future of the discipline and enhancing its impact on society. Key challenges include data limitations, the need for interdisciplinary collaboration, and the integration of emerging technologies.

Data Limitations: One of the primary challenges in applied geology is the availability and quality of geological data. In many regions, geological maps and datasets may be incomplete, outdated, or not readily accessible. This lack of reliable data can hinder effective decision-making and risk assessment, particularly in areas that are prone to natural hazards, such as landslides, earthquakes, or flooding, or where resource extraction is planned. For instance, without accurate geological information, it becomes difficult to assess the stability of slopes or the potential for groundwater contamination, leading to increased risks for communities and the environment.

To overcome these limitations, ongoing data collection and updating are essential. This includes not only conducting geological surveys but also integrating new findings into existing databases to create a more comprehensive understanding of geological conditions. Enhanced data-sharing platforms and collaborative efforts among geological organizations, governmental agencies, and academic institutions can help improve data availability and accessibility. By fostering a culture of data sharing and collaboration, the field of applied geology can better equip itself to address the challenges posed by incomplete or outdated information.

Interdisciplinary Collaboration: Another significant challenge in applied geology is the need for effective collaboration with other disciplines. Complex environmental issues often require insights from various fields, including environmental science, engineering, urban planning, and public policy. Geological issues are rarely isolated; they intersect with social, economic, and environmental factors that necessitate a multidisciplinary approach. For example, addressing groundwater contamination may require not only geological expertise but also an understanding of public health, regulatory frameworks, and community engagement.

Building strong interdisciplinary teams is crucial for addressing these complex challenges. This involves fostering communication and collaboration among professionals from different fields, ensuring that geological insights are integrated into broader environmental and planning frameworks. By working together, interdisciplinary teams can develop more holistic solutions that consider the multifaceted nature of environmental issues. This collaborative approach can lead to more effective land-use planning, resource management, and disaster risk reduction strategies.

Emerging Technologies: The field of applied geology is also being transformed by advances in technology, including machine learning, artificial intelligence, and improved remote sensing techniques. These technologies offer new opportunities for data analysis, modeling, and visualization, enhancing the capabilities of applied geologists. For instance, machine learning algorithms can analyze large datasets to identify patterns and predict geological phenomena, while remote sensing technologies can provide real-time data on land use changes and environmental conditions.

However, with these advancements come the challenge of staying abreast of technological developments. Future practitioners must be proactive in keeping informed about these innovations and be willing to adapt their methodologies accordingly. This may involve ongoing education and training to incorporate new tools and techniques into their work. By embracing emerging technologies, applied geologists can enhance their analytical capabilities and improve the accuracy and efficiency of their assessments.

In conclusion, the challenges facing applied geology are multifaceted and require a proactive approach to address effectively. By focusing on improving data availability, fostering interdisciplinary collaboration, and embracing emerging technologies, the field can enhance its contributions to sustainable development and environmental management. As applied geologists navigate these challenges, they will play a vital role in shaping a more resilient and sustainable future for communities and ecosystems worldwide.

Applications of AI and Machine Learning in Ore Deposit Exploration

The integration of artificial intelligence (AI) and machine learning (ML) into ore deposit exploration is revolutionizing the mining industry by enhancing the efficiency and accuracy of exploration processes while also addressing environmental impacts. Here are several key applications of these technologies:

Data Integration

AI algorithms excel at integrating diverse datasets, which is crucial in ore deposit exploration. Geological, geophysical, and geochemical information often comes from various sources and formats, making it challenging to analyze comprehensively. AI can synthesize this information into cohesive models that provide a holistic view of potential ore deposits. By combining data from satellite imagery, geological surveys, and historical mining records, AI can help geologists identify correlations and trends that may not be apparent when examining datasets in isolation. This comprehensive modeling enables more informed decision-making and prioritization of exploration efforts, ultimately leading to more sustainable practices by reducing unnecessary exploration activities that could harm the environment.

Pattern Recognition

One of the most powerful applications of machine learning in ore deposit exploration is its ability to recognize complex patterns within large datasets. Traditional exploration methods often rely on human intuition and experience, which can lead to missed opportunities. Machine learning techniques, such as neural networks and clustering algorithms, can analyze vast amounts of geological data to identify subtle indicators of mineralization that may be overlooked by conventional methods. By improving the accuracy of mineral deposit identification, these algorithms can help minimize the environmental footprint of exploration by reducing the need for extensive drilling and land disturbance in areas that are less likely to yield valuable resources.

Predictive Modeling

AI models can significantly enhance predictive modeling in ore deposit exploration. By analyzing existing geological data and identifying similar geological features, AI can predict the likelihood of finding ore deposits in unexplored areas. This predictive capability is particularly valuable in regions where traditional exploration methods have been less effective. Machine learning algorithms can be trained on historical data from known deposits to develop models that forecast the potential for mineralization in new areas. By focusing exploration efforts on the most promising sites, companies can reduce the environmental impacts associated with exploratory drilling and land clearing, thereby promoting more responsible resource management.

Automated Analysis

The automation of data analysis through machine learning is transforming the pace of exploration activities. Large datasets generated from exploration activities can be overwhelming for geologists to analyze manually. Machine learning algorithms can process and analyze these datasets quickly, significantly reducing the time required to identify potential mining sites. This automation allows exploration teams to focus on interpreting results and making strategic decisions rather than getting bogged down in data processing. As a result, the overall efficiency of exploration projects is enhanced, leading to faster and more effective identification of viable ore deposits while minimizing the environmental disruption associated with prolonged exploration activities.

Real-time Monitoring

AI systems enable real-time monitoring of exploration activities, providing immediate insights that can inform exploration strategies. By analyzing data from various sources, such as drilling operations, geophysical surveys, and environmental sensors, AI can detect changes or anomalies as they occur. This capability allows exploration teams to make timely adjustments to their strategies based on new findings, optimizing resource allocation and exploration efforts. For example, if real-time data indicates unexpected geological conditions or environmental impacts, teams can quickly adapt their drilling plans or modify their exploration techniques to better suit the situation, thereby reducing potential harm to surrounding ecosystems.

Risk Assessment

Machine learning can play a crucial role in assessing risks associated with exploration projects. By analyzing historical data on geological hazards, operational challenges, and environmental impacts, AI can help identify potential risks before they become significant issues. This proactive approach enables exploration teams to develop risk mitigation strategies and make informed decisions about where and how to explore. For instance, machine learning models can predict the likelihood of encountering geological hazards, such as landslides or groundwater contamination, allowing teams to implement safety measures and minimize environmental impacts. By understanding these risks, companies can better protect local ecosystems and communities from the adverse effects of mining activities.

Conclusion

The applications of AI and machine learning in ore deposit exploration are transforming the mining industry by enhancing data integration, improving pattern recognition, enabling predictive modeling, automating analysis, facilitating real-time monitoring, and assessing risks. Importantly, these technologies also contribute to minimizing environmental impacts by optimizing exploration processes, reducing unnecessary land disturbance, and promoting responsible resource management. As these technologies continue to evolve, they will play an increasingly vital role in ensuring that mineral exploration is conducted sustainably, balancing economic needs with environmental stewardship. By harnessing the power of AI and machine learning, the mining industry can navigate the complexities of exploration more effectively, paving the way for sustainable and responsible resource extraction that protects both the environment and local communities.

2.6 Summary of Chapter 2

Fundamentals of Applied Geology

In this chapter, we explored the fundamentals of applied geology, including its definition, scope, key principles, and methodologies. We discussed the importance of geological mapping, stratigraphy, structural geology, and hydrogeology in informing practical applications. Various methodologies, including field investigations, laboratory analysis, Geographic Information

Systems (GIS), and remote sensing, were highlighted as essential tools for applied geologists.

Real-World Applications

Case studies provided real-world examples of how these principles and methodologies are applied in practice, demonstrating the relevance of applied geology in addressing contemporary challenges. We examined site investigations for urban development, groundwater contamination remediation, and natural hazard mitigation, showcasing the diverse applications of geological expertise. Each case study illustrated not only the technical aspects of applied geology but also the environmental issues associated with these practices, such as habitat disruption from urbanization, the risks of groundwater contamination, and the importance of assessing natural hazards to protect both human and ecological communities.

Integration of Robotic Technologies

We also discussed the integration of robotic technologies in applied geology, which can enhance data collection and analysis while minimizing environmental impacts. Autonomous robotic systems can be employed for remote sensing, automated sampling, and monitoring geological conditions, particularly in hazardous or hard-to-reach areas. This technological advancement not only improves the efficiency of geological assessments but also contributes to more sustainable practices by reducing the need for invasive fieldwork.

Challenges and Future Directions

Finally, we addressed the challenges and future directions in the field, emphasizing the need for ongoing research, interdisciplinary collaboration, and the integration of emerging technologies in applied geology. Environmental issues, including data limitations, the impacts of resource extraction, and the necessity for effective hazard mitigation strategies, were identified as critical areas requiring attention. As environmental challenges become increasingly complex, the role of applied geologists will be crucial in developing sustainable solutions that balance human needs with environmental protection.

Sustainable Resource Management

The sustainable extraction of minerals and resources is a critical area where applied geology plays a significant role. Responsible mining practices, guided by geological assessments, can minimize habitat destruction, soil erosion, and water pollution. However, poor management can lead to significant environmental degradation, including deforestation, loss of biodiversity, and contamination of soil and water resources. As the demand for minerals and resources continues to grow, applied geologists must advocate for and implement sustainable practices that balance economic needs with environmental stewardship.

Groundwater Protection

Groundwater is a vital resource for drinking water, agriculture, and industry. The methodologies discussed in this chapter, such as groundwater contamination assessments, are critical for protecting these water resources. Effective remediation strategies can restore contaminated aquifers, ensuring safe drinking water and preserving aquatic ecosystems. However, inadequate assessments or remediation efforts can lead to prolonged contamination, affecting both human health and the environment. Future directions in groundwater protection will require enhanced monitoring systems, improved data collection methods, and greater public awareness of the importance of safeguarding this essential resource.

Natural Hazard Mitigation

The application of geological principles in assessing natural hazards, such as landslides and flooding, can significantly reduce environmental impacts. By identifying vulnerable areas and implementing mitigation strategies, applied geologists help protect ecosystems and human settlements from the destructive effects of these hazards. Failure to address these risks can result in catastrophic environmental consequences, including habitat destruction and loss of life. As climate change exacerbates the frequency and intensity of natural hazards, the need for robust hazard assessment and mitigation strategies will become increasingly urgent.

Urban Development and Land Use Planning

The impact of urbanization on geological conditions is profound. Site investigations for urban development are essential for understanding the geological conditions of a site. Properly conducted investigations can lead to sustainable urban planning that minimizes environmental impacts, such as soil erosion, water runoff, and habitat fragmentation. However, inadequate geological assessments can result in poorly planned developments that exacerbate environmental degradation. Future urban planning efforts must prioritize geological assessments to ensure that developments are resilient and sustainable.

Climate Change Adaptation

Applied geology plays a vital role in assessing the impacts of climate change, particularly in coastal areas. By understanding geological processes and landforms, geologists can inform strategies for coastal protection and resilience, helping to mitigate the effects of sea-level rise and erosion. Failure to address these issues can lead to significant environmental loss and increased vulnerability of coastal ecosystems. As climate change continues to pose challenges, applied geologists will need to develop innovative solutions that enhance the resilience of both natural and human systems.

Data Limitations and Environmental Decision-Making

The challenges related to data limitations can have profound environmental implications. Incomplete or outdated geological data can lead to poor decision-making regarding land use, resource management, and hazard mitigation. This can result in environmental degradation, increased vulnerability to natural hazards, and unsustainable resource extraction practices. To address these challenges, the field must prioritize the collection and dissemination of high-quality geological data, ensuring that decision-makers have access to the information they need to make informed choices.

Interdisciplinary Collaboration for Environmental Solutions

The need for interdisciplinary collaboration is crucial for addressing complex environmental challenges. By integrating geological insights with environmental science, engineering, and urban planning, applied geologists can contribute to more effective and sustainable solutions that protect ecosystems and human communities. Collaborative efforts can lead to innovative approaches that consider the interconnectedness of geological, environmental, and social factors, ultimately fostering a more holistic understanding of the challenges at hand.

Conclusion

Future of Applied Geology

In conclusion, the future of applied geology hinges on its ability to adapt to emerging challenges and leverage new technologies while fostering interdisciplinary collaboration. By prioritizing data collection, enhancing public awareness, and advocating for sustainable practices, applied geologists can play a pivotal role in shaping a resilient and sustainable future for both human and ecological communities.

Integration of Geological Insights

As we advance, integrating geological insights into broader environmental and societal frameworks will be essential for addressing the pressing challenges of our time. The incorporation of robotic technologies into geological practices not only enhances efficiency but also minimizes environmental impacts, facilitating more responsible resource management.

Environmental Impacts and Sustainable Practices

The environmental impacts of applied geology are significant and multifaceted, influencing resource management, hazard mitigation, urban development, and climate change adaptation. By employing sound geological principles and methodologies, applied geologists can help mitigate negative environmental impacts while promoting sustainable practices.

Role of AI and Machine Learning

The applications of AI and machine learning in ore deposit exploration are transforming the mining industry by enhancing data integration, improving pattern recognition, enabling predictive modeling, automating analysis, facilitating real-time monitoring, and assessing risks. These technologies contribute to minimizing environmental impacts by optimizing exploration processes, reducing unnecessary land disturbance, and promoting responsible resource management.

Collaboration and Innovation

Ultimately, the collaboration between applied geology and innovative technologies, such as robotics and AI, will be instrumental in navigating the complexities of modern environmental challenges. By embracing these advancements and fostering a culture of sustainability, applied geology can continue to evolve and make meaningful contributions to the stewardship of our planet's resources and the well-being of future generations.

Case Study: The Application of Applied Geology in the Sequestration of CO_2 within Coal Bed Methane (CBM) Formations

Introduction

This study aims to investigate the critical role of geology in the sequestration of CO_2 within coal bed methane (CBM) formations, emphasizing a dual-purpose strategy that enhances methane recovery while reducing greenhouse gas emissions. It will assess the annual capture of approximately 44 million metric tons of CO_2 equivalent from various industrial activities, with over 95% of this captured CO_2 being utilized in enhanced oil and gas recovery (EOR).

Geological Sequestration Process

The research will delve into the process of geological sequestration, which involves injecting captured CO_2 into deep underground formations to utilize their natural properties for long-term containment. This approach not only supports sustainable energy practices but also assists industries in meeting global decarbonization targets.

Adsorption Behavior of Coal Seams

Additionally, the study will explore the unique adsorption behavior of coal seams, which effectively trap CO_2 within their structure. This property not only facilitates CO_2 storage but also displaces methane, making it commercially recoverable. The higher affinity of coal for CO_2 compared to methane allows for the storage of double the amount of CO_2, forming the foundation for Enhanced Coal Bed Methane (ECBM) recovery—a technique that serves both carbon sequestration and methane extraction purposes.

Importance of Geological Sequestration

Moreover, the research will highlight the importance of geological sequestration as a key component of carbon capture and storage (CCS) technologies, aimed at mitigating greenhouse gas emissions from major industrial sources, such as coal-fired power plants. The effectiveness of CO_2 storage will be evaluated based on the porosity, permeability, and integrity of geological formations, including the presence of robust cap rocks to prevent leakage. Various geological storage options, such as deep saline formations, enhanced oil recovery (EOR), and depleted oil and gas fields, will be analyzed for their capacity to securely contain CO_2 over extended periods (Figure 4).

Conclusion

Ultimately, this study seeks to demonstrate that integrating CCS technologies into the geological sequestration of CO_2 in CBM formations is essential for addressing climate change and promoting environmental stewardship, thereby contributing to a more sustainable energy future.

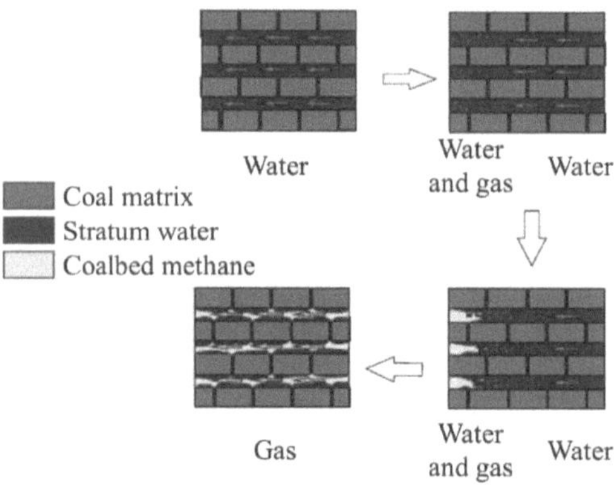

Figure 4: Sample of Coal Seam Formations

Evaluation of CBM Reservoirs

Coal bed methane (CBM) reservoirs are distinct from conventional gas reservoirs, as they function both as the source and storage rock for methane. Efficient extraction of methane from these reservoirs necessitates a thorough understanding of coal's unique dual-porosity system, which consists of (Figure 5):

1. **Macropores (Cleats):** These are the primary conduits for fluid flow within the coal seam. Cleats facilitate the movement of gas and water, playing a crucial role in the extraction process.

2. **Micropores (Matrix):** This is where gas is predominantly stored through adsorption. The micropores provide a significant surface area for methane to adhere to the coal, allowing for effective storage until it is released during extraction.

Understanding the interplay between these two pore systems is essential for optimizing methane recovery from CBM reservoirs and enhancing overall production efficiency.

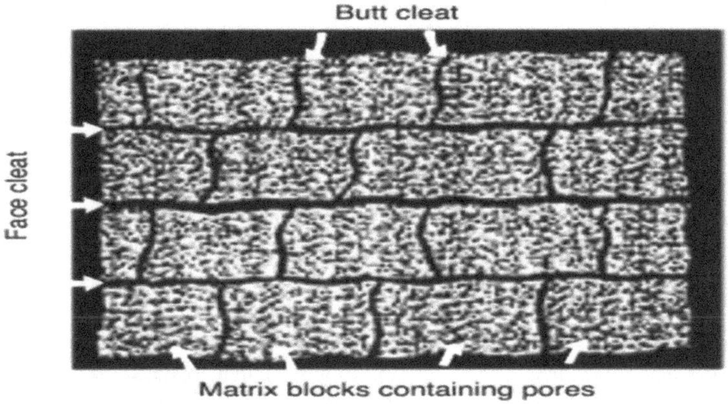

Figure 5: Coal Porosity System

Reservoir engineers must consider several key attributes when evaluating CBM reservoirs to model mechanisms for efficient methane production:

- **Gas Content:** The total amount of methane present in the coal seam.
- **Adsorption Isotherm:** This describes the gas storage capacity relative to pressure, providing insights into how gas is retained within the coal matrix.
- **Permeability Anisotropy:** Variations in permeability across different directions within the reservoir, which can affect fluid flow and extraction efficiency.
- **Relative Permeability:** The relationship between gas and water flow in the reservoir, crucial for understanding how these fluids interact during production.
- **Reservoir Pressure:** This influences gas desorption and overall production rates, making it a critical factor in reservoir management.

The relationship between gas storage capacity and pressure is defined by a sorption isotherm, which is influenced by factors such as coal rank, temperature, and moisture content. The sorption isotherm predicts gas release as reservoir pressure decreases and is essential for estimating recovery factors using desorption data.

CO_2 Storage and Enhanced CBM Recovery (ECBM)

In coal, gas can exist as free gas within the macropores or as an adsorbed layer on the internal surfaces of the micropores. The majority of the gas in coal is stored through adsorption in the coal matrix. Consequently, the pressure-volume relationship is defined by the sorption isotherm, as illustrated in Figure 6. This relationship is critical for understanding how CO_2 can be effectively stored and utilized in Enhanced Coal Bed Methane (ECBM) recovery, where CO_2 injection not only aids in methane extraction but also contributes to carbon sequestration.

The storage capacity of a CBM reservoir can be reliably estimated by analyzing the gas desorption behavior in relation to gas content. If the initial reservoir conditions, including pressure, gas content, desorption isotherm, and abandonment pressure, are known, the recovery factor at the economic limit can be estimated using the following equation:

$$R_f = \frac{G_C - G_{s_a}}{G_C}$$

Where Gc – gas storage capacity at initial pressure

G_{s_a} – gas storage capacity at abandonment pressure

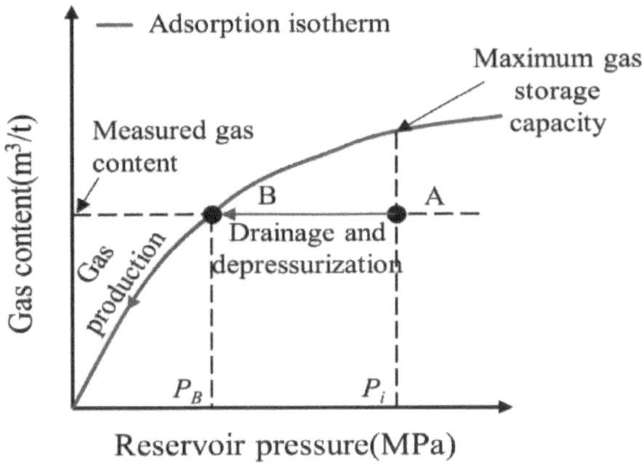

Figure 6 : Sorption Isotherm

From the graph:

The **red curve** represents the maximum gas storage capacity at varying pressures. As pressure increases, the amount of gas adsorbed by the coal or shale also increases, reaching the maximum gas storage capacity. The **measured gas content** reflects the actual amount of gas present in the reservoir under current conditions.

The **blue line**, extending from **Point A** to **Point B**, illustrates the process of gas production through depressurization and drainage. As pressure in the reservoir decreases due to gas extraction, the release of adsorbed gas from the formation occurs. **Point A** signifies the initial reservoir pressure (**P_1**) and the maximum gas storage capacity, while **Point B** indicates the pressure after production (**P_B**), which corresponds to a reduced gas content.

Applicability of CO_2 Storage in Enhanced CBM Recovery

Conventional CBM recovery is achieved through reservoir pressure depletion. This reduction in pressure leads to methane desorption, which follows the sorption isotherm. Notably, the sorption isotherm is non-linear and skewed towards lower pressures, indicating that a significant amount of methane can be released as pressure decreases (Figure 7).

In the context of Enhanced Coal Bed Methane (ECBM) recovery, the injection of CO_2 can further enhance methane production. By introducing CO_2 into the reservoir, the pressure is maintained or increased, which can help to displace adsorbed methane and promote its release. This process not only aids in methane recovery but also facilitates the storage of CO_2, contributing to carbon sequestration efforts.

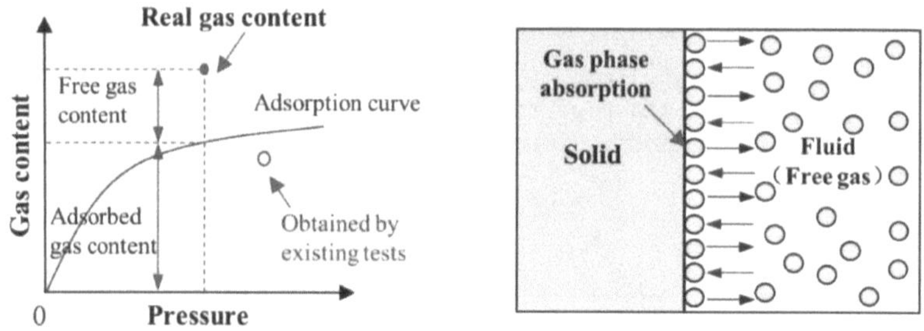

(a) Gas phase adsorption curve **(b) physical model**

Figure 7: CO_2 adsorption curve and physical model

Gas Phase Adsorption Curve Interpretation and Enhanced CBM Recovery

The adsorption curve (blue line) illustrates the relationship between pressure and the amount of gas adsorbed onto a solid surface, following an asymptotic trend that indicates maximum adsorption capacity at higher pressures. The adsorbed gas content increases with pressure until saturation is reached, while free gas content refers to gas present in fluid or pore spaces. The combination of adsorbed and free gas yields the Real Gas Content.

Point **0** represents the gas content measurable through laboratory tests or field sampling. The physical model shows free gas in pore spaces or fractures, which is mobile and not chemically bonded to the solid. The gas phase absorption process involves gas molecules moving from the fluid phase to the solid surface, as indicated by purple arrows, with yellow circles representing gas molecules distinguishing between adsorbed and free gas.

However, a reduction in reservoir pressure diminishes gas production and significantly decreases coal permeability, which is stress-dependent. These challenges have prompted research into Enhanced CBM recovery (ECBM) and CO_2 sequestration in CBM reservoirs.

ECBM has been explored using two methods: nitrogen injection and CO_2 injection. The principle behind ECBM recovery is that the amount of methane stored in coal depends on its partial pressure. Injecting an inert gas like nitrogen lowers methane desorption without reducing reservoir pressure, allowing for methane liberation as nitrogen passes through the reservoir. In contrast, CO_2 flooding reduces methane's partial pressure and adsorbs into the coal, leading to competitive sorption. Since coal seams have a higher sorption capacity for CO_2 than for methane, injected CO_2 molecules replace adsorbed methane, facilitating its release.

Environmental Impacts of CO_2 Sequestration

CO_2 sequestration in coal bed methane formations can help mitigate greenhouse gas emissions by storing carbon dioxide in unminable coal beds, which also enhances methane recovery. However, potential environmental impacts include leakage risks, induced seismic activity, and the need for long-term monitoring to ensure storage integrity.

CO_2 Sequestration Benefits

- Geological sequestration of CO_2 in coal seams can significantly contribute to greenhouse gas mitigation.
- This process not only aids in reducing atmospheric CO_2 levels but also enhances the recovery of coalbed methane, providing dual benefits for energy production.

Potential Environmental Concerns

- **Leakage Risks**: There is a possibility of CO_2 leaking back to the surface, which could negate the benefits of sequestration and pose risks to local ecosystems.
- **Induced Seismic Activity**: The injection of CO_2 into geological formations can potentially trigger seismic events, raising concerns about the stability of the surrounding area.
- **Long-term Monitoring**: Continuous monitoring is essential to verify the integrity of CO_2 storage sites and to ensure that no harmful emissions occur over time.

Mitigation Strategies

- Implementing robust monitoring systems can help detect any leakage early and assess the effectiveness of CO_2 storage.
- Developing best practices for CO_2 injection can minimize the risk of induced seismicity and ensure safe operations.
- Engaging with local communities and stakeholders can enhance transparency and address environmental concerns related to CO_2 sequestration projects.

Conclusion of the Case Study on the Application of Applied Geology in the Sequestration of CO_2 within Coal Bed Methane (CBM) Formations

Summary of Findings

This case study highlights the critical role of applied geology in the effective sequestration of CO_2 within coal bed methane (CBM) formations. The unique geological characteristics of CBM reservoirs, including their dual-porosity system and the interplay between adsorbed and free gas, are essential for optimizing methane recovery and enhancing CO_2 storage.

Influence of Gas Phase Adsorption

The analysis of gas phase adsorption behavior demonstrates that pressure dynamics significantly influence gas desorption and reservoir performance. Enhanced CBM recovery methods, such as nitrogen and CO_2 injection, leverage these geological properties to improve methane extraction while simultaneously facilitating CO_2 sequestration. CO_2 injection not only reduces the partial pressure of methane but also utilizes competitive sorption to enhance methane release, showcasing a sustainable approach to managing greenhouse gas emissions.

Pathway for Climate Change Mitigation

Overall, the integration of geological principles in the design and implementation of CO_2 sequestration strategies within CBM formations presents a promising pathway for addressing climate change while maximizing energy resources. This case study underscores the importance of continued research and innovation in applied geology to advance carbon capture and storage technologies in the context of coal bed methane reservoirs.

CHAPTER 3

GEOLOGICAL HAZARDS AND RISKS

3.0 Introduction to Geological Hazards and Risks

Deformational structures such as faults and folds play a crucial role in shaping the Earth's geology and have significant impacts as major causes of geohazard risks. Faults can lead to the formation of significant topographical features and are indicative of tectonic processes, often resulting in earthquakes, landslides, and other geological hazards that pose risks to human safety and infrastructure. Similarly, folds represent the complex interactions of stress and strain in the Earth's crust, which can also contribute to geohazards, particularly in regions where folded rock layers are subjected to erosion or instability. Understanding these structures is essential for geologists in interpreting geological history, assessing natural resources, and mitigating the risks associated with geohazards. By studying faults and folds, scientists can better predict geological events and develop strategies to protect communities and the environment from their potentially devastating effects.

3.1 Definition of Geological Hazards

Geological hazards are natural events that arise from geological processes, such as tectonic movements, volcanic activity, and erosion, which pose significant risks to human life, property, and the environment. These hazards can manifest in various forms and can lead to catastrophic consequences, including loss of life, destruction of infrastructure, and long-term environmental degradation. The impact of geological hazards can be profound, affecting not only the immediate area but also having far-reaching effects on regional and even global scales.

The consequences of geological hazards can be devastating. For instance, earthquakes can lead to the collapse of buildings, bridges, and roads, resulting in significant casualties and economic losses. Landslides can bury homes and infrastructure under tons of debris, while volcanic eruptions can release ash and lava flows that devastate surrounding areas. Tsunamis, often triggered by underwater earthquakes, can inundate coastal communities, causing widespread destruction and loss of life. Sinkholes can suddenly appear, swallowing structures and creating hazardous conditions for residents.

Understanding geological hazards is essential for effective risk management and disaster preparedness. By studying the geological processes that lead to these hazards, scientists and engineers can develop models to predict when and where such events are likely to occur. This knowledge is crucial for creating early warning systems, establishing building codes, and implementing land-use planning that minimizes risk.

The main types of geological hazards include:

- **Earthquakes**: Earthquakes are caused by the sudden release of energy in the Earth's crust, typically due to tectonic plate movements along fault lines. The impacts of earthquakes can be devastating, leading to structural damage, loss of life, and secondary hazards such as tsunamis and landslides.

- **Landslides**: Landslides occur when there is a sudden and rapid movement of rock, soil, and debris down a slope. They can be triggered by various factors, including heavy rainfall, earthquakes, volcanic activity, and human activities such as excavation and deforestation. Different types of landslides include rockfalls, debris flows, and rotational slides. The impacts of landslides can be severe, particularly in populated areas, where they can destroy homes, roads, and infrastructure.

- **Volcanic Eruptions**: Volcanic eruptions are caused by the movement of magma from beneath the Earth's crust to the surface. Eruptions can be effusive (lava flows) or explosive (ash clouds, pyroclastic flows). The hazards associated with volcanic activity include lava flows, ash fall, volcanic gases, and pyroclastic flows, which can devastate surrounding areas and pose significant risks to human health and safety.

- **Tsunamis**: Tsunamis are large ocean waves typically caused by underwater earthquakes, volcanic eruptions, or landslides. When these events displace a significant volume of water, they can generate waves that travel across oceans at high speeds. Upon reaching coastal areas, tsunamis can cause widespread flooding and destruction, impacting communities, ecosystems, and infrastructure.

- **Sinkholes**: Sinkholes are depressions or holes in the ground caused by the dissolution of soluble rocks, such as limestone, gypsum, or salt. They can form suddenly and pose risks to buildings, roads, and other infrastructure, particularly in urban areas. Factors contributing to sinkhole formation include heavy rainfall, groundwater withdrawal, and changes in land use.

Figure 8: Deformation structures that cause geohazards

By recognizing the relationship between geological structures and these hazards, we can enhance our ability to predict, prepare for, and respond to geological risks. This understanding allows for the development of effective mitigation strategies, such as reinforcing buildings in earthquake-prone areas, implementing proper drainage systems to reduce landslide risks, and establishing evacuation plans for coastal communities at risk of tsunamis. Ultimately, by prioritizing geological hazard awareness and preparedness, we can safeguard communities and ecosystems, reducing the potential for loss and ensuring a more resilient future in the face of natural disasters.

Causes of Geological Hazards: Deformational Structures

Faults

A geological fault is a planar feature formed by the fracturing (breaking) of a rock mass along which movement accompanied by displacement has occurred (Figure 8). Faults can be classified into two main types: normal faults and reverse faults. A normal fault has its hade (θ_1) directed towards the down throw (D), resulting in lateral separation (heave) of the rock layers. In this case, the hanging wall moves downward relative to the footwall, indicative of extensional tectonics. Conversely, a reverse fault has its hade directed towards the up-throw (U), leading to the overlapping of faulted beds. Here, the hanging wall moves upward relative to the footwall, characteristic of compressional tectonics. Each fault has an inclination to the horizontal known as the angle of dip (θ_2), and it causes a vertical displacement in the affected beds referred to as throw.

Strike-Slip Faults

A strike-slip fault, also known as a transcurrent fault, lateral fault, tear fault, or wrench fault, is characterized by horizontal movement of the faulted beds, with minimal vertical displacement. This type of fault is closely related to plate tectonics, where one tectonic plate moves laterally against another. Notable examples of strike-slip faults include the San Andreas Fault in California and the Anatolian Fault in Turkey. These faults often serve as tectonic boundaries, where significant geological activity occurs, leading to earthquakes and other geological phenomena.

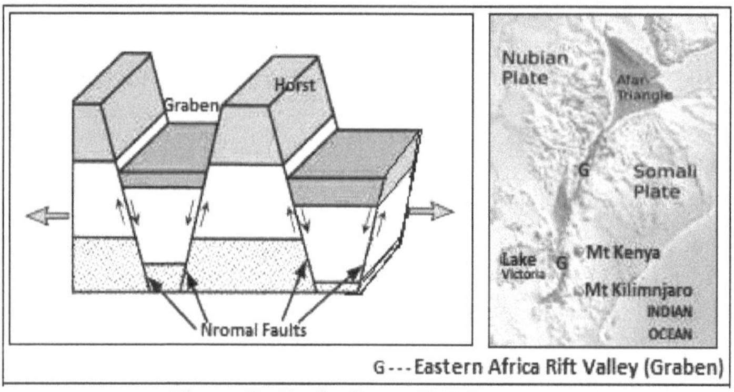

Figure 9: Fault-Controlled Structures

Fault-Controlled Features

Major fault-controlled topographical features include grabens and horsts (Figure 9). These are elongate faulted blocks of stratified rock masses that have been lowered (grabens) or raised (horsts) relative to their surroundings due to normal faulting. Horsts are elevated blocks of rock that have been pushed upward, while grabens are depressed blocks that have been pulled downward. The size of horsts and grabens can range from a few meters to tens of kilometers in width, with vertical movements that can reach several meters or kilometers. Other fault-controlled features include stepped fault scarps, which are prevalent on the floor of the Great Rift Valley, and waterfalls, which are often positioned across active stream channels.

Folds

A geological fold occurs when a planar feature or rock bed becomes bent or curved as a result of permanent deformation. Folds can also form due to the slumping of sedimentary deposits before they become lithified, referred to as syn-sedimentary folds. The two primary types of folds are anticlinal folds and synclinal folds. An anticlinal fold is an upward arching fold where the oldest rock layers are at the core, while a synclinal fold is a downward trough-like fold where the youngest rock layers are at the core. The trace of the fold plane for a syncline illustrates the curvature of the rock layers, highlighting the complex interactions of stress and strain in the Earth's crust.

Summary

Deformational structures such as faults and folds play a crucial role in shaping the Earth's geology and have significant impacts as major causes of geohazard risks. Faults can lead to the formation of significant topographical features and are indicative of tectonic processes, often resulting in earthquakes, landslides, and other geological hazards that pose risks to human safety and infrastructure. Similarly, folds represent the complex interactions of stress and strain in the Earth's crust, which can also contribute to geohazards, particularly in regions where folded rock layers are subjected to erosion or instability. Understanding these structures is essential for geologists in interpreting geological history, assessing natural resources, and mitigating the risks associated with geohazards. By studying faults and folds, scientists can better predict geological events and develop strategies to protect communities and the environment from their potentially devastating effects.

3.2 Environmental Impacts Caused by Geological Hazards

Geological hazards can lead to significant environmental impacts, including habitat destruction, soil erosion, and water contamination. Events like landslides and volcanic eruptions can disrupt ecosystems, while earthquakes may alter landscapes and affect groundwater systems, ultimately threatening biodiversity and natural resources. Additionally, geological hazards can release harmful substances into the environment, such as ash and toxic gases from volcanic eruptions, which can affect air quality and contribute to climate change. The alteration of landforms and drainage patterns due to these events can lead to increased flooding and sedimentation in rivers, further impacting aquatic habitats. The long-term consequences of these hazards can hinder recovery efforts and disrupt the balance of local ecosystems, making it crucial to consider environmental factors in disaster preparedness and response strategies.

3.3 Risk Assessment Methodologies

Effective risk assessment is crucial for understanding and mitigating geological hazards. The first step in this process is **hazard identification**, which involves recognizing geological hazards in a specific area through the analysis of historical data, geological surveys, and field investigations. This foundational step is essential for understanding the types and frequencies of hazards that may affect a region, enabling effective planning and response strategies. By identifying potential hazards, stakeholders can prioritize areas that require further investigation and intervention.

Following hazard identification, the next step is **vulnerability assessment**. This process evaluates the susceptibility of people, infrastructure, and the environment to geological hazards. Factors such as population density, building codes, and the resilience of critical infrastructure are analyzed to determine how vulnerable a community is to specific hazards. Understanding vulnerability helps in tailoring mitigation strategies to protect the most at-risk populations and assets.

Risk analysis combines the insights gained from hazard identification and vulnerability assessments to quantify the likelihood and potential impacts of geological hazards. This analysis can involve both quantitative methods, such as statistical modeling, and qualitative methods, which rely on expert judgment. By quantifying risk, decision-makers can prioritize resources and interventions based on the severity and likelihood of potential hazards.

Finally, **mapping and modeling** play a vital role in visualizing risk areas. Geographic Information Systems (GIS) and remote sensing technologies are employed to create hazard maps and models that illustrate spatial relationships and risk zones. These tools help decision-makers understand the geographical distribution of hazards and prioritize areas for intervention, ultimately enhancing preparedness and response efforts.

3.4 Mitigation Strategies

Mitigation strategies are essential for reducing the impacts of geological hazards. One key strategy is **land use planning**, which involves incorporating geological hazard assessments into zoning regulations and land use policies. By restricting development in high-risk areas or implementing buffer zones around hazardous sites, communities can minimize their exposure to geological threats. This proactive approach helps ensure that future developments are resilient to potential hazards.

Engineering solutions are another critical component of mitigation. Designing infrastructure to withstand geological hazards is vital for protecting communities. This includes constructing earthquake-resistant buildings, implementing retaining walls to stabilize slopes prone to landslides, and developing flood defenses for coastal areas. By investing in resilient infrastructure, communities can significantly reduce the potential damage caused by geological events.

The implementation of **early warning systems** is also crucial for effective hazard mitigation. These systems provide timely alerts for events such as earthquakes, tsunamis, and volcanic eruptions, allowing communities to evacuate or take protective measures. Early warning systems can save lives and reduce property damage by ensuring that residents have the information they need to respond quickly to imminent threats.

Lastly, **public education and preparedness** are vital for enhancing community resilience. Educating residents about geological hazards and promoting disaster preparedness plans can empower communities to respond effectively in emergencies. Public awareness campaigns can inform residents about evacuation routes, emergency kits, and safety protocols, fostering a culture of preparedness that can mitigate the impacts of geological hazards.

3.5 Case Studies of Geological Hazards

Several case studies illustrate the importance of effective geological hazard management. **Case Study 1** focuses on the 2010 Haiti Earthquake, which struck on January 12, resulting in widespread destruction and loss of life. This disaster underscored the critical need for understanding seismic risks and the importance of improved building codes and disaster preparedness in vulnerable regions. The lessons learned from this event have informed subsequent efforts to enhance resilience in earthquake-prone areas.

Case Study 2 examines the 1980 eruption of Mount St. Helens in Washington State, one of the most significant volcanic events in U.S. history. The eruption caused extensive damage to the surrounding area and provided valuable lessons in volcanic monitoring and emergency response. The experiences gained from this event have shaped current practices in volcanic hazard assessment and community preparedness.

Case Study 3 highlights the issue of landslides in the Appalachian region, which is prone to such events due to its steep terrain and geological conditions. This case study explores the factors contributing to landslides in the region and the mitigation efforts implemented to protect communities and infrastructure. By analyzing these factors, stakeholders can develop targeted strategies to reduce the risk of landslides and enhance community safety.

3.6 Future Directions in Geological Hazard Management

Technological Advancements

Looking ahead, several future directions in geological hazard management are emerging. Advancements in technology are playing a pivotal role in enhancing hazard detection, monitoring, and response capabilities. Emerging technologies, such as machine learning and improved remote sensing techniques, can lead to more accurate risk assessments and timely interventions. These innovations have the potential to revolutionize how geological hazards are monitored and managed.

Climate Change Considerations

Climate change considerations are also becoming increasingly important in geological hazard management. Climate change is expected to influence the frequency and intensity of geological hazards, with increased rainfall and extreme weather events exacerbating landslide risks and rising sea levels heightening the impact of tsunamis and coastal flooding. Understanding these interactions is crucial for developing adaptive strategies that consider future climate scenarios and their implications for geological hazards.

Interdisciplinary Approaches

Finally, interdisciplinary approaches are essential for effective geological hazard management. Collaboration among geologists, engineers, urban planners, emergency responders, and policymakers is necessary to create comprehensive risk management plans that address the multifaceted nature of geological hazards. By leveraging diverse expertise, interdisciplinary teams can develop more effective strategies for mitigating risks and enhancing community resilience.

Summary of Methodologies

In summary, the methodologies for risk assessment, mitigation strategies, case studies, and future directions in geological hazard management highlight the critical role of applied geology in understanding and addressing geological hazards. These methodologies not only focus on the technical aspects of hazard identification and risk analysis but also consider the environmental impacts associated with geological events. For instance, effective risk assessment can help identify areas where geological hazards may lead to significant environmental degradation, such as habitat destruction, soil erosion, and water contamination.

Mitigation Strategies

Mitigation strategies, such as land use planning and engineering solutions, are essential for reducing both human and environmental vulnerabilities to geological hazards. By incorporating geological hazard assessments into land use planning, communities can minimize risks while also protecting natural ecosystems from the adverse effects of development. Engineering solutions, such as constructing earthquake-resistant buildings and flood defenses, not only safeguard infrastructure but also help preserve the surrounding environment by reducing the potential for ecological damage during geological events.

Case Studies

The case studies discussed, such as the 2010 Haiti Earthquake and the eruption of Mount St. Helens, illustrate the profound environmental impacts that geological hazards can have on ecosystems and communities. These events underscore the importance of preparedness and response strategies that not only focus on human safety but also consider the long-term health of the environment.

Future Outlook

Looking to the future, advancements in technology and climate change considerations will further shape geological hazard management. As climate change influences the frequency and intensity of geological hazards, understanding these interactions will be crucial for developing adaptive strategies that protect both human populations and the environment. Interdisciplinary approaches that bring together geologists, engineers, urban planners, and environmental scientists will be essential for creating comprehensive risk management plans that address the multifaceted nature of geological hazards and their environmental impacts.

Holistic Approach

By employing a comprehensive approach that integrates risk assessment, engineering solutions, public education, and interdisciplinary collaboration, communities can enhance their resilience to geological threats while also safeguarding the environment. This holistic perspective ensures that both lives and ecosystems are protected, fostering a sustainable future in the face of geological challenges.

3.7 Methods of Prospecting for Minerals

a) Radiometric Method

The radiometric method is a specialized technique primarily employed for prospecting minerals such as uranium and thorium, which are classified as rare earth minerals due to their capacity to emit continuous gamma rays. This method utilizes a sensing device known as a Geiger counter, designed to detect gamma rays within its ionization chamber. When gamma rays are detected, the Geiger counter provides audible feedback, alerting the operator to the presence of radioactive substances. This device quantifies the percentage of radioactivity in rock samples, making it an invaluable tool for identifying potential mineral deposits. In addition to the Geiger counter, fluorescent lamp tubes can be employed to detect fluorescent minerals, such as fluorspar and diamonds, which emit visible light when exposed to ultraviolet radiation.

Environmental Impacts: While the radiometric method is non-invasive and does not directly disturb the environment, the prospecting of radioactive minerals can raise concerns about radiation exposure to workers and nearby communities. Additionally, if mining operations are initiated based on radiometric surveys, there may be significant environmental impacts associated with the extraction and processing of radioactive materials, including potential contamination of soil and water resources.

b) Geochemical Prospecting

Geochemical prospecting is grounded in the principle that any element present in excess quantities compared to the typical composition of a rock type will tend to migrate in an ionic state. Most elements can migrate in various forms and become absorbed in the soil that covers the bedrock. By analyzing soil samples for specific elements, geologists can infer the potential presence of mineral deposits beneath the surface. For instance, if soil samples reveal copper concentrations of 80-90 parts per million (ppm) compared to a normal background level of 2-4 ppm, this indicates a promising area for copper deposits. The results of geochemical analyses are often compiled into geochemical maps, which visually delineate areas of potential mineral concentration.

Environmental Impacts: Geochemical prospecting can lead to soil and water contamination if improper sampling techniques or chemical reagents are used. The use of hazardous chemicals in the analysis process can pose risks to local ecosystems and human health if not managed properly. Additionally, the disturbance of soil during sampling can lead to erosion and habitat disruption.

c) Aerial Survey

Aerial surveys serve as a powerful method for geological mapping, particularly in terrains that are challenging to access by ground. In this approach, aircraft equipped with high-resolution cameras fly at altitudes of approximately 300 meters, capturing detailed images of the landscape below. These aircraft follow a systematic grid pattern, maintaining a fixed speed and automatically taking photographs at regular intervals. The resulting images provide an excellent record of topography, rock type exposures, and structural details.

Environmental Impacts: While aerial surveys are generally non-invasive, they can still have environmental impacts, particularly if low-flying aircraft disturb wildlife or sensitive habitats. Noise pollution from aircraft can also affect local fauna. Additionally, the use of drones for aerial surveys, while more efficient, may raise concerns about privacy and the potential for disturbance to wildlife.

d) Prospecting and Exploration

Objective of Prospecting: The primary aim of prospecting is to identify areas that exhibit signs of economic importance based on various survey methods previously mentioned. This initial assessment is crucial for determining which locations warrant further investigation.

Initial Assessment Techniques: Following the initial assessment, a limited amount of pitting, trenching, and drilling may be conducted to ascertain the extent of mineralization and its grade. These preliminary techniques help geologists gather essential data about the mineral deposit.

Transition to Exploration: This phase is succeeded by detailed prospecting, known as exploration, where geologists strive to obtain a three-dimensional understanding of a deposit. This involves determining both the quantitative and qualitative features of the mineral resource.

Economic Viability Assessment: Economic viability is then assessed from both mining and investment perspectives. Understanding the potential profitability of a mineral deposit is essential for making informed decisions about further development.

Method Selection Based on Deposit Type: The type of mineral deposit and the nature of mineralization will ultimately dictate the methods employed and the evaluation process during exploration. For example, in the case of bedded deposits, it is common practice to create pits at regular intervals or to conduct drilling at intervals of 300 to 500 meters apart. This spacing is generally adequate for obtaining a representative understanding of the deposit's characteristics.

Intensive Drilling for Vein Deposits: Conversely, for vein deposits, such as those containing non-ferrous minerals, a more intensive drilling approach is required, with drill holes spaced at a maximum of 50 to 100 meters apart. This closer spacing is necessary due to the typically more localized nature of vein deposits, which can exhibit significant variability in mineral content over short distances.

Validation of Drilling Data: In both scenarios, the data obtained from drilling are often validated through experimental underground mining. This correlation between surface data and underground results is crucial for ensuring the accuracy of estimates regarding the grade and reserves of the mineral deposit.

Environmental Impacts: The processes of pitting, trenching, and drilling can lead to significant environmental disturbances. These activities can result in soil erosion, habitat destruction, and disruption of local ecosystems. Additionally, the disposal of drill cuttings and other waste materials can lead to soil and water contamination if not managed properly. The use of heavy machinery can also contribute to soil compaction and degradation of the land.

Importance of Accurate Grade Estimation: Accurate estimation of the mineral grade is of paramount importance in exploration, as it directly influences the economic viability of a mining project. To achieve this, systematic and scientific sampling methods are employed, which help to minimize bias and ensure that the samples collected are representative of the overall deposit.

Environmental Impacts: While the focus on accurate grade estimation is essential for economic reasons, it is also important to consider the environmental implications of exploration activities. The extraction of samples can disturb the land and ecosystems, and if the exploration leads to mining, the environmental impacts can be magnified. Sustainable practices must be integrated into exploration to mitigate these effects.

Goals of Mineral Exploration: Ultimately, the goal of mineral exploration is to ascertain the actual tonnage of the mineral resource available within defined dimensions, along with an accurate assessment of its grade. This comprehensive understanding is essential for making informed decisions regarding the feasibility of mining operations and for securing the necessary investments to develop the resource.

Environmental Considerations: As exploration progresses, it is vital to incorporate environmental assessments to evaluate potential impacts on local ecosystems, water resources, and communities. This proactive approach can help identify and mitigate adverse effects before they occur.

Integration of Prospecting Methods: By employing rigorous sampling and drilling techniques, geologists can provide reliable estimates that are critical for the successful exploration and eventual extraction of mineral deposits. The integration of these methods allows for a comprehensive approach to mineral exploration, where each technique complements the others.

Environmental Impacts: The combined use of various prospecting methods can lead to cumulative environmental impacts, particularly if not managed carefully. For instance, the disturbance from drilling and sampling can be exacerbated by the use of aerial surveys and geochemical analyses, leading to broader ecological consequences.

Complementary Techniques: For instance, radiometric surveys can highlight areas of potential radioactivity, which can then be further investigated through geochemical sampling to confirm the presence of specific minerals. Aerial surveys can provide a broad overview of the terrain, helping to identify regions that warrant more detailed ground-based exploration.

Environmental Considerations: While complementary techniques enhance the efficiency of exploration, they also necessitate careful planning to minimize environmental disruption. Each method should be evaluated for its potential impacts, and strategies should be developed to mitigate any negative effects.

Systematic Sampling and Drilling: Moreover, the systematic approach to sampling and drilling ensures that the data collected is not only accurate but also representative of the mineral deposit as a whole. This is particularly important in the context of economic viability, as the correct estimation of mineral grades and reserves directly impacts the feasibility of mining operations.

Environmental Impacts: The systematic nature of these activities can lead to predictable patterns of disturbance, which can be managed through careful planning and environmental monitoring. Implementing best practices in sampling and drilling can help reduce the ecological footprint of exploration activities.

Technological Advancements: As technology continues to advance, the methods of mineral prospecting are likely to evolve, incorporating new tools and techniques that enhance the efficiency and accuracy of exploration efforts. The use of drones for aerial surveys, improved geochemical analysis techniques, and enhanced remote sensing capabilities are just a few examples of how the field is progressing.

Environmental Benefits: Technological advancements can also lead to reduced environmental impacts. For example, drones can minimize the need for ground disturbance, and remote sensing can provide valuable data without direct contact with the land. These innovations can help balance the need for resource exploration with environmental stewardship.

Sustainable Development Goals: Ultimately, the goal of mineral exploration remains the same: to discover and evaluate mineral resources that can be economically extracted, contributing to the sustainable development of natural resources and supporting the needs of society. By employing a combination of traditional and innovative methods, geologists can continue to uncover valuable mineral deposits that play a vital role in various industries, from energy production to technology and beyond.

Environmental Responsibility: As the industry moves forward, it is essential to prioritize sustainable practices that minimize environmental impacts. This includes conducting thorough environmental assessments, engaging with local communities, and implementing strategies to rehabilitate disturbed areas. By doing so, the mineral exploration sector can contribute to a more sustainable future while meeting the growing demand for mineral resources.

Conclusion: In summary, the methods of prospecting for minerals—such as the radiometric method, geochemical prospecting, aerial surveys, and various exploration techniques—each play a crucial role in identifying and evaluating mineral resources. These methods provide valuable insights into the subsurface geology, helping geologists to locate potential deposits and assess their economic viability. However, it is essential to recognize that these activities can have significant environmental impacts, which must be carefully managed.

Integration of Sustainable Practices: As the demand for mineral resources continues to grow, the mineral exploration industry must prioritize sustainable practices that minimize ecological disruption. This includes:

- **Conducting Comprehensive Environmental Assessments:** Before initiating exploration activities, thorough assessments should be conducted to evaluate potential impacts on local ecosystems, water resources, and communities. This proactive approach can help identify risks and develop strategies to mitigate them.

- **Engaging with Local Communities:** Building strong relationships with local communities is vital for ensuring that exploration activities are conducted responsibly. Engaging stakeholders in the decision-making process can foster trust and collaboration, leading to better outcomes for both the industry and the community.

- **Implementing Rehabilitation Strategies:** After exploration activities are completed, it is essential to rehabilitate disturbed areas to restore ecosystems and minimize long-term impacts. This may involve replanting vegetation, restoring soil quality, and monitoring the recovery of local habitats.

- **Adopting Innovative Technologies:** Embracing advancements in technology can enhance the efficiency and accuracy of exploration while reducing environmental footprints. For example, using drones for aerial surveys can minimize ground disturbance, and remote sensing techniques can provide valuable geological data without direct contact with the land.

Future Directions: The future of mineral exploration lies in the balance between resource extraction and environmental stewardship. By integrating traditional prospecting methods with innovative technologies and sustainable practices, geologists can continue to discover and evaluate mineral resources that are essential for modern society. This approach not only supports economic development but also ensures that the exploration and extraction of minerals are conducted in a manner that respects and preserves the environment for future generations.

Ultimately, the goal of mineral exploration is to contribute to the sustainable development of natural resources, meeting the needs of society while safeguarding the planet. By prioritizing environmental responsibility and community engagement, the mineral exploration sector can play a vital role in achieving a sustainable future.

3.8 Four Types of Geophysical Prospecting Methods

1. Magnetic Method

The magnetic method is a geophysical prospecting technique primarily used to locate magnetic ore deposits, such as magnetite, pyrrhotite, and cobalt. This method employs a magnetometer, an instrument designed to detect variations in the Earth's magnetic field caused by the presence of magnetic minerals. By mapping these variations, geologists can identify potential ore deposits and delineate contact zones between different rock types, particularly between sedimentary rocks and basic igneous rocks. The magnetic method is particularly valuable in mineral exploration, as it provides a non-invasive means of assessing subsurface geology and identifying areas of interest for further investigation. Additionally, this method can be applied in archaeological studies to locate buried structures and artifacts, making it a versatile tool in both geological and archaeological prospecting.

2. Electrical Method

The electrical method involves measuring the flow of electric current between an ore body using two electrodes placed several meters apart. When an electric current is passed through the ground, the flow is typically uniform if the Earth's composition is consistent. However, the presence of mineral lodes, which may be good conductors or non-conductors of electricity, can cause the current flow to converge or diverge. A galvanometer is used to measure any changes in electrical potential, indicating anomalies that suggest the possible localization of mineral deposits. For instance, mineral deposits containing sulfides are good conductors of electricity, while quartz reefs are non-conductors. This method is particularly useful for delineating the contours of bedrock and identifying areas with potential mineralization. Furthermore, the electrical method can be adapted for various applications, including groundwater exploration and environmental assessments, making it a valuable technique in geophysical prospecting.

3. Seismic Method

The seismic method is a crucial technique in the prospecting of petroliferous sedimentary beds. This method utilizes artificially generated seismic waves, which are sent into the ground and then reflected back to the surface. By analyzing the time it takes for these waves to return, geologists can infer the thickness and depth of various sedimentary formations. The reflection and refraction of seismic waves provide valuable information about subsurface structures, allowing for the identification of potential hydrocarbon reservoirs. This method is widely used in the oil and gas industry, as it helps in mapping geological formations and assessing their potential for resource extraction. Additionally, advancements in seismic technology, such as 3D seismic imaging, have significantly improved the accuracy and resolution of subsurface mapping, enabling more effective exploration and development of hydrocarbon resources.

4. Gravimetric Method

The gravimetric method is employed to identify areas of high and low gravitational pull, which can indicate the presence of different geological formations. By measuring variations in gravitational intensity compared to the mean Earth's gravity, geologists can select areas for further exploration. Heavy masses, such as dense mineral deposits, will attract a plumb bob more strongly than lighter masses, such as those found in sedimentary basins or acidic igneous rocks. The identification of high or low intensity zones can suggest potential occurrences of minerals that are genetically related to the geological characteristics of those zones. This method is particularly useful in mineral exploration, as it provides insights into the subsurface distribution of various geological materials. Moreover, the gravimetric method can be integrated with other geophysical techniques to enhance the overall understanding of subsurface geology, making it a valuable component of comprehensive exploration strategies.

In summary, these four geophysical prospecting methods—magnetic, electrical, seismic, and gravimetric—each offer unique advantages and applications in the exploration of mineral and hydrocarbon resources. By employing these techniques, geologists can gain valuable insights into subsurface geology, facilitating more effective resource management and exploration efforts.

3.9 Five Successive Stratigraphic Steps Used by Geologists

1. Discrimination of Stratigraphic Units

The first step in stratigraphic analysis involves the discrimination of stratigraphic units, which entails the detailed description and classification of various formations. This process is essential for reducing the complexities inherent in stratigraphic sections, allowing geologists to simplify and organize the data effectively. By clearly defining these units—such as formations, members, and beds—geologists can better understand the relationships between different rock layers and their respective characteristics. This foundational step sets the stage for subsequent analyses, as it provides a structured framework for interpreting the geological history of an area.

2. Determination of Stratigraphic Sequence

Once the stratigraphic units are identified, the next step is to determine the stratigraphic sequence. This involves recognizing that stratified rocks are deposited in an orderly manner over time, reflecting the geological history of the area. Understanding the sequence in which these layers were laid down is crucial for reconstructing past environments and interpreting the geological processes that shaped the region. By analyzing the vertical and lateral relationships of the stratigraphic units, geologists can infer the relative ages of the layers and the environmental conditions that prevailed during their deposition, such as changes in sea level, sediment supply, and tectonic activity.

3. Geologic Mapping

Geologic mapping is a critical step that includes both the areal mapping of outcropping units and the continuous tracing of subsurface units. This process aims to create a spatial distribution of three-dimensional rock units, which is particularly important in engineering geology and resource exploration. Accurate mapping allows geologists to visualize the arrangement of different rock types and their relationships, facilitating further exploration and analysis. By integrating field observations with remote sensing and geophysical data, geologists can produce detailed maps that illustrate the distribution of stratigraphic units, structural features, and potential mineral resources, aiding in land-use planning and environmental assessments.

4. Stratigraphic Correlation

Stratigraphic correlation involves comparing stratigraphic sequences observed at different locations to identify similarities and differences. This step relies on all available evidence, including physical characteristics (such as lithology and sedimentary structures) and biological characteristics (such as fossil content). The application of petrology and paleontology is often necessary to solve correlation problems, as these disciplines provide insights into the composition and age of the rock units being studied. By establishing correlations between stratigraphic units across different geographic areas, geologists can reconstruct regional geological histories and identify widespread depositional environments, enhancing the understanding of sedimentary processes and paleoenvironments.

5. Stratigraphic Interpretation

The final step is stratigraphic interpretation, where geologists analyze the correlated stratigraphic units to derive insights about paleogeography and geological history. This interpretation helps to reconstruct past environments, understand the processes that led to the formation of the current geological landscape, and assess the potential for mineral deposits within the stratigraphic framework. By synthesizing data from the previous steps, geologists can develop models of sedimentary basins, evaluate the impact of tectonic events, and predict the distribution of resources such as hydrocarbons, minerals, and groundwater. This comprehensive understanding is vital for effective resource management, environmental protection, and hazard assessment.

In summary, these five successive stratigraphic steps—discrimination of stratigraphic units, determination of stratigraphic sequence, geologic mapping, stratigraphic correlation, and stratigraphic interpretation—provide a systematic approach for geologists to analyze and understand the complexities of the Earth's geological history. By following these steps, geologists can enhance their ability to interpret past environments, assess geological resources, and inform land-use decisions.

3.10 Six Guides that Characterize the Location of Mineral Deposits by Conventional Exploration Practices

1. Geometry or Symmetry of Geological Features

The geometry or symmetry of geological features is a fundamental aspect in identifying potential mineral deposits. Geologists meticulously analyze the shapes, orientations, and arrangements of rock formations, such as folds, faults, and stratigraphic layers, to discern patterns that may indicate mineralization. For instance, certain geometrical configurations, like synclines and anticlines, can trap mineral-rich fluids, leading to the concentration of valuable minerals. By understanding the spatial relationships and symmetry of these geological features, geologists can effectively target areas that are more likely to host economically viable mineral deposits.

2. Apparent Geological Anomalies

Geological anomalies, which manifest as unexpected changes in rock type, structure, or mineral composition, can serve as critical indicators of potential mineral deposits. These anomalies may include unusual rock formations, unexpected mineral assemblages, or variations in geophysical properties that deviate from the surrounding geology. Such irregularities often warrant further investigation, as they may signal the presence of underlying mineralization. For example, a sudden shift from sedimentary to igneous rock types could suggest the proximity of a mineral deposit, prompting geologists to conduct more detailed surveys and sampling in the area.

3. Regional Trends

Understanding regional geological trends is essential for effective mineral exploration. By studying the broader geological context, including tectonic history, sedimentation patterns, and regional metamorphism, geologists can identify overarching patterns that may suggest the presence of mineral deposits in specific areas. For instance, regions with a history of volcanic activity may be more likely to host certain types of mineral deposits, such as copper or gold. Recognizing these regional trends allows geologists to prioritize exploration efforts in areas with a higher likelihood of mineralization, thereby optimizing resource allocation and increasing the chances of successful discovery.

4. Favourable Factors

Favourable geological factors, such as faulting, intrusion, and the presence of specific lithologies, often intersect to create prospective locales for mineral exploration. Faults can act as conduits for mineral-rich fluids, while intrusions can provide heat and pressure conditions conducive to mineral formation. Additionally, certain rock types, such as those rich in silica or carbonate, may be more receptive to mineralization. By identifying and analyzing these favourable geological features, geologists can enhance the likelihood of finding mineral deposits, guiding exploration efforts toward areas with the greatest potential for economic viability.

5. Relevant Mineralogical or Geochemical Phenomenon Patterns

Patterns in mineralogy or geochemistry can provide invaluable insights into the location of mineral deposits. Geologists often conduct detailed analyses of mineral compositions and geochemical signatures to identify areas with a higher probability of mineralization. For example, the presence of specific trace elements or isotopic ratios can indicate the proximity of a mineral deposit. By mapping these mineralogical and geochemical patterns, geologists can develop targeted exploration strategies, focusing on regions that exhibit promising signs of mineral wealth.

6. Secondary Clues to Geological Phenomena

Secondary clues, such as changes in vegetation, soil composition, and stream patterns, can also serve as important indicators of underlying geological phenomena. For instance, certain plants may thrive in mineral-rich soils, while variations in soil chemistry can signal the presence of nearby mineral deposits. Additionally, stream sediment analysis can reveal concentrations of heavy minerals that have been eroded from upstream deposits. By observing and

interpreting these secondary clues, geologists can guide their exploration efforts, highlighting areas of interest that may warrant further investigation and sampling.

In summary, these six guides—geometry of geological features, geological anomalies, regional trends, favourable factors, mineralogical and geochemical patterns, and secondary clues—provide a comprehensive framework for conventional exploration practices. By leveraging these guides, geologists can enhance their understanding of mineral deposit formation and improve the efficiency and effectiveness of exploration efforts, ultimately leading to more successful discoveries of valuable mineral resources.

3.11 Five Categories of Surface Indicators of Possible Mineralization or Aids to Reconnaissance Mapping Identification by Remote Sensing

1. Topography

Topography serves as a critical surface indicator, as physical features reflect underlying geological structures, stratigraphy, igneous processes, and geomorphological information. These surface expressions are essential for reconnaissance mapping and identifying locations for prospecting, especially when correlated with regional mineralization. Structural expressions such as depressions, subsidence, domes, and folding can provide insights into hidden geology.

2. Igneous and Volcanic Features

Igneous and volcanic features are important indicators of potential mineralization. The presence of volcanic rocks and associated structures can suggest areas where mineral deposits may be concentrated, particularly in regions with a history of volcanic activity.

3. Lineaments and Geological Structure

Lineaments, which are linear features on the Earth's surface, can indicate underlying geological structures such as faults and fractures. These features are often associated with mineral deposits, making them valuable for exploration.

4. Mineralogical-Lithologic Association

The association between different mineralogical and lithologic units can provide clues about potential mineralization. By studying these associations, geologists can identify areas where specific minerals are likely to occur, thereby guiding exploration efforts toward regions with a higher probability of containing valuablmineral deposits.

5. Stratigraphic Sequence

Stratigraphy is often considered one of the more disappointing areas of LANDSAT applications in geology and mineral exploration. Even in regions where stratigraphy is well exposed, the individual formations are typically not thick enough to be distinctly identified, particularly by rock type. As a result, the identification of stratigraphic units largely relies on their expressions as ridge markers or other known characteristics, as well as vegetational guides that can indicate underlying geological features. In this context, the use of aircraft subsystems becomes crucial, as they can provide the essential detail needed to differentiate between various rock types, such as evaporites, limestones, detrital deposits, shales, and sandstones. In fact, airborne radar and aerial photography are significantly more effective than LANDSAT imagery for accurately defining rock units within a stratigraphic sequence. This highlights the limitations of satellite imagery in stratigraphic analysis and underscores the importance of utilizing more advanced aerial techniques to enhance geological understanding and mineral exploration efforts.

3.12 Remote Sensing in Mineral Exploration

Introduction

Remote sensing has become an invaluable tool in mineral exploration, offering a range of advantages that enhance the efficiency and effectiveness of geological surveys.

Advantages of Remote Sensing

1. **Continuous Data Acquisition:** One of the primary benefits is the ability to enable continuous acquisition of data, allowing geologists to monitor changes over time without the need for constant field visits.

2. **Wide Regional Coverage:** Remote sensing provides wide regional coverage, capturing extensive areas in a single sweep, which is particularly useful for identifying mineral deposits across large landscapes.

3. **Good Spectral Resolution:** The technology boasts good spectral resolution, enabling the differentiation of various materials based on their spectral signatures. This results in accurate data that can be utilized for detailed information and analysis, making it easier to identify potential mineralization zones.

4. **Historical Data Archive:** Moreover, remote sensing serves as a large archive of historical data, allowing researchers to compare past and present conditions and track changes in the landscape.

5. **Three-Dimensional Perspective:** The three-dimensional perspective offered by remote sensing technologies enhances the understanding of geological structures and formations, providing insights that are often difficult to obtain through traditional methods.

6. **Rapid Data Collection:** The rapid method of data collection is another significant advantage, as it allows for quick assessments of vast areas, facilitating timely decision-making in exploration efforts.

7. **Enhanced Ground Observation:** Increased spectral resolution further aids in distinguishing between different rock types and mineral deposits, while the ability to extend ground observation enhances the overall understanding of the geological context.

8. **Improved Vantage Point:** Finally, remote sensing provides an improved vantage point, offering a synoptic view of the landscape that can reveal patterns and features not easily discernible from the ground.

Disadvantages of Remote Sensing

Despite its many advantages, remote sensing also has several disadvantages that must be considered:

1. **Limited Subsurface Penetration:** One major limitation is that current remote sensing methods cannot penetrate deeply into the surface of the ground, which restricts their ability to provide information about subsurface mineral deposits.

2. **Vegetation Cover Limitations:** Remote sensing is often not usable in areas with heavy vegetation cover, although it can still be effective in locating altered vegetation that may indicate the presence of ore deposits.

3. **High Equipment Costs:** The equipment used in remote sensing can be very expensive to build and operate, which may pose a barrier to some exploration projects.

4. **Measurement Uncertainty:** Furthermore, measurements obtained through remote sensing can have significant uncertainty, particularly in complex geological settings.

5. **Coarse Resolution:** The resolution of remote sensing data is often coarse, especially in cloudy areas, which can hinder the ability to detect finer geological features.

6. **Data Interpretation Challenges:** Lastly, data interpretation can be challenging, necessitating a solid understanding of the geological phenomena being sampled to ensure accurate conclusions are drawn from the data collected.

Conclusion

In summary, while remote sensing offers numerous advantages that enhance mineral exploration, it is essential to be aware of its limitations and challenges. By understanding both the strengths and weaknesses of remote sensing technologies, geologists can make informed decisions about their application in mineral exploration and effectively integrate them with traditional exploration methods.

3.13 Summary of Chapter 3

Overview of Geological Hazard Management

In this chapter, we explored the critical aspects of geological hazard management, emphasizing the importance of understanding geological hazards, effective risk assessment methodologies, and the implementation of mitigation strategies. We began by discussing the various types of geological hazards, including earthquakes, landslides, volcanic eruptions, and flooding, highlighting their potential impacts on communities and the environment. Understanding these hazards is essential for developing effective response strategies and enhancing community resilience while also considering the environmental consequences that can arise from such events.

Risk Assessment Methodologies

We then delved into the methodologies employed in risk assessment, which are crucial for identifying and quantifying geological hazards. Key processes such as hazard identification, vulnerability assessment, and risk analysis were outlined, demonstrating how these methodologies work together to provide a comprehensive understanding of potential risks. The use of Geographic Information Systems (GIS) and remote sensing technologies for mapping and modeling hazards was also highlighted, showcasing their role in visualizing risk areas and informing decision-making. These tools not only help identify human vulnerabilities but also assess the potential environmental impacts of geological hazards, such as habitat destruction and water contamination.

Mitigation Strategies

The chapter further examined various mitigation strategies that can be implemented to reduce the impacts of geological hazards. Land use planning, engineering solutions, early warning systems, and public education were identified as essential components of effective hazard mitigation. By incorporating geological hazard assessments into land use planning and designing resilient infrastructure, communities can minimize their exposure to risks and protect both lives and the environment. For instance, effective land use planning can prevent development in high-risk areas, thereby preserving natural ecosystems and reducing the likelihood of environmental degradation during geological events.

Case Studies

Case studies, such as the 2010 Haiti Earthquake and the eruption of Mount St. Helens, provided real-world examples of the consequences of geological hazards and the lessons learned from these events. These case studies underscored the importance of preparedness, response, and the need for improved building codes and disaster management practices. They also highlighted the significant environmental impacts that can result from geological hazards, such as soil erosion, loss of biodiversity, and contamination of water resources, emphasizing the need for a holistic approach to hazard management.

Future Directions in Geological Hazard Management

Finally, we discussed future directions in geological hazard management, including advancements in technology, climate change considerations, and the necessity for interdisciplinary approaches. As the frequency and intensity of geological hazards are influenced by climate change, understanding these interactions will be vital for developing adaptive strategies that protect both human populations and the environment. For example, increased rainfall and extreme weather events can exacerbate landslide risks, while rising sea levels may heighten the impact of tsunamis and coastal flooding.

Conclusion

In conclusion, this chapter emphasized that a comprehensive approach to geological hazard management—integrating risk assessment, mitigation strategies, and interdisciplinary collaboration—is essential for enhancing community resilience. By understanding geological hazards and their environmental impacts, and by implementing effective strategies, we can better protect lives, infrastructure, and ecosystems in the face of geological challenges. This holistic perspective ensures that both human safety and environmental health are prioritized, fostering a sustainable future in the context of geological risks.

Environmental Responsibility: The exploration and extraction of minerals must be conducted with a strong commitment to minimizing environmental impacts. This includes implementing best practices in resource management, such as reducing waste, preventing pollution, and rehabilitating disturbed areas. By adopting sustainable practices, the industry can protect ecosystems, preserve biodiversity, and ensure that natural resources are available for future generations.

Community Engagement: Engaging local communities in the exploration process is crucial for fostering trust and collaboration. By involving stakeholders in decision-making and addressing their concerns, the mineral exploration sector can enhance social acceptance and support for projects. This engagement not only helps to mitigate potential conflicts but also ensures that the benefits of mineral development are shared with local populations, contributing to their economic and social well-being.

Innovative Technologies: The adoption of innovative technologies plays a significant role in enhancing the efficiency and effectiveness of mineral exploration. Advanced techniques such as remote sensing, geophysical surveys, and data analytics can provide valuable insights into subsurface geology while minimizing the need for invasive exploration methods. These technologies enable geologists to make informed decisions, optimize resource extraction, and reduce the environmental footprint of exploration activities.

Holistic Approach: A holistic approach that combines geological hazard management with mineral exploration is essential for addressing the complexities of resource development in a changing environment. By understanding the interactions between geological hazards and mineral resources, geologists can develop strategies that not only enhance resource recovery but also protect communities and ecosystems from potential risks. This integrated perspective ensures that both human safety and environmental health are prioritized in the face of geological challenges.

Future Outlook: As the global demand for minerals continues to rise, driven by technological advancements and the transition to renewable energy sources, the mineral exploration sector must adapt to meet these challenges responsibly. By embracing sustainable development principles, the industry can play a pivotal role in supporting economic growth while safeguarding the environment. This commitment to sustainability will be crucial in navigating the complexities of resource management in the 21st century.

In conclusion, the synergy between geological hazard management and mineral exploration practices is essential for fostering a sustainable future. By prioritizing environmental stewardship, engaging communities, and leveraging innovative technologies, the mineral exploration sector can effectively contribute to the responsible development of natural resources, ensuring that the needs of society are met without compromising the health of our planet.

CHAPTER 4

SOIL AND WATER RESOURCES

4.1 Importance of Soil and Water Resources

Soil and water are vital resources that underpin the health of ecosystems and human societies. They play a crucial role in supporting plant growth, regulating water cycles, and providing habitat for a diverse array of organisms. The interdependence of soil and water is evident in their contributions to agriculture, industry, and domestic life. For instance, agriculture relies heavily on soil for crop production and water for irrigation, while industries depend on these resources for various processes.

Moreover, the quality of soil and water directly impacts environmental health. Healthy soils contribute to clean water by filtering pollutants and supporting biodiversity, while contaminated water can lead to adverse effects on ecosystems and human health. Understanding the dynamics of soil and water resources is essential for sustainable management practices that ensure their availability for future generations.

Role in Ecosystems

Soil and water are fundamental components of ecosystems, playing critical roles in maintaining ecological balance and supporting life.

Support for Plant Growth: Soil provides essential nutrients, minerals, and organic matter necessary for plant growth. It serves as a medium for root development, anchoring plants and facilitating the uptake of water and nutrients. Healthy soil is rich in microorganisms that contribute to nutrient cycling, enhancing soil fertility and promoting robust plant growth.

Regulation of Water Cycles: Water is a vital resource that regulates local and global water cycles. It is involved in processes such as precipitation, evaporation, and transpiration. Soil acts as a natural filter, absorbing rainfall and facilitating groundwater recharge. This regulation helps maintain hydrological balance, ensuring that ecosystems receive adequate moisture for sustaining plant and animal life.

Habitat for Organisms: Soil and water provide habitats for a diverse array of organisms, including bacteria, fungi, insects, and larger animals. These organisms play essential roles in nutrient cycling, decomposition, and soil formation. Aquatic ecosystems, such as rivers, lakes, and wetlands, support a variety of species, contributing to biodiversity and ecosystem resilience.

Human Dependency

Human societies are deeply dependent on soil and water resources for various aspects of life, including agriculture, industry, and domestic needs.

- **Agriculture:** Soil is the foundation of agriculture, providing the necessary nutrients and structure for crop production. Healthy soils are crucial for food security, as they support the growth of staple crops and other agricultural products. Irrigation, which relies on water resources, is essential for enhancing agricultural productivity, especially in regions with variable rainfall.

- **Industry:** Many industries depend on soil and water resources for their operations. For example, the agricultural sector relies on water for irrigation and livestock production, while the manufacturing sector may require water for processing and cooling. Additionally, soil is a source of raw materials for construction, mining, and other industrial activities.

- **Domestic Life:** Water is essential for human survival, providing drinking water, sanitation, and hygiene. Access to clean water is critical for public health, as contaminated water can lead to waterborne diseases. Soil also plays a role in urban planning and landscaping, contributing to green spaces that enhance the quality of life in urban environments.

Environmental Health

The quality of soil and water resources is closely linked to overall environmental health, with significant implications for biodiversity and ecosystem services.

- **Soil Quality:** Healthy soils are characterized by good structure, fertility, and biodiversity. Degraded soils, often resulting from erosion, contamination, or unsustainable agricultural practices, can lead to reduced agricultural productivity and increased vulnerability to climate change. Soil health is essential for maintaining ecosystem services, such as carbon sequestration, water filtration, and habitat provision.

- **Water Quality:** Clean water is vital for sustaining aquatic ecosystems and human health. Pollution from agricultural runoff, industrial discharges, and urban waste can degrade water quality, harming aquatic life and making water unsafe for human consumption. Healthy water bodies support biodiversity, providing habitats for fish, amphibians, and other organisms that contribute to ecosystem stability.
- **Biodiversity and Ecosystem Services:** The interdependence of soil and water quality affects biodiversity and the services ecosystems provide. Healthy ecosystems offer services such as pollination, pest control, and climate regulation, which are essential for human well-being. Conversely, degraded soil and water resources can lead to loss of biodiversity, diminished ecosystem resilience, and reduced capacity to provide essential services.

Conclusion

The environmental impacts of groundwater systems and aquifers are multifaceted and interconnected, affecting ecosystems, human health, and economic stability. The challenges posed by aquifer depletion, contamination, and changing hydrological patterns necessitate a comprehensive approach to groundwater management that prioritizes sustainability and resilience.

To mitigate these impacts, it is essential to adopt integrated water resource management practices that consider the interconnectedness of groundwater with surface water, ecosystems, and human activities. This includes promoting conservation, enhancing recharge, preventing contamination, and ensuring equitable access to clean water.

By addressing the environmental impacts associated with groundwater systems and aquifers, we can work towards a future where these vital resources are protected, ensuring their availability and quality for generations to come. Sustainable groundwater management is not only crucial for maintaining ecological balance but also for supporting human well-being and economic development in an increasingly water-scarce world.

Further Considerations for Sustainable Soil and Water Management

As we delve deeper into the sustainable management of soil and water resources, several additional considerations emerge that can enhance our approach and effectiveness in addressing the challenges we face:

Integrated Water Resource Management (IWRM): IWRM is a holistic approach that considers the interconnectedness of water resources across different sectors and scales. By integrating land use planning, agricultural practices, and water management, IWRM promotes the sustainable use of water resources while ensuring that ecological health is maintained. This approach encourages collaboration among various stakeholders, including government agencies, local communities, and private sectors, to develop comprehensive strategies that address water scarcity, quality, and ecosystem health.

Soil Health Management: Emphasizing soil health is critical for sustainable agriculture and ecosystem resilience. Practices such as crop rotation, cover cropping, reduced tillage, and organic amendments can enhance soil structure, increase organic matter, and improve nutrient cycling. Healthy soils are better able to retain water, support diverse plant life, and sequester carbon, contributing to climate change mitigation. Promoting soil health not only benefits agricultural productivity but also enhances the overall functioning of ecosystems.

Ecosystem-Based Approaches: Implementing ecosystem-based approaches to resource management can help restore and maintain the natural functions of ecosystems. This includes protecting and restoring wetlands, forests, and riparian zones, which play critical roles in water filtration, flood regulation, and habitat provision. By prioritizing ecosystem health, we can enhance the resilience of both natural systems and human communities to environmental changes.

Technological Innovations: Advancements in technology can significantly improve the efficiency and sustainability of soil and water management practices. Precision agriculture, for example, utilizes data analytics, remote sensing, and GPS technology to optimize irrigation, fertilization, and pest management. These innovations can reduce resource waste, minimize environmental impacts, and enhance productivity. Additionally, technologies for water recycling and treatment can help address water scarcity and contamination issues.

Community-Led Initiatives: Empowering local communities to take the lead in managing their soil and water resources can yield significant benefits. Community-led initiatives often incorporate traditional knowledge and practices that are well-suited to local conditions. Supporting local stewardship programs, participatory monitoring, and community education can foster a sense of ownership and responsibility, leading to more effective and sustainable resource management.

Global Cooperation and Knowledge Sharing: Addressing soil and water challenges is a global issue that requires international cooperation and knowledge sharing. Collaborative efforts among countries can facilitate the exchange of best practices, technologies, and research findings. Global initiatives, such as the United Nations Sustainable Development Goals (SDGs), emphasize the importance of sustainable water management and soil conservation, providing a framework for collective action.

Long-Term Monitoring and Adaptive Management: Establishing long-term monitoring programs is essential for assessing the health of soil and water resources and the effectiveness of management practices. Adaptive management approaches allow for flexibility in decision-making, enabling stakeholders to respond to changing conditions and new information. By continuously evaluating outcomes and adjusting strategies, we can improve resilience and sustainability over time.

Conclusion

In conclusion, the sustainable management of soil and water resources is a multifaceted challenge that requires a comprehensive and integrated approach. By considering the interconnectedness of these resources and the environmental impacts associated with their use, we can develop strategies that promote ecological health, support human well-being, and ensure the availability of these vital resources for future generations.

The way forward involves collaboration, innovation, and a commitment to stewardship at all levels of society. By embracing sustainable practices, investing in research and technology, and fostering community engagement, we can create a resilient and equitable future where soil and water resources are protected and managed wisely. Together, we can build a sustainable world that honors the intrinsic value of our natural resources and supports the diverse needs of all living beings.

Importance of Fundamentals of Soil Mechanics

Competency and Evaluation-Based Approach in the Study of Soils

Introduction: The study of soils is a critical aspect of civil engineering, geotechnical engineering, and environmental science. Understanding the fundamentals of soil mechanics is essential for evaluating soil behavior, which directly impacts construction, land use, and environmental management. A competency and evaluation-based approach to soil mechanics emphasizes the practical application of theoretical knowledge, ensuring that students and professionals can effectively analyze and address real-world challenges related to soil.

1. Understanding Soil Behavior: A solid foundation in soil mechanics allows students and professionals to understand how soils behave under various conditions. This includes knowledge of soil properties, which encompasses composition, structure, and classification. Understanding properties such as cohesion, friction angle, permeability, and compressibility is crucial for predicting how soils will respond to loads and environmental changes. Additionally, soil mechanics provides insights into the stress-strain behavior of soils, which is essential for designing foundations, retaining structures, and earthworks. A competency-based approach ensures that learners can apply these concepts to real-world scenarios, enhancing their ability to make informed decisions in engineering practices.

2. Application in Engineering Design: The principles of soil mechanics are vital for various engineering applications. For instance, understanding soil mechanics is essential for designing safe and effective foundations for buildings, bridges, and other structures. Evaluating soil bearing capacity, settlement, and lateral earth pressures is critical for ensuring structural stability. Furthermore, a thorough understanding of soil mechanics allows engineers to assess the stability of slopes and embankments, which is crucial for preventing landslides and ensuring the safety of infrastructure. Additionally, soil mechanics principles guide the design of retaining walls, sheet piles, and other structures that support soil, making it essential to evaluate soil pressures and potential failure mechanisms for effective design.

3. Environmental Considerations: Soil mechanics also plays a significant role in environmental engineering and management. Understanding soil properties and behavior is essential for assessing the movement of contaminants through soil and groundwater. This knowledge is critical for designing remediation strategies and protecting water resources. Moreover, a competency-based approach to soil mechanics helps professionals evaluate soil suitability for various land uses, including agriculture, construction, and waste disposal. This evaluation is essential for sustainable land management, ensuring that land use practices do not compromise environmental integrity.

4. Competency Development: A competency and evaluation-based approach to soil mechanics emphasizes the development of practical skills and critical thinking. Engaging in laboratory experiments, fieldwork, and simulations allows students to apply theoretical knowledge to real-world situations. This hands-on experience is crucial for developing a deep understanding of soil behavior. Additionally, evaluating soil conditions and predicting behavior requires analytical thinking and problem-solving skills. A competency-based approach fosters these skills, preparing students for the challenges they will face in their careers. Furthermore, soil mechanics intersects with various disciplines, including geology, hydrology, and environmental science. A comprehensive understanding of these connections enhances the ability to address complex issues related to soil and its interactions with other systems.

Conclusion: The fundamentals of soil mechanics are essential for understanding soil behavior, informing engineering design, and addressing environmental challenges. A competency and evaluation-based approach ensures that students and professionals are equipped with the practical skills and knowledge necessary to analyze and solve real-world problems related to soil. By emphasizing the application of theoretical concepts, this approach prepares individuals to contribute effectively to the fields of civil engineering, geotechnical engineering, and environmental management, ultimately promoting sustainable practices and enhancing public safety. The integration of theoretical knowledge with practical application not only enriches the learning experience but also fosters a generation of professionals capable of tackling the pressing challenges of soil management in a rapidly changing world.

Construction Perception of Soil and Soil Mechanics

(a) Pioneers of Soil Mechanics

The term "soil" encompasses various meanings depending on the context in which it is considered. For construction engineers, soil is viewed as an aggregate of mineral deposits, which may be cemented or un-cemented, and may or may not contain organic particles. The engineering study of soils, including the behavior of soil masses, is known as soil mechanics. This field is one of the youngest disciplines within civil engineering and involves the study of soil properties and their application to construction works.

Soil mechanics applies principles from engineering mechanics—such as kinematics, dynamics, fluid mechanics, and materials science—to predict soil behavior under various conditions. The first scientific study of soil mechanics was conducted by French physicist Charles-Augustin de Coulomb, who published a theory of earth pressure in 1773. A century later, Scottish engineer William Rankine (1857) published his work on the theory of earth masses, which remains foundational in quantifying earth stresses.

In the 20th century, these theories were refined to incorporate cohesion as a significant factor influencing soil behavior under stress. Renowned scientist Karl Terzaghi (1948) defined soil mechanics as the application of the laws of mechanics and hydraulics to engineering problems involving sediments and unconsolidated accumulations of solid particles, regardless of the presence of organic materials. Since then, soil mechanics has played a crucial role in construction, guiding engineers in the design and analysis of structures.

(b) Essential Soil Properties

Understanding soil properties is vital for designing stable, durable, and safe engineering structures that impose stresses on the soils beneath them. The response of soils to imposed loading during and after construction depends on several key properties:

1. **Soil Shear Strength:** This property is characterized by internal friction and cohesion, which are critical shear strength parameters. Internal friction, the resistance of a soil mass to sliding, is typically greater in sands and gravels than in clays. Cohesion, which refers to the molecular attraction between soil particles, is significantly higher in clays compared to sands or silts. Both parameters reduce the tendency of soils to shear or slide along failure planes and are influenced by moisture content. Soil shear strength is primarily determined in laboratory settings while considering field moisture conditions.

2. **Compressibility:** Compressibility refers to the degree to which soil can be densified through mechanical means, such as tamping and vibration, to support greater loads. This property is crucial for understanding

settlement in construction works and requires thorough investigation in sites with compressible soils.

3. **Elasticity:** Elasticity describes the ability of soil to re-expand after being compressed. Compressible soils exhibit this property, particularly those that have experienced higher stress in the geological past and are now undergoing expansion due to stress release from regional erosion and melting ice cover.

4. **Permeability:** Soil permeability is the ability of soil to allow water to seep through it, represented by a coefficient that can be determined under both laboratory and field conditions (hydraulic conductivity). Coarse soils typically exhibit higher permeability than fine-grained soils.

5. **Capillarity:** Capillarity refers to the ability of water to be drawn upward from the water table into the soil. Water held above the water table exists at negative pressure, filling voids between solid particles or being adsorbed onto them. The quantity of water held increases with higher clay mineral content, affecting the overall effective stress distribution within the soil mass. Effective stress in the field increases with depth.

(c) Typical Applications

A clear understanding of the aforementioned soil properties is paramount for ensuring the provision of safe and durable infrastructure. Throughout history, civilization has been enhanced by engineers who construct structures based on a thorough evaluation of the geotechnical properties of construction sites. Most construction projects focus on providing modern buildings and transportation networks that comply with established codes of practice (see fig. 1). Special attention is given to regions prone to geo-disasters, which pose risks to human safety and can cause significant property damage.

In addition to traditional construction, the principles of soil mechanics are increasingly applied in innovative ways, such as in the design of green infrastructure and sustainable urban development (Figure 10). For instance, engineers are now utilizing soil properties to create permeable pavements that enhance stormwater management and reduce urban flooding. Furthermore, understanding soil behavior is crucial in the implementation of earth-retaining structures, such as retaining walls and slopes, which are essential in preventing landslides and ensuring the stability of roadways and buildings. By integrating these advanced applications into engineering practices, professionals can address contemporary environmental challenges while promoting sustainability. In summary, the construction perception of soil and soil mechanics is essential for the successful design and implementation of engineering projects. By understanding the fundamental properties of soil and applying this knowledge effectively, engineers can create structures that are not only functional but also resilient to environmental challenges. This knowledge is critical for advancing infrastructure development and ensuring the safety and well-being of communities.

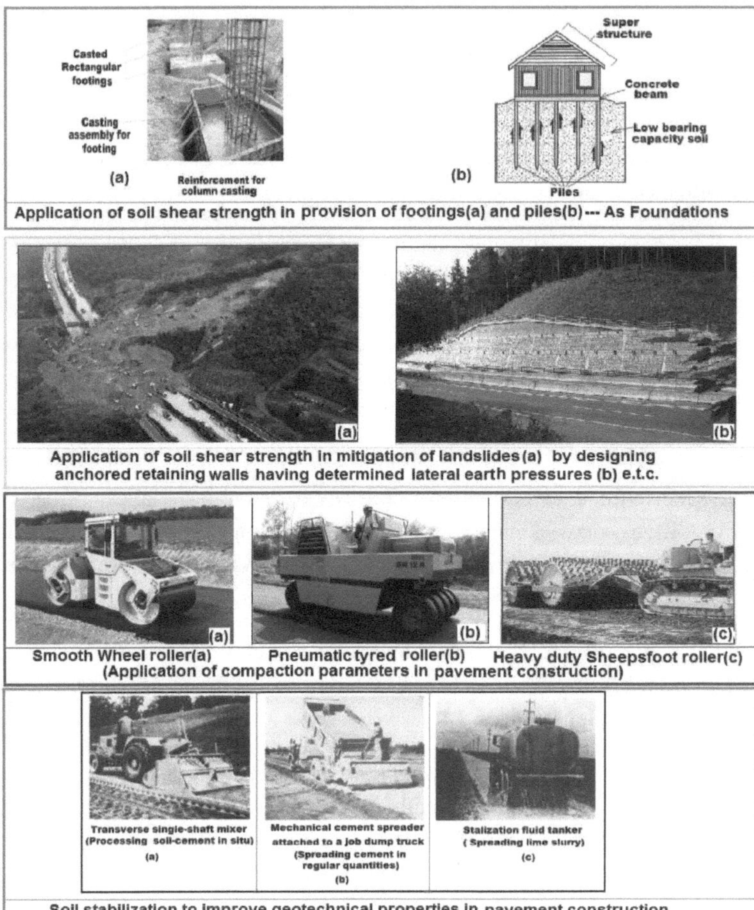

Fig.10: Typical landslide and infrastructural works

4.2 Soil Formation and Types

Soil Formation Processes

Soil formation is a complex process influenced by several factors, including parent material, climate, topography, organisms, and time. These five soil-forming factors interact to create diverse soil types across different landscapes. The processes involved in soil formation include weathering of rocks, leaching of minerals, and the accumulation of organic matter.

Soils are typically organized into distinct layers, known as horizons. The O horizon consists of organic material, while the A horizon is rich in nutrients and minerals. The E horizon is characterized by leaching, and the B horizon accumulates minerals leached from above. The C horizon consists of weathered parent material, and the R horizon is unweathered bedrock. Understanding these layers is crucial for assessing soil fertility and water retention capabilities.

Soil types vary widely, with sandy, clayey, and loamy soils being the most common. Sandy soils drain quickly but may lack nutrients, while clayey soils retain water but can become compacted. Loamy soils, a mixture of sand, silt, and clay, are often considered ideal for agriculture due to their balanced properties. Understanding these factors is essential for comprehending how different soils develop and their characteristics.

Parent Material: The mineral composition and characteristics of the parent material significantly influence soil formation. Parent material can be derived from various sources, including igneous, metamorphic, or sedimentary rocks. The type of rock determines the minerals available for soil development, which affects soil texture, fertility, and drainage properties.

For example, soils derived from granite may be sandy and well-drained, while those from basalt may be richer in nutrients due to the presence of minerals like iron and magnesium.

Climate: Climate plays a crucial role in soil formation through temperature and precipitation patterns. Weathering processes, such as physical and chemical weathering, are accelerated in warm, moist climates, leading to the breakdown of parent material and the development of soil horizons.

In arid climates, soils may develop more slowly due to limited moisture, resulting in less weathering and lower organic matter accumulation.

Conversely, tropical climates often produce highly weathered soils, such as laterites, which are rich in iron and aluminum oxides.

Topography: The landscape's slope, elevation, and drainage patterns affect soil formation. Steep slopes may experience erosion, while flat areas may accumulate sediments. Topography influences water movement, which in turn affects soil moisture and nutrient availability.

For instance, soils on hilltops may be thinner and less fertile due to erosion, while valley bottoms may accumulate rich alluvial soils from flooding rivers.

Organisms: Living organisms, including plants, animals, and microorganisms, contribute to soil formation through organic matter accumulation and bioturbation. Plant roots help break down rocks, while decomposing organic matter enriches the soil with nutrients.

Soil fauna, such as earthworms and insects, play a vital role in aerating the soil and mixing organic matter, enhancing soil structure and fertility.

Time: Soil formation is a gradual process that occurs over long periods. The length of time that soil has been developing influences its characteristics, including horizon development, nutrient content, and overall fertility.

Older soils may exhibit more developed horizons and greater leaching of nutrients, while younger soils may retain more of their parent material characteristics.

Processes of Soil Formation

Weathering: The breakdown of rocks into smaller particles through mechanical, chemical, and biological processes. Mechanical weathering involves physical forces, such as freeze-thaw cycles and abrasion, while chemical weathering alters the mineral composition of rocks through reactions with water and acids.

Leaching: The process by which water-soluble substances are washed out of the soil, often leading to nutrient depletion. Leaching can significantly affect soil fertility, particularly in areas with high rainfall, where essential nutrients are washed away from the root zone.

Organic Matter Accumulation: The buildup of decomposed plant and animal material contributes to soil fertility and structure. Organic matter enhances soil's water retention capacity, improves aeration, and provides essential nutrients for plant growth.

Soil Horizons

Soil is organized into distinct layers known as horizons, each with unique characteristics that influence soil fertility and water retention:

O Horizon (Organic Layer): Composed mainly of organic matter, such as decomposed leaves, plant material, and microorganisms. This layer is rich in nutrients and plays a vital role in soil fertility. The O horizon is typically dark in color due to the high organic content and is crucial for supporting plant life.

A Horizon (Topsoil): The uppermost mineral layer, rich in organic material and nutrients. It is crucial for plant growth and is typically darker in color due to organic matter content. The A horizon is where most biological activity occurs, and it is the layer that farmers and gardeners often amend with fertilizers and organic matter to enhance fertility.

E Horizon (Eluviation Layer): A leached layer where minerals and nutrients are washed out, resulting in a lighter color. This horizon is often found in forested areas and indicates the movement of materials downward. The E horizon is significant for understanding nutrient dynamics and the overall health of the soil profile.

B Horizon (Subsoil): Accumulates minerals and nutrients leached from the upper layers. It is denser and often has a higher clay content, affecting water retention and root penetration. The B horizon is critical for plant growth, as it serves as a reservoir of nutrients and minerals.

C Horizon (Parent Material): Composed of weathered rock and partially disintegrated material. This layer serves as the source of minerals for soil formation and development. The C horizon is less affected by biological activity and weathering processes compared to the upper horizons, but it provides essential minerals that contribute to the soil's overall fertility.

R Horizon (Bedrock): The unweathered rock layer beneath the soil profile. It is the source of the parent material for soil formation and can vary widely in composition, depending on the geological history of the area. The R horizon is typically not directly involved in soil processes but serves as the foundation upon which soil develops.

Impact on Soil Fertility and Water Retention: The arrangement and characteristics of these horizons significantly affect soil fertility and water retention. The O and A horizons are critical for nutrient availability, while the B horizon influences water movement and retention. The presence of clay in the B horizon can enhance water retention, while sandy soils may drain quickly, affecting plant growth.

The E horizon, with its leached materials, can indicate nutrient deficiencies, which may require amendments to support healthy plant growth. Understanding the composition and characteristics of each horizon allows land managers, farmers, and engineers to make informed decisions about soil management, crop selection, and construction practices.

Soil Types

Soils can be classified into various types based on their texture, structure, and formation processes. Major soil types include:

Sandy Soils: Composed of large particles, sandy soils have high permeability and low water retention. They warm quickly in the spring but may require more frequent irrigation due to rapid drainage. Sandy soils are often low in nutrients, making them less suitable for certain crops without proper fertilization.

Clayey Soils: Characterized by fine particles, clayey soils have low permeability and high water retention. They can become compacted and may pose challenges for drainage but are often rich in nutrients. Clayey soils can support a wide range of crops but may require careful management to prevent waterlogging and improve aeration.

Loamy Soils: A balanced mixture of sand, silt, and clay, loamy soils are ideal for agriculture due to their good drainage, nutrient-holding capacity, and ability to retain moisture. Loamy soils provide an optimal environment for root growth and are often considered the best soil type for farming.

Silty Soils: Silty soils consist of fine particles that are smaller than sand but larger than clay. They have good water retention and nutrient-holding capacity, making them suitable for agriculture. However, silty soils can be prone to erosion if not managed properly.

Peaty Soils: Rich in organic matter, peaty soils are formed in wetland areas where plant material decomposes slowly. They have high water retention and can be very fertile, but their acidity and low nutrient availability can limit their agricultural use.

Saline Soils: These soils contain high levels of soluble salts, which can inhibit plant growth. Saline soils are often found in arid regions and can result from poor drainage or excessive irrigation. Management practices, such as leaching and the use of salt-tolerant crops, are necessary to mitigate their effects.

Significance of Soil Texture and Structure

Soil texture and structure are critical factors in agricultural productivity and water management. Texture affects water retention, drainage, and aeration, while structure influences root penetration and the movement of air and water within the soil. Understanding these aspects is essential for optimizing crop growth and ensuring sustainable land use.

Soil Texture: The relative proportions of sand, silt, and clay determine the soil's texture, which significantly influences its physical properties. For example, sandy soils drain quickly but may not retain enough moisture for crops, making them less suitable for certain agricultural practices without adequate irrigation. In contrast, clayey soils retain water well but can become compacted, limiting root growth and reducing aeration. Understanding soil texture helps farmers select appropriate crops and irrigation methods, ensuring that they can maximize yields while minimizing water use.

Soil Structure: The arrangement of soil particles into aggregates affects how water and air move through the soil. Well-structured soils promote good drainage and aeration, which are essential for healthy root development and overall plant health. Poorly structured soils can lead to waterlogging, reduced plant growth, and increased susceptibility to erosion. Soil management practices, such as cover cropping, reduced tillage, and the addition of organic matter, can improve soil structure over time, enhancing its capacity to support agricultural activities.

4.3 Types of Soils Based on Mode of Formation

Soils can also be categorized based on their mode of formation, which includes transported soils and residual soils. Transported soils are carried away from their original location by agents such as gravity, water, wind, and glaciers, and they can be further classified into several types:

Alluvial Soils: Formed along stream beds, alluvial soils consist mainly of gravels and sands, with some clay. They are well-sorted based on grain size and are fertile due to nutrient deposition during flooding. Alluvial soils are often found in floodplains and deltas, making them ideal for agriculture due to their rich nutrient content.

Aeolian Soils: These soils are transported and deposited by wind, prevalent in arid regions. They often consist of fine quartz and clay minerals, forming deposits like loess, which can pose construction challenges due to their tendency to liquefy when saturated. While aeolian soils can be highly fertile, they require careful management to prevent erosion and maintain their productivity.

Marine and Lacustrine Soils: Deposited in seas and lakes, respectively, marine soils typically have higher salt content and different characteristics compared to lacustrine soils, which are usually less saline and more fertile. These soils can support unique ecosystems and agricultural practices, particularly in coastal and wetland areas.

Glacial Soils: Resulting from glacial activity, these soils are unsorted and heterogeneous, with two main types: glacial till (boulder clay) and fluvio-glacial soils, which are better sorted and stratified. Glacial soils can vary widely in composition and are often found in regions previously covered by ice, presenting both challenges and opportunities for land use.

Residual Soils: These soils form in place from the weathering of the underlying bedrock. They retain many characteristics of the parent material and are often less fertile than transported soils. The fertility of residual soils can vary significantly depending on the type of rock from which they are derived, influencing their agricultural potential.

Organic Soils: Formed primarily from the accumulation of organic matter in wetland areas, organic soils are rich in nutrients but can be acidic. They are often found in peat bogs and marshes, supporting unique plant communities and providing important ecological functions.

Volcanic Soils: Formed from volcanic ash and lava, these soils are often very fertile due to the presence of minerals released during volcanic eruptions. Volcanic soils can support diverse agricultural practices and are typically found in regions with active or dormant volcanoes, contributing to local food security.

Conclusion

In summary, understanding soil formation processes, horizons, and types is essential for effective land management, agricultural productivity, and construction practices. By recognizing the characteristics and behaviors of different soils, engineers, farmers, and land planners can make informed decisions that promote sustainability and resilience in the face of environmental challenges. The interplay of soil properties, formation processes, and human activities underscores the importance of soil as a vital resource for sustaining life and supporting economic development. Proper management and conservation of soil resources are crucial for ensuring food security, environmental health, and the overall well-being of future generations.

Soil of Mixed Composition

In nature, soils occur as mixtures, and it is the silica-rich soils that are widespread. Thus, it is common to encounter soils such as clayey sand, sandy gravel, silty sand, and very clayey organic silt. The occurrence of calcareous soils also needs to be highlighted. Calcareous material can be incorporated into transported soils, represented in the form of calcite minerals. Percolating groundwater often introduces calcareous grains into the soils through a recrystallization process, further influencing the soil's chemical properties and fertility. The presence of calcareous materials can enhance soil alkalinity and provide essential nutrients, particularly calcium, which is vital for plant growth.

The diversity of soil types and their mixed compositions underscores the complexity of soil ecosystems. For instance, soils with a combination of sand, silt, and clay can exhibit unique characteristics that affect their behavior in terms of water retention, drainage, and nutrient availability. This variability is crucial for agricultural practices, as it allows for the cultivation of a wide range of crops suited to different soil conditions.

Moreover, understanding the mixed composition of soils is essential for effective land management and conservation strategies. For example, recognizing the presence of calcareous materials can inform decisions regarding fertilization and soil amendments, ensuring that agricultural practices are tailored to the specific needs of the soil. Additionally, mixed soils may require different irrigation strategies to optimize water use and prevent issues such as salinization or erosion.

In conclusion, the significance of soil texture, structure, and composition cannot be overstated. These factors play a critical role in determining soil fertility, water management, and overall agricultural productivity. By understanding the complexities of soil formation and the interactions between various soil types, stakeholders can implement more effective land-use practices that promote sustainability and resilience. This knowledge is vital for addressing the challenges posed by climate change, urbanization, and other environmental pressures, ensuring that soil resources are managed wisely for future generations. Proper soil management not only supports agricultural productivity but also contributes to environmental health and the sustainability of ecosystems, highlighting the integral role of soil in our lives.

Some Effects of Clay Minerals

Surface Charge and Absorption

Clay minerals play a crucial role in soil chemistry and physics due to their unique structural properties. The platy surface of a clay mineral layer possesses a fairly high net negative charge. This negative charge arises from the presence of free O^{2-} and OH^{-} ions on surfaces that are not covalently linked to any other layer. These ions readily form bonds with positively charged ions when they come into contact, leading to significant implications for soil behavior and nutrient availability.

For instance, when water dissociates, it produces positively charged hydronium ions (H_3O^+). These ions are firmly held onto the surface of the clay mineral layers, resulting in the absorption of water (Figure 11). This process is not limited to hydronium ions; other cations such as sodium (Na^+), potassium (K^+), and calcium (Ca^{2+}), which are derived from salts that dissolve in water, are absorbed in a similar manner. The capacity of clay minerals to absorb these cations is essential for soil fertility, as it influences the availability of nutrients to plants.

The amount of water that is absorbed onto the mineral flakes depends on several factors, including the magnitude of the net negative charge, the specific surface area of the clay minerals, and the concentration and type of salts dissolved in the water. Clay minerals with a higher surface area and greater negative charge can hold more water and cations, enhancing their ability to retain moisture and nutrients.

However, the water that is strongly held onto clay minerals is not easily removed. This characteristic significantly lowers the rate of water movement within the soil. Reduced water movement can lead to slower drainage rates and hinder soil consolidation, which is critical for maintaining soil structure and health.

Moreover, the presence of excess water in clay-rich soils can adversely affect their mechanical properties. The shear strength of soils is influenced by the amount of water present and the drainage conditions at any given time. When soils become saturated, the excess water reduces their shear strength, diminishing the soil's ability to resist deformation when subjected to stress. This can lead to issues such as soil slippage, erosion, and instability, particularly in construction and agricultural contexts.

In summary, the effects of clay minerals on surface charge and absorption are fundamental to understanding soil behavior. Their ability to retain water and nutrients not only impacts agricultural productivity but also plays a critical role in soil stability and health. Recognizing these effects is essential for effective land management and sustainable agricultural practices.

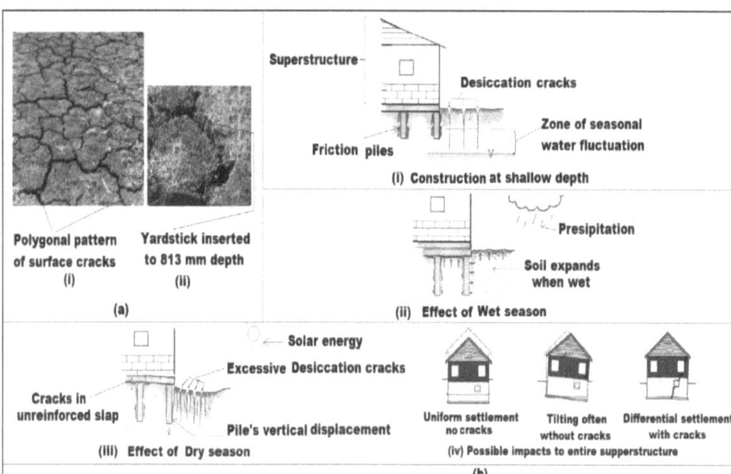

Fig.11: Some negative construction effects of clay soils

Plasticity and Liquefaction

Soils that exhibit plasticity have the ability to undergo deformation over a wide range of moisture content without crumbling. This characteristic is particularly prevalent in soils with a high concentration of clay minerals. The plastic behavior of these soils is influenced by several factors, including the specific surface area of the clay minerals and the types of clay flakes present in the soil matrix.

Plasticity in Clay Soils

1. **Specific Surface Area:** The specific surface area of clay minerals plays a crucial role in determining the plasticity of the soil. Soils containing minerals with high specific surface areas, such as montmorillonite, tend to exhibit greater plasticity. This is because the larger surface area allows for more significant interaction between the clay particles and water, enhancing the soil's ability to deform without cracking. The increased surface area facilitates the adsorption of water molecules, which contributes to the soil's plastic behavior.

2. **Attractive Forces:** The attractive forces that exist between clay mineral flakes also influence the soil's behavior. When the moisture content is low, strong attractive forces develop, resulting in a binding effect that creates internal tension known as cohesion. This cohesion can lead to the development of cracks if the moisture content is not sufficient to maintain plasticity. Conversely, when moisture levels are adequate, these attractive forces allow the soil to maintain its plasticity, enabling it to be molded and shaped without fracturing.

3. **Plastic Range:** The moisture content that allows for deformation without the development of cracks falls within the plastic range of the soil. This range is critical for various engineering applications, as it indicates the moisture levels at which the soil can be manipulated without compromising its structural integrity. Understanding the plastic range is essential for construction activities, as it helps engineers determine the optimal moisture content for soil compaction and stabilization.

4. **Water Absorption Capacity:** The presence of minerals with high absorption capacity, such as montmorillonite, increases the soil's ability to retain water. Strong bonding develops between the clay minerals and

the absorbed water, which helps to maintain the soil's plasticity. However, if the moisture content becomes excessively high, particularly in fine soils subjected to dynamic loading, the soil can undergo liquefaction.

Liquefaction

Liquefaction occurs when saturated soils lose their strength and behave like a liquid due to sudden dynamic loading, such as during an earthquake. In this state, the forces between clay mineral flakes and water molecules, as well as the effective unit weight of the soil, are significantly reduced. The implications of liquefaction are critical:

- **Shear Strength Reduction:** The shear strength of the soil, which is dependent on cohesion, is compromised during liquefaction. This loss of strength can lead to catastrophic failures in structures built on or within the affected soil. Buildings, bridges, and other infrastructure can experience severe damage or collapse if the underlying soil liquefies during seismic events.
- **Mechanical Stabilization:** Cohesion is also vital for the mechanical stabilization of soils. It increases the binding forces that develop between solid grains during the compaction process, enhancing the overall stability of the soil structure. When liquefaction occurs, the loss of cohesion disrupts these binding forces, leading to a significant reduction in the soil's mechanical stability and increasing the risk of ground deformation and settlement.

In summary, understanding the concepts of plasticity and liquefaction is essential for geotechnical engineering and construction practices. The ability of clay soils to exhibit plastic behavior under varying moisture conditions is crucial for effective soil management and engineering design. Additionally, recognizing the conditions that lead to liquefaction allows engineers to implement appropriate measures to mitigate risks associated with saturated soils during dynamic loading events. This knowledge is vital for ensuring the safety and stability of structures built on or within these soils.

Flocculation and Dispersion

Platy surfaces of clay minerals usually carry negative charge. Cations absorbed by a clay mineral get distributed near and around the mineral surface to form a positively charged layer. When two mineral flakes come into contact, repulsion may occur to make the flakes free and dispersed. The magnitude of repulsion that occurs depends on the concentration of the ions present in the absorbed water and other factors. If the absorbed layer is thin enough then the net attractive forces may dominate. The flaky particles then form an edge to face contact (Figure 12). The behavior of clay minerals in soil is also influenced by the processes of flocculation and dispersion, which are critical for understanding soil structure and stability.

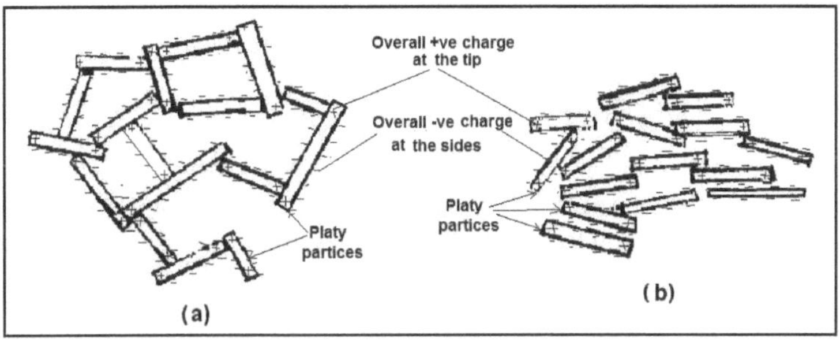

Fig. 12: Schematic diagram of deposits containing salty water (a) and fresh water (b)

Flocculation

Flocculation is a critical process in soil science, particularly concerning clay minerals. The platy surfaces of clay minerals typically carry a negative charge, which plays a significant role in their behavior and interactions. When cations, which are positively charged ions, are absorbed by the clay minerals, they create a positively charged layer around the mineral surface. This charge distribution is essential for the interactions between clay particles, as it influences how they behave in the soil matrix.

When two mineral flakes come into contact, repulsion may occur due to the negative charges on their surfaces. This repulsion can cause the flakes to remain dispersed within the soil matrix, preventing them from aggregating. The magnitude of this repulsion depends on several factors, including the concentration of ions present in the absorbed water and other environmental conditions. Higher concentrations of cations can reduce the repulsive forces, allowing for closer interactions between clay particles.

If the absorbed layer of cations is thin enough, the net attractive forces between the clay particles may dominate over the repulsive forces. In this scenario, the flaky particles can form an edge-to-face contact, leading to flocculation. This arrangement enhances soil structure and stability, as the particles are held together more firmly. Flocculation is beneficial for soil health, as it improves aeration, water retention, and nutrient availability, ultimately contributing to better plant growth.

Conclusion: Environmental Impacts Summary

Importance of Clay Soil Behaviors

The behaviors of swelling, shrinkage, plasticity, liquefaction, flocculation, and dispersion in clay-rich soils have significant environmental and engineering implications. Understanding these characteristics is essential for effective soil management and construction practices.

Swelling and Shrinkage

The swelling and shrinkage behavior of clay soils can lead to structural issues in construction, such as excessive settlement and damage to foundations. When clay soils absorb water, they can swell significantly, and when they dry out, they can shrink, leading to cracks and instability. Proper investigation and management of these properties are crucial to mitigate risks associated with these phenomena, ensuring that structures remain stable and safe over time.

Plasticity and Liquefaction

Plasticity is another important characteristic of clay soils, allowing for deformation without cracking, which can be beneficial in various applications. However, the risk of liquefaction under dynamic loading conditions poses a significant threat to infrastructure stability. Liquefaction occurs when saturated soil temporarily loses strength and stiffness, leading to potential failure during events such as earthquakes. Engineers must account for these risks when designing structures in areas with clay-rich soils, implementing appropriate measures to enhance stability and resilience.

Flocculation and Dispersion

The processes of flocculation and dispersion also influence soil structure and stability. Understanding these interactions can help in managing soil health and fertility, as well as in preventing erosion and maintaining water quality. Flocculation contributes to the formation of stable soil aggregates, which are essential for preventing soil erosion and promoting healthy ecosystems. Conversely, excessive dispersion can lead to soil degradation and reduced agricultural productivity. Therefore, effective management of these processes is vital for sustainable land use and environmental protection.

Summary of Implications

In summary, the complex behaviors of clay minerals in soils have profound implications for environmental management, agricultural practices, and civil engineering. By recognizing and addressing these characteristics, we can develop more sustainable practices that enhance soil health, improve water retention, and support ecosystem resilience in the face of changing environmental conditions.

Swelling and shrinkage

Water that is absorbed by a soil can be retained to cause volume increase. The amount of water that is absorbed varies with the type of clay mineral content present in a soil material. Clay minerals of large specific surface and strong net surface charge absorb more water. Such minerals swell considerable. The swelling characteristic is typical of expansive soils such as Bentonite clay. Marine soils also exhibit a high swelling capacity. They contain a high concentration of expansive minerals such as illite and montmorillonite. Reduction in moisture content can occur due to excessive drainage or rapid surface evaporation at high environmental temperatures, the reduction in moisture content may lead to shrinkage.

Residual soils

Residual soils refer to soils that are deposited in situ without undergoing transportation. They are formed by chemical weathering processes that involve precipitation of hydroxides in regions which experience rapid evaporation of the ground. Typical examples of residual soils are laterite and calcrete. The laterite consists of mainly insoluble hydroxides of iron and aluminum. The soil is derived from volcanic lavas such as basalts and granitic intrusions as well as gneissic rocks while calcrete soil is derived from limestone deposits. Calcrete soils that are rich in calcarious material consist of mainly calcite minerals which usually provide whitish coloration to the soils. Lateritic soils vary in colour from deep red to pink depending on the amount of iron oxide present. The soils are generally referred to as "red tropical soils" and are wide spread in the tropical regions. In these regions, intense chemical weathering processes are predominant.

Lateritization processes involve leaching of silica and re-deposition of hydrous iron oxide constituents. The constituents are inhomogeneously distributed in different zones. The top zone consists of a residue of loose cohesionless sandy and silty material. The zone is underlain by a reddish crust that is enriched with sesqui oxides mixed with kaolin clay. In some case, with increased drainage and prolonged weathering, the iron may be leached to leave a pale enrichment of aluminum known as bauxite. Concentration of the enriched aluminum may be of economic significance so as to warrant its exploitation. For instance, in Guyana bauxite deposits are exploited as they occur on large scale leading their economic viability.

Grain Size Classification

Several schemes have been developed for classification of soils based on a particular purpose. In construction engineering, grain size consideration is emphasized thereby giving five main classes. However, this is a physical approach, which has no genetic bearing. In this format clay represents finest soil material while boulders were coarsest. Normally, laboratory and field description of a soil for construction needs also to consider other properties such as consistency etc. standard particle sizes are as follows in Table 1.1:

Table 1.1: Grain size classification in construction engineering

FINE GRAINED				COARSE GRAINED							
CLAY	**SILT**			**SAND**			**GRAVEL**			**STONE**	
	F	**M**	**C**	**F**	**M**	**C**	**F**	**M**	**C**	**Cobbles**	**boulders**
0.002	0.006	0.02	0.06	0.2	0.6	2	6	20	60	200	
——— Particle size diameter in mm ——→											

Cohesive Soils

Cohesive soils are those that contain sufficient amounts of clay and silt, which collectively contribute to their unique engineering behavior. The characteristics of the clay and silt-sized particles present in these soils significantly influence their properties, particularly plasticity and cohesion. Plasticity refers to the ability of the soil to deform without cracking, while cohesion is the internal binding force that holds soil particles together. These properties are crucial for various engineering applications, including foundation design and slope stability analysis. The presence of clay minerals, such as montmorillonite, enhances the plasticity of cohesive soils, allowing them to retain moisture and maintain their structural integrity under varying conditions.

Granular Soils

Granular soils, on the other hand, are characterized by coarse-grained fragments, primarily composed of gravels and sands. In these soils, the influence of surface forces is considered negligible, and the concentration of clay and silt-sized particles is minimal. Granular soils typically exhibit good drainage properties and are less prone to plastic deformation compared to cohesive soils. Their behavior under load is primarily governed by the intergranular friction between the larger particles, making them suitable for applications such as road construction and drainage systems. The lack of cohesion in granular soils means that they rely on the mechanical interlocking of particles to maintain stability, which is essential for ensuring the integrity of structures built on or within these soils.

Soil of Mixed Composition

In nature, soils often occur as mixtures, leading to a wide variety of soil types. Silica-rich soils are particularly prevalent, resulting in the common occurrence of various soil types such as clayey sand, sandy gravel, silty sand, and very clayey organic silt. The presence of calcareous soils, which contain calcareous materials represented in the form of calcite minerals, is also noteworthy. These materials can be incorporated into transported soils, and percolating groundwater often introduces calcareous grains into the soil through a recrystallization process. The mixed composition of soils can significantly influence their physical and chemical properties, affecting their behavior in engineering and environmental contexts. Understanding these mixtures is essential for effective land use planning and management, as they can impact soil fertility, drainage, and overall soil health.

Environmental Impact of Soil Formation and Types

Introduction

The formation and types of soil have significant environmental impacts that influence ecosystems, agriculture, and human activities. Understanding these impacts is crucial for sustainable land management and environmental conservation.

Role of Soil in Ecosystems

Soil is a critical component of terrestrial ecosystems, providing habitat for a diverse range of organisms, including plants, animals, and microorganisms. Healthy soils support biodiversity, which is essential for ecosystem resilience and function. Soil degradation, such as erosion and nutrient depletion, can lead to loss of biodiversity and disrupt ecosystem services, including water filtration, carbon storage, and nutrient cycling.

Impact on Agriculture

The type and quality of soil directly affect agricultural productivity. Fertile soils with good structure and nutrient content support healthy crop growth, while poor soils can lead to reduced yields and increased reliance on fertilizers and pesticides. Unsustainable agricultural practices, such as over-tillage and monocropping, can degrade soil health, leading to erosion, compaction, and loss of organic matter. This degradation not only impacts food security but also contributes to environmental issues such as water pollution from runoff.

Water Management Considerations

Additionally, soil types influence water retention and drainage, which are critical for effective water management. Sandy soils drain quickly, requiring more frequent irrigation, while clayey soils retain water but can lead to waterlogging. Understanding soil properties helps in designing effective irrigation systems and managing water resources sustainably.

Soil and Carbon Sequestration

Soils play a vital role in carbon sequestration, acting as a significant carbon sink. Healthy soils with high organic matter content can store carbon, helping to mitigate climate change. However, soil degradation and erosion can release stored carbon back into the atmosphere, contributing to greenhouse gas emissions. Sustainable soil management practices, such as cover cropping and reduced tillage, can enhance soil carbon storage and improve overall soil health.

Contamination and Remediation

Furthermore, soil types and their formation processes can influence the susceptibility of soils to contamination from pollutants, such as heavy metals, pesticides, and industrial waste. Contaminated soils can pose risks to human health and the environment, affecting food safety and water quality. Remediation efforts are often necessary to restore contaminated soils, which can be costly and time-consuming.

Land Use Planning

Understanding soil types and their properties is essential for effective land use planning. Different soils have varying capabilities for supporting infrastructure, agriculture, and natural habitats. Poor land use decisions can lead to soil degradation, increased erosion, and loss of arable land. Sustainable land management practices that consider soil characteristics can help balance development needs with environmental protection.

Conclusion

In summary, understanding soil formation processes, horizons, and types is essential for effective land management, agricultural productivity, and construction practices. The interplay of soil properties, formation processes, and human activities underscores the importance of soil as a vital resource for sustaining life and supporting economic development.

4.3 Groundwater Systems and Aquifers

Groundwater Importance

Groundwater is a critical resource, providing drinking water for billions of people and supporting agricultural irrigation and industrial processes. Aquifers, which are underground layers of water-bearing rock or sediment, play a vital role in storing and transmitting groundwater. There are two main types of aquifers: unconfined aquifers, which are directly recharged by precipitation, and confined aquifers, which are separated from the surface by impermeable layers. The hydrological cycle describes the movement of water through the environment, including the interaction between groundwater and surface water bodies. The water table represents the upper surface of the saturated zone in an aquifer, and its level can fluctuate based on factors such as precipitation, evaporation, and human extraction.

However, groundwater resources face significant challenges, including over-extraction and contamination. Over-extraction can lead to aquifer depletion, land subsidence, and reduced water quality. Contamination from agricultural runoff, industrial discharges, and septic systems poses serious risks to groundwater quality, necessitating effective monitoring and management strategies. The importance of groundwater can be categorized into several key areas.

Drinking Water Supply

Groundwater serves as a primary source of drinking water for many communities, particularly in rural and semi-urban areas where surface water sources may be limited or contaminated. Wells and boreholes are commonly used to access groundwater, providing a reliable and often cleaner alternative to surface water. The quality of groundwater is generally higher than that of surface water due to natural filtration processes that occur as water percolates through soil and rock layers.

Irrigation

Agriculture is one of the largest consumers of groundwater, accounting for approximately 70% of global freshwater use. Groundwater irrigation is essential for food production, especially in arid and semi-arid regions where rainfall is insufficient to support crops. The ability to access groundwater allows farmers to maintain crop yields and sustain agricultural productivity, contributing to food security and economic stability.

Industrial Use

Many industries rely on groundwater for various processes, including manufacturing, cooling, and cleaning. Industries such as food processing, textiles, and pharmaceuticals utilize groundwater as a critical resource. The availability of groundwater can influence industrial location decisions and economic development in regions where surface water is scarce.

Ecosystem Support

Groundwater contributes to the health of ecosystems by maintaining base flows in rivers and streams, supporting wetlands, and providing habitat for aquatic life. The interaction between groundwater and surface water is essential for sustaining biodiversity and ecological balance.

Aquifer Types

Aquifers are geological formations that can store and transmit groundwater. They are classified into two main types based on their confinement and hydrological characteristics.

Unconfined Aquifers

Unconfined aquifers are those in which the water table is open to the atmosphere and is directly recharged by precipitation and surface water infiltration. The upper boundary of an unconfined aquifer is the water table, which fluctuates based on seasonal variations in rainfall and groundwater extraction. Unconfined aquifers are typically more susceptible to contamination due to their direct exposure to surface activities. They often have higher recharge rates compared to confined aquifers, as water can easily infiltrate through the overlying soil and rock layers. The hydraulic conductivity of unconfined aquifers can vary significantly depending on the soil and rock types, influencing the rate at which groundwater can flow.

Confined Aquifers

Confined aquifers are situated beneath a layer of impermeable rock or clay, which restricts the flow of water into and out of the aquifer. The water in confined aquifers is under pressure, and when tapped by wells, it can rise above the top of the aquifer, sometimes resulting in artesian wells. Confined aquifers typically have lower recharge rates compared to unconfined aquifers due to the presence of the impermeable layer. Recharge occurs through specific areas known as recharge zones, where the confining layer is absent or thin. The water quality in confined aquifers is often better than in unconfined aquifers, as the overlying layers can filter out contaminants. However, confined aquifers can still be vulnerable to contamination from deep sources.

Groundwater Flow

The movement of groundwater is a critical component of the hydrological cycle, which describes the continuous circulation of water within the Earth and atmosphere. Groundwater flow is influenced by various factors, including hydraulic gradients, soil and rock permeability, and the presence of aquifers.

Hydrological Cycle

The hydrological cycle consists of several processes, including evaporation, condensation, precipitation, infiltration, and runoff. Groundwater plays a significant role in this cycle, as it is recharged by precipitation that infiltrates the soil and percolates down to the water table. Groundwater can also discharge into surface water bodies, such as rivers, lakes, and wetlands, contributing to their flow and maintaining ecological health. This interaction is crucial for sustaining aquatic ecosystems, especially during dry periods when surface water levels may drop.

Water Table

The water table is the upper surface of the saturated zone in an unconfined aquifer, where the soil and rock are fully saturated with water. The depth of the water table can vary significantly based on factors such as rainfall, evaporation, and groundwater extraction. Seasonal variations in precipitation can lead to fluctuations in the water table. During wet seasons, the water table rises due to increased recharge, while it may drop during dry periods or droughts. Human activities, such as excessive groundwater extraction for irrigation or urban development, can also lower the water table, leading to potential long-term impacts on water availability and ecosystem health.

Aquifer Depletion and Contamination

The sustainability of groundwater resources is threatened by various challenges, including over-extraction and contamination. Addressing these issues is critical for ensuring the long-term availability of groundwater.

Aquifer Depletion

Over-extraction of groundwater occurs when the rate of withdrawal exceeds the natural recharge rate of the aquifer. This can lead to a decline in water levels, reduced water availability, and negative impacts on ecosystems. The consequences of depletion can be severe. For instance, land subsidence can occur as a result of aquifer depletion, where the ground sinks due to the compaction of soil and rock layers. This phenomenon can damage infrastructure, reduce land usability, and increase the risk of flooding. Additionally, depletion can lead to the deterioration of water quality, as the concentration of contaminants may increase when water levels drop.

Groundwater Contamination

Groundwater contamination can arise from various sources, including agricultural runoff, industrial discharges, and improper waste disposal. Contaminants such as pesticides, fertilizers, heavy metals, and pathogens can infiltrate groundwater, posing risks to human health and the environment.

Sources of Contamination: Agricultural runoff is a significant contributor to groundwater contamination. The use of fertilizers and pesticides in agriculture can lead to nutrient loading and chemical contamination of groundwater. Nitrates from fertilizers are particularly concerning, as they can cause health issues such as methemoglobinemia (blue baby syndrome) in infants. Industrial discharges also pose a threat, as industries may release pollutants into the environment that can leach into groundwater. Contaminants such as solvents, heavy metals, and hydrocarbons can persist in groundwater and pose significant health risks. Additionally, improperly managed landfills and waste disposal sites can lead to leachate formation, which can contaminate groundwater with harmful substances.

Impacts on Water Quality

Contaminated groundwater can have severe implications for drinking water supplies, agricultural practices, and ecosystem health. The presence of contaminants can render groundwater unsafe for human consumption and necessitate costly treatment processes. Furthermore, contaminated groundwater can affect surface water bodies when it discharges into rivers and lakes, leading to broader environmental impacts. The degradation of water quality can disrupt aquatic ecosystems, harm wildlife, and compromise the health of communities that rely on these water sources.

Conclusion

Groundwater systems and aquifers are essential components of the Earth's hydrological cycle, providing critical resources for drinking water, irrigation, and industrial use. Understanding the characteristics of different aquifer types, groundwater flow dynamics, and the challenges of aquifer depletion and contamination is vital for sustainable water management.

As global populations continue to grow and water demands increase, the need for effective groundwater management practices becomes increasingly important. Strategies such as responsible water extraction, pollution prevention, and the implementation of sustainable agricultural practices are essential to protect and preserve groundwater resources for future generations. By prioritizing the sustainable management of groundwater, we can ensure that this vital resource remains available for drinking, agriculture, and ecosystem health, ultimately supporting the well-being of both human and natural communities.

Sustainable Groundwater Management Strategies

Monitoring and Assessment

Regular monitoring of groundwater levels and quality is crucial for understanding the status of aquifers and detecting changes over time. This can involve the use of piezometers to measure water levels and sampling wells to assess water quality. The data collected can inform management decisions and help identify areas at risk of depletion or contamination. By establishing a comprehensive monitoring network, stakeholders can gain insights into groundwater dynamics, enabling proactive measures to protect this vital resource.

Recharge Enhancement

Enhancing groundwater recharge can help maintain aquifer levels and improve water availability. Techniques such as rainwater harvesting, constructing recharge basins, and implementing permeable surfaces in urban areas can facilitate the infiltration of rainwater into the ground. Additionally, restoring wetlands and natural landscapes can enhance natural recharge processes. These strategies not only increase groundwater supplies but also contribute to improved water quality and ecosystem health by filtering pollutants and providing habitat for wildlife.

Water Conservation Practices

Promoting water conservation in agricultural, industrial, and domestic settings can significantly reduce groundwater extraction. Implementing efficient irrigation techniques, such as drip irrigation and scheduling irrigation based on soil moisture levels, can minimize water use in agriculture. Public awareness campaigns can encourage households to adopt water-saving practices, such as fixing leaks, using water-efficient appliances, and practicing mindful water use. By fostering a culture of conservation, communities can help ensure the sustainability of groundwater resources.

Pollution Prevention

Preventing groundwater contamination is critical for protecting water quality. This can involve implementing best management practices (BMPs) in agriculture, such as integrated pest management (IPM) and the use of organic fertilizers. Industries should adhere to regulations regarding waste disposal and emissions to minimize the risk of contaminating groundwater. Additionally, promoting the use of environmentally friendly products and practices can further reduce the likelihood of groundwater pollution, safeguarding this essential resource for future generations.

Regulatory Frameworks

Establishing and enforcing regulations related to groundwater extraction and protection is essential for sustainable management. This can include setting limits on groundwater withdrawals, requiring permits for new wells, and designating protected areas around sensitive aquifers. Collaborative governance involving stakeholders, including local communities, farmers, and industries, can enhance the effectiveness of regulatory measures. By fostering cooperation and shared responsibility, regulatory frameworks can better address the complexities of groundwater management.

Public Education and Engagement

Educating the public about the importance of groundwater and the challenges it faces can foster a culture of stewardship. Community engagement initiatives can involve local stakeholders in groundwater management decisions, ensuring that diverse perspectives are considered and that solutions are tailored to local needs. By raising awareness and encouraging active participation, communities can work together to protect and manage their groundwater resources effectively.

Research and Innovation

Continued research into groundwater systems, aquifer behavior, and contamination sources is vital for developing effective management strategies.

Innovations in technology, such as remote sensing and modeling tools, can enhance our understanding of groundwater dynamics and improve decision-making processes. By investing in research and fostering collaboration between academic institutions, government agencies, and industry, we can develop innovative solutions to address groundwater challenges.

Conclusion

Groundwater systems and aquifers are indispensable resources that support human life, agriculture, and industry. Their sustainable management is critical for ensuring water security and environmental health in the face of growing demands and climate change. By implementing comprehensive strategies that encompass monitoring, recharge enhancement, pollution prevention, and public engagement, we can protect these vital resources for future generations.

The challenges of aquifer depletion and contamination underscore the need for a proactive approach to groundwater management. As we strive to balance the competing demands for water, it is essential to recognize the interconnectedness of groundwater with surface water, ecosystems, and human activities. Through collaborative efforts and informed decision-making, we can safeguard groundwater resources, ensuring their availability and quality for years to come.

In summary, the importance of groundwater cannot be overstated. It is a critical resource that requires careful stewardship to maintain its availability and quality. By understanding the complexities of groundwater systems and implementing sustainable practices, we can work towards a future where groundwater continues to support both human needs and ecological health.

Summary of Environmental Impacts of Groundwater Systems and Aquifers

Groundwater systems and aquifers play a crucial role in sustaining ecosystems, human populations, and agricultural practices. However, their environmental impacts can be significant, particularly when considering issues related to depletion and contamination. Here are the key environmental impacts associated with groundwater systems and aquifers:

Ecosystem Health

Groundwater is essential for maintaining the health of ecosystems, particularly wetlands, rivers, and lakes. It supports base flows in surface water bodies, which are critical for aquatic habitats. Over-extraction of groundwater can lead to reduced stream flows, negatively impacting aquatic life and biodiversity.

Aquifer Depletion

Over-extraction of groundwater can result in aquifer depletion, where the rate of water withdrawal exceeds the natural recharge rate. This can lead to a decline in water levels, reduced water availability for ecosystems and human use, and land subsidence, which can damage infrastructure and alter landscapes.

Water Quality Degradation

Groundwater contamination from agricultural runoff, industrial discharges, and improper waste disposal can severely impact water quality. Contaminants such as nitrates, heavy metals, and pathogens can infiltrate aquifers, posing risks to human health and the environment. Contaminated groundwater can also affect surface water quality when it discharges into rivers and lakes.

Land Subsidence

Excessive groundwater extraction can lead to land subsidence, where the ground sinks due to the compaction of soil and rock layers. This phenomenon can cause structural damage to buildings, roads, and other infrastructure, as well as increase the risk of flooding in low-lying areas.

Altered Hydrological Cycles

Groundwater systems are integral to the hydrological cycle. Changes in groundwater levels can disrupt natural water flow patterns, affecting precipitation, evaporation, and surface water interactions. This alteration can lead to changes in local climate conditions and affect agricultural practices.

Impact on Agriculture

While groundwater is vital for irrigation, over-reliance on this resource can lead to unsustainable practices that degrade soil health and reduce agricultural productivity. Additionally, groundwater contamination from agricultural inputs can compromise food safety and ecosystem health.

Climate Change Vulnerability

Groundwater systems are increasingly vulnerable to the impacts of climate change, including altered precipitation patterns and increased evaporation rates. These changes can exacerbate existing challenges related to groundwater availability and quality, making sustainable management even more critical.

Biodiversity Loss

The depletion and contamination of groundwater can lead to habitat loss for various species that depend on consistent water availability. Aquatic and semi-aquatic ecosystems, such as wetlands and riparian zones, are particularly vulnerable. As groundwater levels decline, these habitats may shrink or disappear, leading to a decline in biodiversity and the potential extinction of sensitive species.

Salinization

In coastal areas, excessive groundwater extraction can lead to saltwater intrusion, where seawater encroaches into freshwater aquifers. This salinization of groundwater can render it unsuitable for drinking and irrigation, posing significant challenges for agriculture and drinking water supplies. Salinization can also affect soil health and crop yields, further exacerbating food security issues.

Public Health Risks

Contaminated groundwater poses direct risks to public health. The presence of pathogens, heavy metals, and chemical pollutants can lead to serious health issues, including gastrointestinal diseases, neurological disorders, and reproductive problems. Communities relying on contaminated groundwater for drinking water may face increased healthcare costs and reduced quality of life.

Economic Impacts

The degradation of groundwater resources can have significant economic consequences. Reduced water availability can lead to decreased agricultural productivity, impacting food supply and farmer livelihoods. Additionally, the costs associated with treating contaminated water or restoring degraded aquifers can be substantial, placing financial burdens on communities and governments.

Social and Equity Issues

Access to clean and sufficient groundwater is often uneven, leading to social and equity issues. Vulnerable communities, particularly in rural and marginalized areas, may face greater challenges in accessing safe drinking water. Groundwater depletion and contamination can exacerbate existing inequalities, leading to conflicts over water resources and increased social tensions.

Conclusion

The environmental impacts of groundwater systems and aquifers are multifaceted and interconnected, affecting ecosystems, human health, and economic stability. The challenges posed by aquifer depletion, contamination, and changing hydrological patterns necessitate a comprehensive approach to groundwater management that prioritizes sustainability and resilience.

To mitigate these impacts, it is essential to adopt integrated water resource management practices that consider the interconnectedness of groundwater with surface water, ecosystems, and human activities. This includes promoting conservation, enhancing recharge, preventing contamination, and ensuring equitable access to clean water.

By addressing the environmental impacts associated with groundwater systems and aquifers, we can work towards a future where these vital resources are protected, ensuring their availability and quality for generations to come. Sustainable groundwater management is not only crucial for maintaining ecological balance but also for supporting human well-being and economic development in an increasingly water-scarce world.

4.4 Water Quality and Contamination

Assessing water quality is essential for ensuring safe drinking water and protecting aquatic ecosystems. Key parameters for evaluating water quality include pH, turbidity, dissolved oxygen, and the presence of contaminants such as heavy metals, pathogens, and nutrients. Monitoring these parameters helps identify potential health risks and informs management decisions. Water contamination can arise from both point sources, such as industrial discharges, and non-point sources, such as agricultural runoff. Agricultural practices, including the use of fertilizers and pesticides, can lead to nutrient runoff, resulting in eutrophication of water bodies. This process depletes oxygen levels and harms aquatic life.

The impacts of contaminated water extend beyond human health, affecting ecosystems and soil quality. Contaminated water can lead to the degradation of habitats, loss of biodiversity, and reduced agricultural productivity. Effective monitoring and assessment methods, including sampling techniques and laboratory analysis, are crucial for identifying contamination sources and implementing remediation strategies.

Water Quality Parameters

Assessing water quality is essential for ensuring safe drinking water, protecting aquatic ecosystems, and maintaining overall environmental health. Key parameters for evaluating water quality include:

pH

pH measures the acidity or alkalinity of water on a scale from 0 to 14, with 7 being neutral. Values below 7 indicate acidic conditions, while values above 7 indicate alkalinity. pH affects the solubility and availability of nutrients and contaminants, influencing aquatic life and the effectiveness of water treatment processes.

Turbidity

Turbidity refers to the cloudiness or haziness of water caused by suspended particles, such as sediments, algae, and organic matter. High turbidity can reduce light penetration, affecting photosynthesis in aquatic plants and disrupting the habitat for fish and other organisms. It can also indicate the presence of pollutants.

Dissolved Oxygen (DO)

Dissolved oxygen is crucial for the survival of aquatic organisms. It is the amount of oxygen available in water for fish, invertebrates, and microorganisms. DO levels can be affected by temperature, organic matter decomposition, and nutrient levels. Low DO can lead to hypoxic conditions, harming aquatic life and disrupting ecosystems.

Presence of Contaminants

Contaminants in water can significantly impact both human health and the environment. Heavy metals, such as lead, mercury, cadmium, and arsenic, can enter water bodies through industrial discharges, mining activities, and urban runoff. These metals can accumulate in aquatic organisms and pose serious health risks to humans and wildlife. Pathogens, including bacteria, viruses, and parasites, can contaminate water sources through sewage discharge, agricultural runoff, and wildlife. These pathogens can cause waterborne diseases, posing significant health risks to humans. Additionally, excessive nutrients, particularly nitrogen and phosphorus, can lead to eutrophication, resulting in algal blooms that deplete oxygen levels and harm aquatic life. Nutrient pollution often originates from agricultural runoff and wastewater discharges.

Sources of Contamination

Water pollution can originate from various sources, which are generally categorized into point and non-point sources:

Point Sources

Point sources are identifiable and discrete sources of pollution. Examples include industrial discharges, where factories and manufacturing plants may release pollutants directly into water bodies through pipes or outfalls. Wastewater treatment plants can also introduce contaminants into rivers, lakes, and oceans, whether through treated or untreated sewage. Additionally, leachate from landfills can contaminate groundwater and surface water if not properly managed.

Non-Point Sources

Non-point sources are diffuse and harder to trace. Agricultural runoff is a significant contributor, as fertilizers, pesticides, and sediments from agricultural fields can wash into nearby water bodies during rainfall or irrigation. Urban runoff is another concern, as rainwater can carry pollutants from streets, parking lots, and construction sites into storm drains and waterways. Atmospheric deposition also plays a role, with pollutants being deposited into water bodies through precipitation, including acid rain and airborne heavy metals.

Impact of Contamination

Contaminated water can have severe effects on human health, aquatic ecosystems, and soil quality:

Human Health

Contaminated water can lead to a range of health issues, including gastrointestinal diseases, neurological disorders, and reproductive problems. Vulnerable populations, such as children and the elderly, are particularly at risk. Waterborne pathogens can cause outbreaks of diseases like cholera, dysentery, and hepatitis A, highlighting the critical need for safe drinking water.

Aquatic Ecosystems

Water contamination can disrupt aquatic ecosystems by harming fish and other organisms, reducing biodiversity, and altering food webs. Eutrophication can lead to hypoxic zones, where oxygen levels are too low to support most marine life, resulting in fish kills and loss of biodiversity. The overall health of aquatic ecosystems is jeopardized, affecting not only the organisms that inhabit them but also the communities that rely on these ecosystems for their livelihoods.

Soil Quality

Contaminants can leach into the soil, affecting its quality and health. Heavy metals and chemicals can accumulate in the soil, impacting plant growth and entering the food chain. Contaminated soil can also affect groundwater quality, creating a cycle of pollution that is difficult to remediate. This degradation of soil quality can have long-term implications for agricultural productivity and ecosystem health.

Monitoring and Assessment

Monitoring water quality is essential for detecting contamination, assessing ecosystem health, and ensuring compliance with regulatory standards. Key methods for monitoring water quality include:

Sampling Techniques

Water samples can be collected from various locations and depths to provide a representative assessment of water quality. Sampling methods may include grab sampling, which involves collecting a single sample at a specific time, and composite sampling, which combines multiple samples over time to provide a more comprehensive view of water quality.

Laboratory Analysis

Collected water samples are analyzed in laboratories to measure various parameters, including pH, turbidity, dissolved oxygen, and the presence of contaminants. Common analytical techniques include spectrophotometry, which is used to measure the concentration of specific contaminants, microbiological testing to detect pathogens, and chromatography to separate and identify complex mixtures of contaminants.

Field Testing

Portable water quality testing kits can be used for on-site assessments, providing immediate results for parameters like pH, turbidity, and dissolved oxygen. These kits are valuable for rapid screening and preliminary assessments, allowing for quick decision-making in response to potential contamination events.

Remote Sensing

Satellite and aerial imagery can be used to monitor large water bodies and assess changes in water quality over time. Remote sensing can help identify algal blooms, sedimentation patterns, and other indicators of water quality, providing valuable data for managing water resources effectively.

Conclusion

Understanding water quality parameters, sources of contamination, and the impacts of pollution is essential for protecting human health and aquatic ecosystems. The environmental impacts of water contamination are far-reaching, affecting not only the immediate water bodies but also the surrounding ecosystems, soil quality, and overall biodiversity. Contaminated water can lead to the degradation of aquatic habitats, resulting in loss of species and disruption of food webs. Additionally, pollutants can accumulate in sediments, affecting organisms that rely on these habitats and potentially entering the food chain, which poses risks to both wildlife and human health.

Effective monitoring and assessment methods are crucial for identifying contamination issues and implementing appropriate management strategies. By prioritizing water quality and addressing contamination sources, we can safeguard this vital resource for current and future generations. The consequences of neglecting water quality can lead to severe ecological imbalances, increased health risks, and economic burdens associated with water treatment and ecosystem restoration.

Sustainable practices, public awareness, and regulatory frameworks will play a key role in ensuring clean and safe water for all. Promoting responsible agricultural practices, reducing urban runoff, and enforcing regulations on industrial discharges are essential steps in mitigating contamination. Furthermore, fostering community engagement and education about the importance of water conservation and pollution prevention can empower individuals to take action in protecting their local water resources.

In summary, the health of our water systems is intricately linked to the well-being of both human populations and the environment. By recognizing the environmental impacts of water contamination and committing to sustainable management practices, we can work towards a future where clean, safe water is accessible to all, supporting thriving ecosystems and healthy communities.

4.5 Sustainable Management Practices

Sustainable management practices are essential for preserving soil and water resources. Soil conservation techniques, such as crop rotation, cover cropping, and reduced tillage, enhance soil health and prevent erosion. These practices improve soil structure, increase organic matter content, and promote biodiversity. Water management strategies, including rainwater harvesting, efficient irrigation techniques, and watershed management, help optimize water use and reduce waste. Integrated Water Resources Management (IWRM) is a holistic approach that considers the interconnectedness of water, land, and ecosystems. IWRM promotes collaboration among stakeholders to balance water supply, demand, and environmental sustainability. Community engagement is also vital for successful soil and water management. Involving local communities in decision-making processes fosters a sense of ownership and responsibility, leading to more effective and sustainable practices. This section discusses various techniques and strategies that contribute to sustainable soil and water management.

Soil Conservation Techniques

Soil conservation is critical for maintaining soil health, preventing erosion, and enhancing agricultural productivity. Key practices include:

Crop Rotation

Crop rotation involves alternating the types of crops grown in a particular field over different seasons or years. This practice helps break pest and disease cycles, improves soil fertility, and enhances biodiversity. By varying root structures and nutrient demands, crop rotation can also reduce soil erosion and improve soil structure, leading to healthier soils.

Cover Cropping

Cover cropping entails planting specific crops, such as legumes or grasses, during the off-season or between main crops. These cover crops protect the soil from erosion, improve soil structure, and enhance organic matter content. They also help suppress weeds, reduce nutrient leaching, and promote beneficial soil microorganisms. Additionally, cover crops can fix nitrogen in the soil, reducing the need for synthetic fertilizers.

Reduced Tillage

Reduced tillage practices minimize soil disturbance, preserving soil structure and organic matter. By limiting the frequency and intensity of tillage, farmers can reduce erosion, enhance water retention, and promote soil health. No-till or conservation tillage methods allow for the maintenance of soil cover, which protects against erosion and improves moisture retention.

Agroforestry

Integrating trees and shrubs into agricultural landscapes can enhance soil conservation. Agroforestry practices improve soil structure, reduce erosion, and increase biodiversity. Trees provide shade, reduce wind speed, and contribute organic matter through leaf litter, enhancing soil fertility.

Water Management Strategies

Sustainable water management practices are essential for optimizing water use, conserving resources, and protecting water quality. Key strategies include:

Rainwater Harvesting

Rainwater harvesting involves collecting and storing rainwater for later use. This practice can reduce reliance on groundwater and surface water sources, especially in arid regions. By capturing rainwater from rooftops or other surfaces, communities can supplement their water supply for irrigation, drinking, and domestic use, thereby enhancing water security.

Efficient Irrigation Techniques

Implementing efficient irrigation methods, such as drip irrigation and sprinkler systems, can significantly reduce water waste. These techniques deliver water directly to the plant roots, minimizing evaporation and runoff. Additionally, scheduling irrigation based on soil moisture levels and crop needs can optimize water use and improve crop yields.

Watershed Management

Watershed management involves the coordinated management of land and water resources within a specific watershed area. This approach aims to protect water quality, enhance water availability, and restore ecosystems. Effective watershed management includes practices such as reforestation, wetland restoration, and erosion control, which help maintain the natural hydrological cycle and improve water quality.

Integrated Water Resources Management (IWRM)

Integrated Water Resources Management (IWRM) is a comprehensive approach that seeks to balance water supply, demand, and environmental sustainability. Key aspects of IWRM include:

Holistic Approach

IWRM recognizes the interconnectedness of water resources with land use, ecosystems, and human activities. By considering the entire water cycle and the various demands on water resources, IWRM promotes a more sustainable and equitable allocation of water.

Stakeholder Involvement

IWRM emphasizes the importance of involving all stakeholders, including government agencies, local communities, and the private sector, in decision-making processes. Engaging diverse perspectives ensures that water management strategies are inclusive, equitable, and responsive to local needs.

Balancing Competing Uses

IWRM aims to balance competing demands for water, such as agricultural, industrial, and domestic uses, while also considering environmental needs. This approach helps prevent over-extraction and degradation of water resources, ensuring their sustainability for future generations.

Adaptive Management

IWRM incorporates adaptive management principles, allowing for flexibility in response to changing conditions, such as climate variability and population growth. By continuously monitoring and evaluating water resources, stakeholders can adjust management strategies to address emerging challenges effectively.

Community Engagement

Community involvement is crucial for the success of sustainable soil and water management practices. Key aspects of community engagement include:

Education and Awareness

Educating communities about the importance of soil and water conservation fosters a sense of responsibility and stewardship. Workshops, training sessions, and informational campaigns can empower individuals to adopt sustainable practices in their daily lives and agricultural activities.

Participatory Decision-Making

Involving community members in decision-making processes ensures that local knowledge and needs are considered. Participatory approaches, such as community forums and stakeholder consultations, can lead to more effective and culturally appropriate management strategies.

Local Initiatives

Supporting local initiatives and grassroots organizations can enhance community engagement in sustainable practices. These initiatives often leverage local knowledge and resources, leading to innovative solutions tailored to specific environmental challenges.

Collaboration with Local Governments

Collaborating with local governments and agencies can strengthen community efforts in soil and water management. By aligning community goals with governmental policies and programs, communities can access resources, technical assistance, and funding for sustainable projects.

Conclusion

Importance of Sustainable Management Practices

Sustainable management practices for soil and water resources are essential for ensuring the health of ecosystems, supporting agricultural productivity, and safeguarding human well-being. By implementing soil conservation techniques such as crop rotation, cover cropping, and reduced tillage, we can enhance soil health, prevent erosion, and promote biodiversity. These practices not only improve agricultural yields but also contribute to the resilience of ecosystems in the face of climate change. The environmental impacts of soil degradation, such as loss of fertility, increased erosion, and reduced water retention, can be mitigated through these sustainable practices, ultimately leading to healthier ecosystems and improved food security.

Effective Water Management Strategies

In addition, effective water management strategies, including rainwater harvesting, efficient irrigation techniques, and comprehensive watershed management, are vital for optimizing water use and protecting water quality. These strategies help mitigate the impacts of water scarcity, reduce reliance on finite resources, and enhance the sustainability of agricultural and urban water supplies. Poor water management can lead to significant environmental consequences, such as habitat destruction, reduced biodiversity, and increased pollution from runoff. By adopting sustainable water management practices, we can protect aquatic ecosystems and ensure the availability of clean water for all.

Integrated Water Resources Management (IWRM)

The concept of Integrated Water Resources Management (IWRM) plays a crucial role in balancing water supply and demand while ensuring environmental sustainability. By adopting a holistic approach that involves all stakeholders, IWRM fosters collaboration and promotes equitable access to water resources. This integrated framework is essential for addressing the complex challenges posed by competing water demands, climate variability, and population growth. Failure to implement IWRM can result in over-extraction of water resources, degradation of aquatic habitats, and increased vulnerability to climate change impacts, further exacerbating environmental challenges.

Community Engagement

Community engagement is a cornerstone of successful sustainable management practices. By involving local communities in education, participatory decision-making, and grassroots initiatives, we can empower individuals to take ownership of their natural resources. This engagement not only enhances the effectiveness of management strategies but also fosters a culture of stewardship and responsibility toward soil and water conservation. Communities that are actively involved in managing their resources are more likely to adopt sustainable practices that protect their environment and promote resilience against environmental degradation.

Summary of a Multifaceted Approach

In summary, the sustainable management of soil and water resources requires a multifaceted approach that combines innovative practices, community involvement, and integrated management frameworks. By prioritizing these sustainable practices, we can protect and enhance our vital natural resources, ensuring their availability and health for current and future generations. The commitment to sustainable soil and water management is not only an environmental imperative but also a pathway to achieving social equity, economic resilience, and ecological integrity in our communities.

A Vision for the Future

Together, we can build a sustainable future that honors the interconnectedness of our ecosystems and the essential resources they provide, ultimately safeguarding the environment for generations to come.

4.6 Case Studies in Soil and Water Management

Case studies provide valuable insights into successful practices and strategies for managing soil and water resources. By examining real-world examples, we can learn from the experiences of different regions and apply these lessons to other contexts. This section presents three case studies that highlight innovative approaches to sustainable soil and water management.

Case Study 1: Sustainable Agriculture in the Midwest

The Midwest region of the United States, often referred to as the "Corn Belt," is a critical area for agricultural production. However, intensive farming practices have historically led to soil degradation, nutrient runoff, and water quality issues. In response, various sustainable agriculture initiatives have been implemented to enhance soil health and improve water quality. Farmers in the Midwest have adopted practices such as cover cropping, no-till farming, and crop rotation to improve soil health. Cover crops, such as clover and rye, are planted during the off-season to protect the soil from erosion, enhance organic matter, and improve nutrient cycling. No-till farming reduces soil disturbance, preserving soil structure and moisture retention.

Precision agriculture techniques have been employed to optimize fertilizer application, reducing nutrient runoff into nearby water bodies. By using soil testing and data analytics, farmers can apply fertilizers more efficiently, ensuring that crops receive the necessary nutrients while minimizing excess application. Collaborative efforts among farmers, conservation organizations, and government agencies have led to the establishment of watershed partnerships. These partnerships focus on implementing best management practices (BMPs) to reduce nutrient runoff and improve water quality in local rivers and lakes. Educational programs and financial incentives encourage farmers to adopt sustainable practices.

As a result of these initiatives, many Midwest farmers have reported improved soil health, increased crop yields, and enhanced water quality in surrounding ecosystems. The adoption of sustainable practices has also contributed to greater resilience against climate variability, helping farmers adapt to changing weather patterns.

Case Study 2: Groundwater Management in California

California faces significant challenges in managing its groundwater resources, particularly in the context of prolonged droughts and over-extraction. The state relies heavily on groundwater for agricultural irrigation and urban water supply, making effective management critical for sustainability. Over-extraction of groundwater has led to declining water tables, land subsidence, and reduced water quality. Drought conditions exacerbate these issues, placing additional stress on already limited water resources. The reliance on groundwater for agriculture has resulted in unsustainable pumping rates, threatening the long-term viability of this resource.

In response to these challenges, California enacted the Sustainable Groundwater Management Act (SGMA) in 2014. This legislation requires local agencies to develop and implement Groundwater Sustainability Plans (GSPs) for high-priority basins. The plans must address issues such as water quality, groundwater levels, and land subsidence. Local agencies are employing various strategies to manage groundwater sustainably, including enhanced monitoring of groundwater levels and quality to inform management decisions and track progress toward sustainability goals. Initiatives to enhance groundwater recharge, such as creating recharge basins and restoring wetlands, are being implemented to increase the amount of water that infiltrates into aquifers. Promoting water conservation practices among agricultural and urban users is essential for reducing demand on groundwater resources.

While challenges remain, the implementation of SGMA has led to increased awareness of groundwater issues and greater collaboration among stakeholders. Local agencies are working to develop sustainable management practices that balance agricultural needs with environmental protection, ultimately aiming to restore and protect California's vital groundwater resources.

Case Study 3: Urban Water Management in Singapore

Singapore is a global leader in innovative urban water management, employing a comprehensive approach to ensure a sustainable and resilient water supply in a densely populated city-state. Faced with limited natural water resources, Singapore has developed a range of strategies to manage water effectively. Singapore's approach to water management is characterized by an integrated framework that combines various sources of water, including rainwater, recycled water, and desalinated water. This holistic approach ensures a reliable and diverse water supply.

The city-state has invested heavily in advanced technologies for water treatment and management. The NEWater initiative involves the treatment and recycling of wastewater into high-quality reclaimed water, which is then used for industrial processes and as a supplementary water source. Additionally, smart water management systems utilize sensors and data analytics to monitor water quality and optimize distribution. Singapore has implemented extensive rainwater harvesting systems, including green roofs and rain gardens, to capture and utilize rainwater. These systems help reduce stormwater runoff and enhance water availability during dry periods.

Public education and community engagement are integral to Singapore's water management strategy. The government conducts outreach programs to raise awareness about water conservation and the importance of sustainable practices. Initiatives such as the "Water Wally" mascot promote water-saving behaviors among residents. Singapore's innovative water management practices have resulted in a highly efficient and resilient water supply system. The country has successfully reduced its reliance on imported water and has become a model for urban water management worldwide. By integrating technology, community involvement, and sustainable practices, Singapore has demonstrated that effective water management is achievable even in resource-limited environments.

Conclusion

Diverse Approaches to Sustainable Management

These case studies illustrate the diverse approaches to sustainable soil and water management across different contexts. From the agricultural practices in the Midwest to groundwater management in California and urban water innovations in Singapore, each example highlights the importance of tailored strategies that consider local conditions and stakeholder involvement.

Agricultural Practices in the Midwest

In the Midwest, the adoption of sustainable agricultural practices such as cover cropping, no-till farming, and precision nutrient management has not only improved soil health and increased crop yields but has also significantly reduced nutrient runoff into local waterways. This reduction in runoff helps mitigate the environmental impacts of eutrophication, which can lead to harmful algal blooms and degraded water quality in rivers and lakes. By enhancing soil structure and organic matter content, these practices contribute to greater biodiversity in agricultural ecosystems, promoting resilience against pests and diseases while also improving carbon sequestration capabilities.

Groundwater Management in California

In California, the Sustainable Groundwater Management Act (SGMA) represents a proactive approach to addressing the critical challenges of groundwater depletion and land subsidence. By requiring local agencies to develop Groundwater Sustainability Plans, California is fostering a collaborative framework that encourages the responsible use of water resources. The implementation of recharge projects and water conservation strategies not only helps to restore groundwater levels but also protects surface water bodies from the adverse effects of over-extraction. This is particularly important in maintaining aquatic habitats and ensuring the availability of clean water for both human and ecological needs. The increased awareness and stakeholder engagement fostered by SGMA also promote a culture of sustainability that can lead to long-term environmental benefits.

Urban Water Innovations in Singapore

Singapore's innovative urban water management practices serve as a model for cities facing similar challenges of limited natural water resources. By integrating various sources of water, including rainwater harvesting and wastewater recycling, Singapore has significantly reduced its reliance on imported water. This approach not only conserves water but also minimizes the environmental impacts associated with water extraction and transportation, such as habitat destruction and pollution. The use of advanced technologies for water treatment and distribution enhances the efficiency of water use, while community engagement initiatives ensure that residents are informed and active participants in water conservation efforts. This holistic approach contributes to the overall resilience of the urban ecosystem, allowing Singapore to adapt to climate change and urbanization pressures.

Learning from Successful Initiatives

By learning from these successful initiatives, we can identify best practices and adapt them to address the unique challenges faced in other regions. The environmental impacts of effective soil and water management are profound, as sustainable practices not only enhance soil health and water quality but also contribute to the overall resilience of ecosystems. For instance, the adoption of soil conservation techniques in the Midwest has led to reduced erosion, improved nutrient retention, and enhanced biodiversity, which are critical for maintaining healthy ecosystems. Similarly, California's groundwater management strategies aim to mitigate the adverse effects of over-extraction, such as land subsidence and declining water quality, thereby protecting vital habitats and ensuring the sustainability of water resources.

The Path Forward

As we move forward, it is crucial to continue sharing knowledge and experiences from various regions to foster collaboration and innovation in soil and water management. The environmental challenges we face, such as climate change, habitat loss, and water scarcity, require collective action and adaptive strategies. By learning from one another and adapting successful practices to local contexts, we can work towards a more sustainable and resilient future for our soil and water resources, ultimately benefiting both people and the environment.

A Vision for the Future

Together, we can build a world where soil and water are managed sustainably, ensuring their availability for generations to come.

4.7 Future Directions in Soil and Water Resource Management

As climate change continues to impact soil and water resources, it is essential to adapt management practices accordingly. The effects of climate change, including changes in precipitation patterns, increased evaporation rates, and altered growing seasons, pose significant challenges to agricultural productivity and water availability. For instance, regions that traditionally relied on consistent rainfall may experience prolonged droughts or intense flooding, which can disrupt planting schedules and reduce crop yields. Understanding these impacts will be crucial for developing resilient management strategies that can withstand the uncertainties of a changing climate.

Adapting Management Practices

Climate-Resilient Agriculture

As climate change continues to impact agricultural systems, farmers will need to adopt climate-resilient agricultural practices that can adapt to changing conditions. This may include selecting drought-resistant crop varieties that can withstand periods of low rainfall and implementing agroecological practices that promote biodiversity and ecosystem health. For instance, farmers can choose native or locally adapted crops that are better suited to the specific climate conditions of their region, thereby enhancing their chances of successful harvests even in adverse weather. Additionally, integrating agroforestry—where trees and shrubs are planted alongside crops—can provide shade, reduce soil erosion, and improve microclimates, further supporting crop resilience.

Diversifying cropping systems is another effective strategy to reduce vulnerability to climate extremes. Practices such as crop rotation and intercropping can enhance soil health by improving nutrient cycling and reducing pest and disease pressures. For example, rotating legumes with cereal crops can naturally replenish soil nitrogen levels, reducing the need for synthetic fertilizers. Intercropping, where two or more crops are grown together, can create a more complex ecosystem that supports beneficial insects and reduces the likelihood of pest outbreaks. These practices not only optimize water use but also contribute to the overall resilience of agricultural systems, allowing farmers to better cope with the uncertainties posed by climate variability.

The environmental impacts of adopting climate-resilient agriculture are significant. By improving soil health and reducing reliance on chemical inputs, these practices can enhance biodiversity, promote beneficial soil organisms, and improve carbon sequestration. Healthier soils are better at retaining moisture, which can mitigate the effects of drought and reduce the need for irrigation. Furthermore, diverse cropping systems can lead to reduced pest outbreaks, minimizing the need for pesticides that can harm non-target species and degrade water quality. By fostering a more balanced ecosystem, climate-resilient agriculture not only supports food security but also contributes to the overall health of the environment.

Water Conservation Techniques

As water scarcity becomes more prevalent, implementing water conservation techniques will be critical for sustainable resource management. Strategies such as rainwater harvesting, greywater recycling, and the use of water-efficient irrigation systems, such as drip irrigation, can help maximize water use efficiency. Rainwater harvesting involves collecting and storing rainwater from rooftops or other surfaces for later use, which can significantly reduce dependence on municipal water supplies. Greywater recycling, which involves reusing water from sinks, showers, and washing machines for irrigation, can further decrease overall water consumption. Additionally, adopting water-efficient irrigation systems, like drip irrigation, allows for targeted watering directly at the plant roots, minimizing evaporation and runoff.

These methods not only allow for the collection and reuse of water but also reduce reliance on traditional water sources and minimize waste. Restoring wetlands and natural water bodies can enhance groundwater recharge and improve overall water availability. Wetlands act as natural sponges, absorbing excess rainwater and slowly releasing it into the groundwater system, which is crucial for maintaining water levels during dry periods. Furthermore, the restoration of these ecosystems can provide additional benefits, such as improved water quality and enhanced biodiversity.

The environmental benefits of these water conservation techniques are profound. By capturing and utilizing rainwater, communities can reduce surface runoff, which often carries pollutants into waterways, thereby improving water quality. Restoring wetlands not only enhances groundwater recharge but also provides critical habitats for wildlife, supports biodiversity, and improves flood resilience. These practices contribute to healthier ecosystems and can help mitigate the impacts of climate change by maintaining hydrological balance in the landscape. By prioritizing water conservation, we can ensure that this vital resource is available for future generations while also protecting the ecosystems that depend on it.

Technological Innovations

Technological innovations present new opportunities for improving soil and water management. One key advancement is precision agriculture, which utilizes data analytics, GPS technology, and remote sensing to optimize farming practices. By monitoring soil moisture levels, nutrient content, and crop health in real-time, farmers can make informed decisions about irrigation, fertilization, and pest management. This targeted approach reduces waste, minimizes environmental impacts, and enhances productivity, ultimately leading to more sustainable agricultural practices.

Remote sensing technologies, such as satellite imagery and drones, allow for large-scale monitoring of soil and water resources. These tools can provide valuable insights into land use changes, soil degradation, and water quality, enabling timely interventions and more effective resource management. Furthermore, innovations in water purification, such as membrane filtration, advanced oxidation processes, and bioremediation, can improve water quality and make previously unusable water sources viable for agricultural and urban use. These technologies can help address contamination issues and ensure a safe and reliable water supply, contributing to the overall sustainability of water resources.

The environmental impacts of these technological innovations are significant. Precision agriculture reduces the over-application of fertilizers and pesticides, which can lead to nutrient runoff and water pollution. By optimizing resource use, these technologies help protect aquatic ecosystems and improve overall water quality. Additionally, advanced water purification techniques can mitigate the effects of industrial and agricultural contaminants, ensuring that water resources remain safe for both human and ecological health.

Effective Policies and Governance Frameworks

Effective policies and governance frameworks are vital for promoting sustainable soil and water resource management. Policymakers must consider the interconnectedness of these resources and engage stakeholders in decision-making processes to ensure long-term sustainability. One key consideration is the promotion of integrated resource management, which recognizes the interdependence of soil and water resources. This approach encourages collaboration among various sectors, including agriculture, urban planning, and environmental conservation, to develop comprehensive strategies that address multiple challenges simultaneously.

Engaging stakeholders, including local communities, farmers, and industry representatives, in the policymaking process is essential for creating effective and equitable management strategies. Participatory approaches can help ensure that diverse perspectives are considered and that policies are tailored to local needs and conditions. Additionally, policymakers can implement financial incentives, such as subsidies or grants, to encourage the adoption of sustainable practices among farmers and landowners. These incentives can support investments in technology, conservation practices, and infrastructure improvements that enhance soil and water management.

The environmental impacts of effective policies and governance frameworks are profound. By promoting integrated resource management, these policies can lead to more holistic approaches that protect both soil and water resources. Engaging stakeholders fosters a sense of ownership and responsibility, which can enhance compliance with sustainable practices and improve environmental outcomes. Financial incentives can accelerate the adoption of practices that reduce pollution, enhance soil health, and improve water quality, ultimately leading to healthier ecosystems and more resilient communities.

Summary of Chapter 4

In this chapter, we explored the significance of soil and water resources, their formation, and the challenges associated with their management. We discussed the importance of soil health and water quality, as well as the impacts of contamination. Sustainable management practices, including soil conservation techniques and integrated water resources management, were highlighted as essential for preserving these vital resources. Through case studies, we illustrated successful approaches to soil and water management, emphasizing the importance of community engagement and innovative solutions.

Additionally, we examined the critical role of education and awareness in promoting sustainable practices among various stakeholders. By fostering a deeper understanding of the interconnectedness of soil and water systems, we can empower communities to take proactive measures in resource management. Educational initiatives that focus on the benefits of sustainable practices can lead to increased participation and commitment from local populations, ultimately enhancing the effectiveness of management strategies.

Finally, we addressed future directions in soil and water resource management, emphasizing the need for adaptive practices in response to climate change, the role of technological innovations in enhancing efficiency, and the importance of effective policies and stakeholder engagement. As we move forward, it is essential to remain flexible and responsive to emerging challenges, ensuring that our approaches to soil and water management are both resilient and sustainable. By integrating diverse perspectives and leveraging advancements in technology, we can create a more sustainable future for these critical resources.

Conclusion: Environmental Impacts Summary

The environmental impacts of soil and water resource management are profound and far-reaching. Sustainable practices not only enhance soil health and water quality but also contribute to the overall resilience of ecosystems. Effective management can mitigate the adverse effects of soil degradation, such as erosion and nutrient loss, while improving water quality by reducing contamination from agricultural runoff. As we face the challenges posed by climate change, the need for adaptive management strategies becomes increasingly critical.

By embracing technological innovations and fostering collaborative governance frameworks, we can ensure that soil and water resources are managed sustainably. This commitment not only addresses immediate environmental concerns but also lays the groundwork for long-term ecological health, social equity, and economic resilience. Ultimately, prioritizing the health of our soil and water resources is essential for creating a balanced relationship between human activities and the natural environment.

Moreover, the role of education and community involvement cannot be overstated in the pursuit of sustainable soil and water management. Engaging local communities in decision-making processes fosters a sense of ownership and responsibility towards natural resources. Educational initiatives that raise awareness about the importance of soil and water conservation can empower individuals to adopt sustainable practices in their daily lives. By cultivating a culture of stewardship, we can inspire collective action that leads to more effective resource management and environmental protection.

In addition, the integration of innovative technologies, such as precision agriculture and water-efficient irrigation systems, can significantly enhance resource management efforts. These technologies allow for more efficient use of inputs, reducing waste and minimizing environmental impacts. By leveraging data-driven approaches and real-time monitoring, we can optimize soil and water management practices, ensuring that resources are used judiciously and sustainably. This technological advancement, combined with community engagement and education, will be crucial in navigating the complexities of resource management in a changing climate.

In summary, the future of soil and water resource management hinges on our ability to adapt to environmental changes, leverage technological advancements, and foster collaborative governance. By prioritizing sustainability, we can build a resilient future that not only meets the needs of current populations but also safeguards the health of our planet for generations to come. Together, we can create a world where soil and water are managed responsibly, ensuring their availability and vitality as essential resources for life.

CHAPTER 5

ENVIRONMENTAL IMPACT ASSESSMENT (EIA)

5.1 Introduction to Environmental Impact Assessment

Environmental Impact Assessment (EIA) is a systematic process used to evaluate the potential environmental effects of proposed projects or actions before they are implemented. The primary purpose of EIA is to inform decision-makers and the public about the potential environmental consequences of a project, allowing for informed choices that promote sustainable development. The EIA process is designed to identify, predict, and evaluate the environmental impacts of proposed actions, ensuring that environmental considerations are integrated into the decision-making process.

The legal and regulatory framework for EIA varies by country, with many nations requiring EIA for specific types of projects, such as infrastructure development, mining, and industrial activities. In some jurisdictions, EIA is mandated by law, while in others, it may be a voluntary process. International guidelines, such as those established by the World Bank and the United Nations, also provide a framework for conducting EIAs, emphasizing the importance of environmental protection and public participation. These guidelines help ensure that EIA processes are consistent, transparent, and effective in addressing environmental concerns.

5.2 Importance of EIA

EIA serves as a critical decision-making tool for assessing the environmental implications of proposed projects. By identifying potential impacts early in the planning process, EIA allows decision-makers to consider alternatives and implement measures to mitigate adverse effects. This proactive approach not only helps protect the environment but also reduces the risk of costly delays and legal challenges that can arise from unanticipated environmental issues.

For instance, projects that undergo EIA are less likely to face opposition from the public or regulatory bodies, as potential concerns are addressed upfront.

Public participation is a fundamental aspect of the EIA process. Engaging stakeholders, including local communities, environmental organizations, and government agencies, fosters transparency and accountability. It allows for diverse perspectives to be considered, ensuring that the concerns of affected parties are addressed. This collaborative approach enhances the legitimacy of the EIA process and promotes community support for projects. Moreover, public involvement can lead to better project outcomes, as local knowledge and insights can inform the assessment and mitigation strategies.

Furthermore, EIA contributes to sustainable development by promoting environmentally responsible practices. By assessing potential impacts and encouraging mitigation measures, EIA helps ensure that projects align with broader sustainability goals, balancing economic growth with environmental protection. This alignment is crucial in an era where environmental degradation and climate change pose significant threats to ecosystems and human well-being.

5.3 EIA Process

The EIA process typically involves several key steps:

Screening

The first step in the EIA process is screening, which determines whether a proposed project requires an EIA. This assessment is based on the project's potential environmental impacts and the regulatory requirements in place. Projects that are likely to have significant environmental effects typically undergo a full EIA. Screening helps to streamline the process by identifying projects that may have minimal impacts and do not require extensive assessment.

Scoping

Once a project is determined to require an EIA, the scoping phase identifies the key issues and concerns that need to be addressed. This involves consultations with stakeholders and regulatory authorities to outline the scope of the assessment, including the specific environmental aspects to be studied.

Scoping is essential for focusing the assessment on the most significant impacts, ensuring that resources are allocated effectively and that the assessment is relevant to stakeholders.

Impact Assessment

During this phase, methodologies are employed to assess the potential environmental impacts of the project. This can include both qualitative and quantitative approaches, such as modeling, field studies, and expert evaluations. The assessment considers direct, indirect, cumulative, and long-term impacts on various environmental components, including air, water, soil, flora, and fauna. A comprehensive impact assessment provides a detailed understanding of how a project may affect the environment and helps identify potential mitigation measures.

Mitigation Measures

Based on the findings of the impact assessment, the EIA report outlines strategies to avoid, minimize, or compensate for adverse environmental impacts. Mitigation measures may include design modifications, implementation of best management practices, and restoration efforts. The effectiveness of these measures is critical for ensuring that the project can proceed without causing significant harm to the environment.

Reporting

The results of the EIA are compiled into an Environmental Impact Statement (EIS) or report, which summarizes the assessment findings, proposed mitigation measures, and public comments. The EIS serves as a key document for regulatory review and public scrutiny. It is essential that the EIS is clear, comprehensive, and accessible to stakeholders, as it forms the basis for decision-making.

Review and Decision-Making

Regulatory authorities review the EIS to ensure compliance with legal requirements and assess the adequacy of the proposed mitigation measures. Based on this review, authorities make decisions regarding project approval, modification, or rejection. This step is crucial for ensuring that projects meet environmental standards and that potential impacts are adequately addressed.

Monitoring and Follow-Up

After project approval, monitoring is essential to ensure that mitigation measures are implemented effectively and to assess the actual environmental impacts of the project. Follow-up assessments can inform future EIAs and improve the overall process. Monitoring helps to ensure accountability and provides valuable data for evaluating the effectiveness of mitigation strategies.

5.4 EIA Methodologies

EIA methodologies can vary widely, but they generally fall into two categories: qualitative and quantitative methods.

Qualitative Methods

These methods involve descriptive assessments of potential impacts based on expert judgment, stakeholder input, and case studies. Qualitative assessments are useful for understanding complex interactions and contextual factors that may not be easily quantifiable. They can provide insights into social, cultural, and economic impacts that are critical for comprehensive decision-making.

Quantitative Methods

Quantitative assessments use numerical data and statistical analyses to evaluate potential impacts. This can include modeling techniques, such as air quality modeling or hydrological modeling, to predict changes resulting from the project. Quantitative methods provide a more objective basis for assessing impacts and can help to identify thresholds beyond which significant effects may occur.

Public participation techniques are also integral to the EIA process. Engaging stakeholders through public meetings, surveys, and workshops allows for the collection of diverse perspectives and concerns. This input can enhance the quality of the assessment and foster community support. Risk assessment is another critical component of EIA. It involves evaluating potential hazards and uncertainties associated with the project, helping to identify areas where additional mitigation measures may be necessary.

5.5 Challenges in EIA

Despite its importance, the EIA process faces several challenges:

Data Limitations

Conducting a comprehensive EIA requires access to high-quality data on environmental conditions and potential impacts. In many cases, data may be incomplete, outdated, or unavailable, hindering the assessment process. This lack of data can lead to uncertainties in the assessment and may result in inadequate mitigation measures.

Political and Economic Influences

The EIA process can be influenced by political and economic factors, including pressure from project proponents, regulatory agencies, and interest groups. These influences can affect the objectivity of the assessment and the implementation of mitigation measures. Ensuring that the EIA process remains transparent and independent is essential for maintaining public trust and confidence.

Public Engagement

Effectively engaging the public and stakeholders in the EIA process can be challenging. Barriers such as lack of awareness, mistrust, and limited access to information can hinder meaningful participation. Ensuring that all voices are heard and considered is essential for a successful EIA. Public engagement is not just a regulatory requirement; it is a vital component that enhances the quality of the assessment and fosters community support for projects. Strategies to improve public engagement may include using accessible language in communications, providing multiple platforms for feedback, and ensuring that engagement activities are inclusive and representative of the affected communities.

5.6 Case Studies in EIA

Case Study 1: Infrastructure Development in Coastal Areas

An EIA conducted for a coastal infrastructure project assessed the potential impacts on marine ecosystems, local fisheries, and coastal communities. The assessment identified key concerns related to habitat loss and water quality, leading to the implementation of mitigation measures such as habitat restoration and monitoring programs. The EIA process highlighted the importance of considering ecological connectivity and the potential long-term effects of construction activities on marine biodiversity. By engaging local fishermen and community members, the project team was able to incorporate traditional ecological knowledge into the assessment, enhancing the relevance and effectiveness of the proposed mitigation strategies.

Case Study 2: Mining Project EIA

The EIA process for a proposed mining project involved extensive assessments of land, water, and air quality impacts. Stakeholder consultations revealed community concerns about potential contamination and displacement. The final Environmental Impact Statement (EIS) included commitments to implement best practices for waste management and community engagement. The EIA also emphasized the need for ongoing monitoring of water quality and the establishment of a community advisory board to facilitate communication between the mining company and local residents. This case study illustrates how effective stakeholder engagement can lead to more socially responsible mining practices and improved community relations.

Geological and Environmental Impacts of Mining Sites

In addition to the broader environmental impacts associated with mining, it is crucial to recognize the specific geological environmental impacts that arise from mining activities, particularly in the context of Artisanal Small-Scale Mining (ASM). Understanding these impacts is essential for developing effective Environmental Impact Assessment (EIA) processes and mitigation strategies (see below section 5.8).

Alteration of Landforms

Mining activities, especially in ASM, often lead to significant alterations in landforms. The extraction of minerals can create open pits, trenches, and waste dumps, which can dramatically change the topography of an area. These alterations can disrupt natural drainage patterns, leading to increased surface runoff and erosion. The resulting changes in landforms can also affect local ecosystems, as species that depend on specific habitats may be displaced or face increased competition for resources.

Soil Compaction and Degradation

The heavy machinery used in mining operations can lead to soil compaction, which reduces the soil's permeability and its ability to support vegetation. Compacted soils are less capable of retaining moisture, which can hinder plant growth and reduce biodiversity. Additionally, the removal of topsoil during mining can lead to soil degradation, making it difficult for ecosystems to recover after mining activities cease.

Ground Instability and Erosion

Mining can create ground instability, particularly in areas where steep slopes are involved. The removal of material can increase the risk of landslides and subsidence, posing hazards to both the environment and local communities. Erosion is another significant concern, as disturbed land is more susceptible to soil loss, which can lead to sedimentation in nearby water bodies. This sedimentation can degrade water quality and disrupt aquatic habitats.

Hydrological Changes

Mining activities can significantly alter hydrological systems. The excavation of minerals can disrupt groundwater flow and surface water drainage, leading to changes in water availability for both ecosystems and local communities. These hydrological changes can result in the drying up of springs and streams, affecting the flora and fauna that depend on these water sources. Additionally, altered water flow can lead to increased flooding in some areas and reduced water supply in others.

Contamination of Geological Resources

The use of chemicals in mining processes, such as cyanide in gold extraction or mercury in artisanal mining, can lead to the contamination of geological resources. These pollutants can leach into the soil and groundwater, posing serious health risks to local communities and wildlife. The contamination of geological resources can also have long-term effects on the environment, making areas uninhabitable and reducing their ecological value.

Impact on Geological Features

Mining can lead to the destruction of significant geological features, such as unique rock formations, fossil beds, and mineral deposits. The loss of these features not only diminishes the geological heritage of an area but also reduces opportunities for scientific research and education. Preserving geological features is essential for maintaining biodiversity and the ecological integrity of mining regions.

Conclusion

Incorporating an understanding of geologica andl environmental impacts into ASM appraisal programs focused on Environmental Impact Assessment is vital for promoting sustainable mining practices in Kenya. By systematically evaluating the potential geological impacts of mining activities, proposing effective mitigation measures, and addressing training needs, these programs can significantly reduce the adverse effects of ASM on the environment.

Case Study 3: Renewable Energy Project

An EIA for a wind energy project evaluated the potential impacts on local wildlife, including bird and bat populations. The assessment included public consultations to address community concerns about visual impacts and noise. Mitigation measures included siting adjustments to avoid critical habitats and the implementation of monitoring programs to minimize impacts on wildlife. This case study demonstrates the importance of balancing renewable energy development with ecological considerations, highlighting how EIA can facilitate the transition to sustainable energy sources while protecting biodiversity.

5.7 Future Directions in EIA

As environmental challenges evolve, the EIA process must adapt to incorporate new considerations:

Integration of Technology

Advancements in technology, such as Geographic Information Systems (GIS) and remote sensing, can enhance the EIA process by providing more accurate data and visualizations of potential impacts. These tools can improve the efficiency and effectiveness of assessments by allowing for real-time monitoring and analysis of environmental conditions. The integration of technology can also facilitate better stakeholder engagement through interactive platforms that allow communities to visualize potential impacts and provide feedback.

Climate Change Considerations

Incorporating climate change impacts into EIA frameworks is essential for addressing the long-term sustainability of projects. Assessing how projects may contribute to or be affected by climate change can inform better decision-making. This includes evaluating the resilience of proposed projects to extreme weather events and changing climatic conditions. By integrating climate considerations into the EIA process, decision-makers can ensure that projects are not only environmentally sustainable but also adaptable to future challenges.

Strengthening Public Participation

Improving public engagement strategies is crucial for enhancing the EIA process. Innovative approaches, such as online platforms for information sharing and feedback, can facilitate broader participation and ensure that diverse perspectives are considered. Additionally, providing training and resources to community members can empower them to engage more effectively in the EIA process. Strengthening public participation not only enhances the legitimacy of the EIA but also leads to more informed and equitable decision-making.

5.8 Environmental Impact Assessment of a Project and Measures to Reduce the Adverse Impact of Mining Activities on the Environment

1. Definition of Environmental Impact Assessment (EIA) and Its Importance

Environmental Impact Assessment (EIA) is a critical process defined by Kenya's National Environment Management Authority (NEMA) as a systematic and thorough examination of the potential effects that a proposed project may have on the environment. This process encompasses the collection and analysis of data regarding the environmental impacts of a project, predicting the likely changes to the environment as a result of the development, evaluating the impacts of various alternatives, and proposing mitigation measures for any significant negative impacts identified. The information gathered during the EIA is made available for decision-making by NEMA, which serves as the designated environmental management and regulatory agency in Kenya.

The importance of conducting an EIA is multifaceted. Firstly, it plays a crucial role in identifying potential impacts, allowing stakeholders—including project developers, government agencies, and local communities—to recognise the likely negative effects of a project on the environment. This identification is essential for informed decision-making, as it provides a foundation for understanding the environmental consequences of proposed actions. Secondly, EIA facilitates informed decision-making by providing comprehensive analyses of potential environmental impacts, thereby assisting decision-makers in determining whether to grant or deny a licence for the proposed project. This process ensures that environmental considerations are integrated into the planning and approval stages of development projects.

Moreover, EIA promotes sustainable development by encouraging the consideration of environmental factors in project planning and implementation. By identifying potential adverse effects and proposing mitigation measures, EIA helps to ensure that economic growth does not come at the expense of environmental degradation. This alignment with sustainable development principles is particularly important in a country like Kenya, where natural resources are vital for the livelihoods of many communities.

2. Major Objectives of an Environmental Impact Assessment

The three major objectives of an EIA are as follows:

1. **To Identify Likely Negative Impacts**: This objective focuses on recognising potential adverse effects that a proposed project may have on the environment, encompassing ecological, social, and economic impacts. By identifying these impacts early in the project planning process, stakeholders can develop strategies to address them effectively.

2. **To Assist Decision Makers**: EIA provides essential information that aids decision-makers in determining whether to approve or reject a project based on its environmental implications. This objective ensures that decision-makers are equipped with the necessary data to make informed choices that balance development needs with environmental protection.

3. **To Identify and Plan Mitigation Measures**: This objective involves proposing strategies to mitigate identified negative impacts, ensuring that any adverse effects on the environment are minimised or eliminated. By planning for mitigation measures, project developers can enhance

the sustainability of their projects and reduce the likelihood of environmental harm.

These objectives serve as a guiding framework for community reviews of EIA reports. They enhance community understanding of the key steps in the EIA process and the potential impacts of the project. Furthermore, by comprehending these objectives, community members can provide systematic feedback—both oral and written—to NEMA regarding the EIA report, ensuring that their concerns are adequately addressed. This engagement aligns with the principles of public service outlined in the Constitution of Kenya, emphasising the importance of community involvement in the decision-making process.

3. Negative Consequences of Mining Activities on the Geo-Environment

Introduction

Mining activities can have several detrimental consequences on the geo-environment, which can be broadly categorized into ecological, social, and economic impacts.

Land Degradation

One of the most significant impacts is land degradation, which occurs as mining operations lead to substantial land disturbance. This disturbance can result in habitat destruction, soil erosion, and a loss of biodiversity, as ecosystems are disrupted and species are displaced. The alteration of landscapes can also affect local climate conditions and water cycles, further exacerbating environmental challenges.

Water Pollution

Water pollution is another critical concern associated with mining activities. The extraction and processing of minerals often involve the use of hazardous chemicals, which can contaminate local water sources through the discharge of pollutants, heavy metals, and sediments. This pollution poses significant risks to aquatic ecosystems, threatening the health of fish and other wildlife, as well as the safety of drinking water for local communities. The long-term effects of water pollution can be devastating, leading to the degradation of water quality and the loss of livelihoods for those who depend on these resources.

Air Pollution

Air pollution is also a significant issue linked to mining operations. Dust and emissions from mining activities can degrade air quality, leading to respiratory problems for nearby communities and wildlife. The release of particulate matter and toxic gases can contribute to a range of health issues, including asthma and other chronic respiratory conditions.

Social Displacement

Furthermore, mining can result in social displacement, as local communities may be forced to relocate due to the encroachment of mining activities on their land. This displacement disrupts local communities and their livelihoods, leading to social tensions and economic instability.

Aesthetic Degradation

The aesthetic degradation of landscapes due to mining operations can diminish their visual appeal, negatively impacting tourism and recreational opportunities. The visual scars left by mining can alter the character of natural landscapes, reducing their attractiveness and value for both residents and visitors.

Comprehensive Approach to Mitigation

To effectively address these challenges, a comprehensive approach is necessary. This approach should begin with a geological-economic appraisal of mineral deposits to ensure sustainable extraction practices.

1. **Improving Extraction Methods:** Adopting advanced techniques that minimize environmental disturbance and enhance resource recovery can significantly mitigate negative impacts. For instance, the implementation of low-impact mining technologies can reduce the physical footprint of mining operations and limit the generation of waste.

2. **Water Management Strategies:** Effective water management is essential for preventing water pollution. This includes treating mine effluents and managing drainage systems to protect groundwater resources. Constructing treatment facilities that ensure any effluents discharged into local water bodies meet regulatory standards for water quality is crucial. Additionally, implementing rainwater harvesting and

recycling systems can help reduce the demand on local water sources and mitigate the impact of mining on water availability.

3. **Rehabilitation of Degraded Lands:** Restoring mined areas to their natural state or repurposing them for beneficial uses can enhance environmental conditions and economic value. Rehabilitation efforts may include replanting native vegetation, restoring soil quality, and creating wildlife habitats. Such initiatives not only help to restore ecological balance but also provide opportunities for community engagement and economic development through eco-tourism and sustainable land use practices.

Community Engagement

Community engagement is paramount in this process. Involving local communities in decision-making and ensuring they are informed about the potential impacts of mining activities fosters a sense of ownership and responsibility towards environmental stewardship. Community participation can take various forms, including public consultations, workshops, and collaborative planning sessions, where local knowledge and concerns are integrated into the Environmental Impact Assessment (EIA) process. This participatory approach not only enhances the legitimacy of the EIA but also empowers communities to advocate for their rights and interests.

Education and Awareness

International organizations, such as UNESCO, have recognized the urgent need for education and awareness in the artisanal and small-scale mining (ASM) sector. By integrating ASM practices into formal education systems, these organizations aim to promote responsible mining practices and environmental protection among operators. Educational programs can focus on sustainable mining techniques, environmental management, and the importance of rehabilitation, thereby equipping ASM operators with the knowledge and skills necessary to minimize their environmental footprint.

This holistic approach not only addresses the economic aspects of mining but also considers social, environmental, and aesthetic factors, ultimately contributing to sustainable development and the preservation of natural resources. By fostering a culture of responsibility and sustainability within the mining sector, it is possible to mitigate the adverse impacts of mining activities on the environment while still supporting economic growth and community development.

Conclusion

In conclusion, the Environmental Impact Assessment (EIA) process is a vital tool for ensuring that the potential impacts of proposed projects, particularly in the mining sector, are thoroughly evaluated and addressed. By identifying likely negative impacts, assisting decision-makers, and planning for mitigation measures, EIA plays a crucial role in promoting sustainable development. The negative consequences of mining activities on the geo-environment are significant and multifaceted, encompassing ecological, social, and economic dimensions. However, through comprehensive approaches that include improved extraction methods, effective water management, land rehabilitation, and community engagement, it is possible to reduce these adverse impacts. Furthermore, the integration of education and awareness initiatives within the ASM sector can foster responsible practices and enhance environmental stewardship. Ultimately, a collaborative effort among stakeholders, including government agencies, local communities, and international organisations, is essential for achieving a balance between economic development and environmental protection in the mining industry.

5.9 Effects of Quarrying and Mining Impacts, Rehabilitation, and Environmental Protection in a Mine Operation

Introduction

The quarrying and mining sectors are integral to the economic development of many nations, supplying essential raw materials for construction, manufacturing, and energy production. However, these activities can have profound and often detrimental impacts on the environment, local communities, and ecosystems. The appraisal program on Quarrying and Mining Impacts, Rehabilitation, and Environmental Protection is designed to equip Artisanal and Small-Scale Mining (ASM) operators with the knowledge and practical skills necessary to mitigate these impacts effectively. This program also aims to establish a criterion for assessing ASM operators' compliance with mining regulations, particularly regarding the safe use of explosives and the implementation of rehabilitation measures in accordance with existing legal frameworks.

Justification: Need to Develop Appraisal Training Programs for ASMs

The protection of quarrying and mining environments from unchecked ASM activities is crucial for sustainable resource management and environmental conservation. Unregulated mining practices can lead to significant environmental degradation, including habitat destruction, soil erosion, deforestation, and water pollution. These impacts not only threaten local biodiversity but also compromise the health and livelihoods of communities that depend on these ecosystems for their survival.

One of the most pressing issues in the ASM sector is the negligence in implementing rehabilitation requirements, which are intended to restore ecosystems to their original state after mining activities cease. Without proper rehabilitation, mined areas can become barren and unproductive, leading to long-term ecological damage. This appraisal program addresses these unchecked activities by highlighting the environmental concerns associated with quarrying and mining and proposing effective mitigation measures.

The program aims to disseminate adequate awareness among ASM operators regarding the environmental impacts of their activities. By enhancing their practical skills and understanding of legal requirements, ASM operators can engage in safer practices that protect the environment while sustainably harnessing natural resources. This approach not only benefits the environment but also promotes the long-term viability of mining operations, ensuring that communities can continue to rely on these resources for their economic well-being.

Program Objectives

The short appraisal program for ASM operators is designed to achieve the following objectives:

1. **Empower Trainees with Knowledge**: Trainees will gain a comprehensive understanding of the impacts caused by hazardous chemicals used in Artisanal and Small-Scale Gold Mining (ASGM) extraction processes. This knowledge is essential for recognising the potential risks associated with chemical use, such as mercury and cyanide, and implementing safer alternatives that minimise health risks to both miners and local communities.

2. **Demonstrate Understanding of Environmental Impacts**: Trainees will learn about the environmental impacts caused by the use of explosives and ASM quarrying and mining practices. This understanding is critical for developing strategies to minimise these impacts, such as adopting best practices for blasting that reduce noise and vibration, and ensuring compliance with environmental regulations that protect surrounding ecosystems.

3. **Understand Rehabilitation Methods and Regulations**: The program will provide trainees with knowledge of the methods and techniques used to rehabilitate quarries and mines that are no longer economically viable. This includes learning about soil restoration, re-vegetation techniques, and the importance of creating wildlife corridors. Additionally, trainees will learn about the regulations that apply to quarrying and mining engagements based on the Mining Act 2016 of Kenya, ensuring they are well-versed in the legal framework governing their activities.

Specific Competences

By the end of this program, trainees will have acquired specific competencies that will enable them to:

1. **Use Explosives Safely**: Trainees will learn how to use explosives in rock blasting while considering safety protocols and related environmental impacts. This knowledge is vital for preventing accidents and minimising harm to the surrounding environment. Emphasis will be placed on the importance of proper training in explosive handling, storage, and disposal to ensure the safety of both workers and local communities.

2. **Identify Mitigation Measures**: Trainees will be equipped to identify and implement mitigation measures that adequately address the environmental impacts of quarrying and mining. This includes understanding best practices for waste management, such as proper disposal of tailings and hazardous materials, water conservation techniques to prevent contamination, and habitat restoration strategies that promote biodiversity.

3. **Rehabilitate Quarries and Mines**: Trainees will gain the skills necessary to rehabilitate quarries and mines in accordance with the

Mining Act. This includes developing comprehensive rehabilitation plans that restore ecosystems, improve land usability, and enhance biodiversity. Trainees will learn how to assess the ecological status of mined areas and implement restoration techniques that promote the recovery of native flora and fauna.

Conclusion

The appraisal program on Quarrying and Mining Impacts, Rehabilitation, and Environmental Protection is a vital initiative aimed at fostering responsible mining practices among ASM operators. By providing knowledge and practical skills, the program empowers trainees to understand the environmental impacts of their activities and implement effective mitigation measures. The focus on compliance with mining regulations, particularly regarding the safe use of explosives and rehabilitation, ensures that ASM operators can contribute to sustainable resource management.

ME 01: ASM Chemical Hazards and Silica Rich Dust

Purpose

The purpose of this module course unit is to provide comprehensive training for Artisanal and Small-Scale Gold Mining (ASGM) operators regarding the chemical hazards associated with their activities. The module aims to equip trainees with essential knowledge and practical skills that are critical for ensuring their safety and the protection of the environment.

Firstly, the module seeks to enhance trainees' understanding of the hazardous chemicals commonly encountered in ASGM activities. These chemicals, which include substances such as mercury and cyanide, pose significant health risks to miners and can lead to severe environmental degradation if not managed properly. By understanding these hazards, trainees will be better prepared to implement safety measures and reduce their exposure to harmful substances.

Secondly, the module is designed to develop the skills necessary for protection against gaseous and hazardous chemicals associated with artisanal mining and mineral processing. This includes practical training on the use of personal protective equipment (PPE), safe handling practices, and emergency response protocols.

Lastly, the module aims to foster a positive attitude among trainees towards the necessity of taking protective measures against hazardous chemicals used in ASGM mineral processing activities, thereby promoting a culture of safety and environmental responsibility.

Specific Objectives

By the end of this module course unit, trainees should be able to achieve several specific objectives that are crucial for their development as responsible ASGM operators.

Firstly, trainees will be able to explain the human health and environmental impacts caused by silica-rich dust and the hazardous chemicals used in mineral processing. Silica dust, which is generated during mining activities, can lead to serious respiratory issues, including silicosis and other chronic lung diseases. Additionally, the improper use of hazardous chemicals can result in soil and water contamination, posing risks not only to miners but also to surrounding communities and ecosystems.

Secondly, trainees will be equipped to discuss the recommended mitigation measures for protection against these hazardous chemicals. This includes understanding best practices for chemical handling, storage, and disposal, as well as implementing effective dust control measures to minimise exposure to silica-rich dust.

Geological Environmental Impacts

In addition to the health risks associated with chemical exposure, ASGM activities can have significant geological environmental impacts. The extraction processes often involve the removal of large volumes of soil and rock, leading to land degradation and alteration of the natural landscape. This disruption can result in soil erosion, loss of topsoil, and changes in land use, which can affect local agriculture and biodiversity.

Moreover, the excavation of mineral deposits can lead to the destabilisation of geological formations, increasing the risk of landslides and other geological hazards. The removal of vegetation during mining operations further exacerbates these issues, as plants play a crucial role in stabilising soil and maintaining the integrity of the landscape.

Water resources are also affected by ASGM activities. The use of water for mineral processing can lead to the contamination of local water bodies with heavy metals and other pollutants, impacting aquatic ecosystems and the health of communities that rely on these water sources for drinking and irrigation. Additionally, the alteration of natural water flow patterns due to mining can lead to changes in local hydrology, affecting both surface and groundwater availability.

Specific Competences

Upon completion of this module course unit, trainees will have acquired specific competencies that will enable them to operate safely and responsibly within the ASGM sector.

1. **Understanding Exposure Risks**: Trainees will demonstrate an understanding of the various ways in which ASGM operators become exposed to silica-rich dust and hazardous chemicals. This includes recognising the medical implications of such exposures, such as acute and chronic health effects, and the importance of implementing mitigation measures to reduce these risks.

2. **Applying Best Practices in Mineral Processing**: Trainees will learn to apply best ASGM gold mineral processing techniques that minimise exposure to hazardous chemicals. This may involve adopting safer processing methods, such as gravity separation or using non-toxic alternatives to mercury, and ensuring proper ventilation in processing areas to reduce inhalation risks.

3. **Preventing Environmental Pollution**: The module will equip trainees with the knowledge to apply best precautions to avoid environmental pollution when exposed to silica-rich dust and when using hazardous chemicals. This includes implementing waste management strategies, such as proper disposal of tailings and ensuring that chemicals are handled and disposed of responsibly to prevent contamination of soil and water sources.

4. **Selecting Protective Gear**: Trainees will learn to identify and select suitable protective gear when engaged in quarrying or mining activities, as well as during the mineral processing of gold-bearing ores. This includes understanding the types of personal protective equipment (PPE) necessary to safeguard against chemical exposure and dust inhalation, such as respirators, gloves, and protective clothing.

5. **Storing Hazardous Chemicals Safely**: The module will provide trainees with the knowledge to understand and apply recommended practices for the safe storage of hazardous chemicals used in ASGM gold extraction. This includes knowledge of proper labelling, containment measures, and storage conditions to prevent leaks and spills, thereby reducing the risk of accidents.

6. **Providing Emergency Response Facilities**: Finally, trainees will develop the capability to provide medical, safety, and transport facilities for emergency response in the event of an accident involving ASGM gaseous and hazardous chemicals. This includes learning about the importance of having emergency plans in place, including first aid measures, evacuation procedures, and access to medical assistance, ensuring that they are prepared to respond effectively to any incidents.

Conclusion

The ME 01 module on ASM Chemical Hazards and Silica Rich Dust is a vital component of training for Artisanal and Small-Scale Gold Mining operators. By equipping trainees with the knowledge and skills necessary to understand and mitigate the risks associated with hazardous chemicals and silica-rich dust, this module aims to promote safer mining practices and protect both human health and the environment.

Through comprehensive training, community engagement, and ongoing education, the ASGM sector can move towards a more sustainable future. By prioritising safety and environmental stewardship, miners can ensure that their activities contribute positively to their communities and the ecosystems they depend on. Ultimately, this training not only empowers individual miners but also fosters a collective commitment to responsible mining practices that benefit current and future generations.

ME 01/M1: Mercury in ASGM

Mercury in the Environment

Mercury is a highly toxic heavy metal that poses significant health risks, particularly in the context of Artisanal and Small-Scale Gold Mining (ASGM). The use of mercury in gold extraction processes can lead to its release into the environment, where it can contaminate soil, water, and air. This contamination occurs through various pathways, including direct discharge of mercury into water bodies, atmospheric emissions from processing activities, and improper disposal of mercury-laden waste.

Environmental Impacts: The environmental impacts of mercury contamination are profound. Mercury can bioaccumulate in aquatic ecosystems, leading to toxic effects on fish and other wildlife. As mercury enters the food chain, it can magnify in concentration, posing risks to both aquatic life and humans who consume contaminated fish. The contamination of water bodies can disrupt local ecosystems, leading to declines in fish populations and affecting the livelihoods of communities that rely on fishing. Additionally, mercury can persist in the environment for long periods, leading to long-term ecological damage.

Health Effects of Mercury Exposure

The health effects of mercury exposure are profound and varied. Acute elemental mercury intoxication can occur when individuals are exposed to high levels of elemental mercury, often through inhalation of mercury vapors. Symptoms of acute exposure can include tremors, insomnia, memory loss, and neurological damage. On the other hand, moderate and severe chronic elemental mercury intoxication can result from prolonged exposure to lower levels of mercury, leading to cognitive deficits, mood changes, and motor dysfunction. Severe cases can result in irreversible damage to the nervous system. Furthermore, chronic methyl mercury intoxication, which can occur through the consumption of contaminated fish, poses significant health risks, particularly to developing fetuses, infants, and children. Exposure during pregnancy can lead to developmental delays, cognitive impairments, and other serious health issues in children.

Treatment and Prevention

Treatment of Mercury Intoxication: The treatment for mercury poisoning typically involves the use of chelating agents, which bind to mercury and facilitate its excretion from the body. Common chelating agents include dimercaprol and DMSA (dimercaptosuccinic acid). Supportive medical care is also essential to manage symptoms and complications, including neurological and respiratory support.

Prevention: Effective prevention strategies at mining sites are crucial for reducing mercury exposure. These strategies include the implementation of safer mining practices, education on the risks of mercury use, and the promotion of alternative gold extraction methods that do not rely on mercury. Community awareness programs can also play a vital role in educating miners and their families about the dangers of mercury exposure and the importance of adopting safer practices.

Risk Pyramid of Mercury Use in Artisanal Gold Mining: The risk pyramid outlines a hierarchy of strategies for managing mercury use. At the base of the pyramid is **Prevention**, which focuses on implementing measures to avoid mercury use altogether. The next level is **Protection**, which involves using protective gear and equipment to minimize exposure. The third level is **Reduction**, which aims to reduce the amount of mercury used in processes. Finally, at the top of the pyramid is **Elimination**, which advocates for the complete phase-out of mercury use in gold extraction.

Mitigation Measures

To mitigate the risks associated with mercury exposure, several measures can be implemented. **Protective gear** is essential; miners should wear appropriate personal protective equipment (PPE), including gloves, masks, and protective clothing, to reduce exposure. **Ventilation** in work areas is also critical, as adequate airflow can help disperse mercury vapors and reduce inhalation risks. The **use of standard retort equipment** for mercury recovery can minimize emissions during gold processing, while proper **tailings management** can prevent mercury from contaminating surrounding environments. Safe **handling and storage** practices for mercury are crucial to prevent spills and accidental exposure. Finally, clear **warning signs** indicating the presence of mercury and associated hazards can help raise awareness and promote safety among miners and their communities.

ME 01/M2: Cyanide in ASGM

Cyanide Exposure in Artisanal Gold Mining

Cyanide is a highly toxic chemical commonly used in gold extraction processes. Its presence in the environment can lead to severe ecological and health consequences. Cyanide can contaminate soil and water sources, posing risks to aquatic life and human health. The chemical can enter the environment through various pathways, including spills during transportation, improper disposal of cyanide-laden tailings, and runoff from mining sites. Natural biodegradation processes can mitigate some of its effects, but improper handling can lead to significant environmental damage.

Environmental Impacts: The environmental impacts of cyanide contamination are severe. Cyanide is highly toxic to aquatic organisms, and even low concentrations can lead to fish kills and the destruction of aquatic ecosystems. Contaminated water sources can affect not only wildlife but also communities that rely on these water bodies for drinking and irrigation. The long-term effects of cyanide pollution can lead to the degradation of local ecosystems, loss of biodiversity, and disruption of food chains.

Effects of Cyanide on Health

The health effects of cyanide exposure can vary based on the level and duration of exposure. **Low acute exposure** to cyanide can cause symptoms such as headache, dizziness, and nausea, which may be mistaken for less serious conditions. In contrast, **high acute exposure** can be life-threatening, leading to respiratory failure, loss of consciousness, and death within minutes. Additionally, **low chronic exposure** to cyanide can result in long-term health issues, including respiratory problems, neurological effects, and potential damage to the cardiovascular system. The cumulative effects of chronic exposure can significantly impact the overall health and well-being of miners and surrounding communities.

Treatment of Cyanide Intoxication

Antidotes for cyanide poisoning are critical for effective treatment. The primary antidote is hydroxocobalamin, which binds to cyanide and forms a non-toxic compound that can be excreted by the body. Sodium thiosulfate is another antidote that can be used in conjunction with hydroxocobalamin to enhance treatment efficacy. In addition to administering antidotes, **supportive**

medical care is essential in managing cyanide poisoning. This care may include oxygen therapy, monitoring of vital signs, and addressing any complications that arise during treatment.

Mitigation Measures

To reduce the risks associated with cyanide exposure, several mitigation measures should be implemented. **Protective gear** is essential; miners should wear appropriate PPE, including gloves, goggles, and respirators, to minimize exposure. **Ventilation** in work areas is critical to ensure that cyanide vapors are adequately dispersed, reducing inhalation risks. The **use of standard methods and equipment** for cyanide application can enhance safety and reduce the likelihood of accidents. Proper **tailings management** is also vital to prevent contamination of surrounding environments, as improperly managed tailings can leach cyanide into soil and water sources. Safe **handling and storage** practices for cyanide are crucial to prevent spills and accidental exposure, including using secure containers and clearly labeling hazardous materials. Finally, clear **warning signs** indicating the presence of cyanide and associated hazards can help raise awareness and promote safety among miners and their communities.

ME 01/M4: Borax in ASGM

Introduction

Borax, also known as sodium borate, is a naturally occurring mineral with various industrial uses, including its role in gold extraction. It is often used as a safer alternative to mercury in the gold refining process. Borax is valued for its ability to lower the melting point of gold and improve the efficiency of the extraction process. Its chemical properties make it an effective flux, helping to separate gold from impurities during smelting.

Environmental Impacts

While borax is considered less toxic than mercury and cyanide, its use in ASGM is not without environmental consequences. The mining and processing of borax can lead to habitat destruction, soil degradation, and water pollution. When borax is used in gold extraction, any residual borax that is not properly managed can leach into the soil and water systems, potentially affecting local flora and fauna. Additionally, the mining of borax itself can result in the disruption of ecosystems, leading to loss of biodiversity and alteration of land use patterns.

Health Risks

Borax poses certain health risks, particularly with improper handling. **Acute toxicity** can occur through manual handling, with symptoms including skin irritation, respiratory distress, and gastrointestinal issues. The lethal dose of borax is relatively high, but fatalities can occur in cases of severe exposure. **Chronic toxicity** is also a concern, particularly with prolonged inhalation or ingestion of borax dust. Studies have indicated potential carcinogenicity, raising concerns about long-term exposure among miners. Additionally, there is evidence suggesting that borax may have reproductive effects, which necessitates caution in its use, especially among women of childbearing age.

ME 01/M1: Silica Rich Dust

Fine Silica in Mining Environment

Silica-rich dust is a significant hazard in the mining environment, particularly in Artisanal and Small-Scale Gold Mining (ASGM). Fine silica particles are prevalent in various areas, including mining, construction, and agriculture. The primary source mineral for silica dust is quartz, which is commonly found in many types of ore. Another source is chalcedony, a form of silica that can also contribute to the generation of fine dust during mining activities. The inhalation of silica dust poses serious health risks to miners and surrounding communities, making it essential to understand the sources and impacts of this hazard.

Environmental Impacts

The environmental impacts of silica dust are multifaceted. When silica dust is released into the air, it can settle on vegetation and water bodies, leading to soil and water contamination. This contamination can affect plant growth and aquatic ecosystems, disrupting local food chains. Additionally, the erosion of silica-rich soils can lead to increased sedimentation in rivers and streams, negatively impacting water quality and aquatic habitats. The long-term effects of silica dust pollution can contribute to the degradation of ecosystems, loss of biodiversity, and reduced agricultural productivity in surrounding areas.

Mining Exposure to Silica

Miners are often exposed to silica dust during various stages of mining, including drilling, blasting, and crushing of ore. The inhalation of fine silica particles can lead to significant health issues, particularly respiratory diseases. Regulatory agencies have established exposure limits to protect workers, but many artisanal miners may not be aware of these guidelines or may lack the means to implement them effectively. Continuous exposure to silica dust can lead to chronic health conditions, making it crucial for miners to understand the risks associated with their work environment.

Typical Silicosis Infections

Silicosis is a lung disease caused by the inhalation of silica dust, leading to inflammation and scarring of lung tissue. There are several types of silicosis, each with distinct characteristics and symptoms. **Pneumoconiosis** is a general term for lung diseases caused by inhaling various types of dust, including silica. **Pneumonitis** refers to inflammation of the lung tissue, which can occur as a result of silica exposure.

The types of silicosis include:

- **Chronic Silicosis**: This form develops after prolonged exposure to low levels of silica dust, typically over many years. Symptoms may include cough, shortness of breath, and fatigue.

- **Accelerated Silicosis**: This occurs with higher levels of exposure over a shorter period, leading to more rapid disease progression. Symptoms can appear within 5 to 10 years of exposure.

- **Acute Silicosis**: This rare form can develop after very high levels of silica exposure over a short time, leading to severe respiratory distress and rapid deterioration of lung function.

Medical Interventions

Medical interventions for silicosis primarily focus on easing symptoms and improving the quality of life for affected individuals. While there is no cure for silicosis, treatments may include bronchodilators to help open airways, corticosteroids to reduce inflammation, and supplemental oxygen for those with severe respiratory issues. Regular medical check-ups, including periodic diagnosis and lung functionality tests, are essential for early detection and management of silicosis. Miners should be encouraged to seek medical attention if they experience respiratory symptoms, as early intervention can help mitigate the progression of the disease.

Mitigation Measures

To reduce the risks associated with silica exposure, several mitigation measures should be implemented in the mining environment. **Protective gear** is essential; miners should wear appropriate personal protective equipment (PPE), including respirators, dust masks, and protective clothing, to minimize inhalation of silica dust. **Ventilation** in work areas is critical to ensure that dust is adequately dispersed, reducing the risk of inhalation. Implementing effective **warning signs** in mining areas can help raise awareness about the dangers of silica dust and encourage miners to take necessary precautions.

Additionally, training programs should be established to educate miners about the risks of silica exposure and the importance of using protective measures. Regular monitoring of air quality in mining areas can help identify hazardous conditions and prompt timely interventions. By adopting these mitigation measures, miners can significantly reduce their risk of developing silicosis and other respiratory diseases associated with silica exposure.

Conclusion

The modules on mercury, cyanide, borax, and silica-rich dust in Artisanal and Small-Scale Gold Mining (ASGM) highlight the critical health and environmental risks associated with these substances. Understanding the hazards, implementing effective treatment and prevention strategies, and adopting mitigation measures are essential for promoting safer mining practices.

By equipping miners with the knowledge and skills necessary to recognize and manage these risks, we can foster a culture of safety and environmental responsibility within the ASGM sector. Ultimately, these efforts will contribute to the well-being of miners, their communities, and the ecosystems they depend on, ensuring a more sustainable future for all involved in the mining industry.

ME 02: Explosives Storage, Rock Blasting Safety, and Environmental Impacts

Purpose

The purpose of this module course unit is to equip trainees with essential knowledge and skills related to the safe handling and storage of explosives, as well as the environmental impacts associated with rock blasting activities. The module aims to provide trainees with a thorough understanding of blasting design techniques, the importance of safe storage practices, and the potential environmental consequences of using explosives in mining and quarrying operations.

By the end of the course, trainees will not only possess the technical skills necessary for implementing safety measures but will also develop a positive attitude towards the significance of mitigation measures in the workplace. This includes fostering a culture of safety and preparedness for rapid response to emergencies involving explosives, thereby enhancing overall workplace safety and environmental stewardship.

Specific Objectives

By the end of this module course unit, trainees should be able to:

1. Demonstrate an understanding of the dynamics involved in the use of energetic materials to weaken rock masses, including the principles of rock fragmentation and the mechanics of explosive detonation.
2. Demonstrate an understanding of the functions and facilities that are specified for the storage of explosives used in rock blasting, including safety protocols and regulatory requirements.
3. Demonstrate an understanding of the impacts caused by explosives in rock blasting activities, including both immediate and long-term environmental effects.

Specific Competences

By the end of this module course unit, trainees will have acquired the abilities to:

1. Identify basic types of energetic materials used in rock blasting, including their properties and applications.

2. Apply recommended safety measures for the safe storage of explosives meant for rock blasting, ensuring compliance with legal and regulatory standards.

3. Apply appropriate mitigation measures for impacts related to rock blasting, including environmental protection strategies.

4. Provide emergency facilities at the workplace for rapid response to incidents associated with the use of explosives, ensuring preparedness for potential accidents.

ME 02/M1: Energetic Materials and Storage of Explosives

Introduction

Energetic materials are substances that undergo rapid chemical reactions, releasing energy in the form of gas and heat. In the context of rock blasting, these materials are crucial for fragmenting rock masses to facilitate mining and construction activities. Common types of energetic materials include nitroglycerin-based explosives, which are known for their high sensitivity and power, as well as single molecule and composite explosives that offer varying degrees of stability and performance. Other formulations, such as slurries and emulsions, combine solid and liquid components to enhance safety and effectiveness. Heavy ANFO (Ammonium Nitrate Fuel Oil) is another widely used explosive that combines ammonium nitrate with fuel oil, providing a cost-effective solution for large-scale blasting operations.

Storage of Explosives

The safe storage of explosives is paramount to prevent accidents and ensure the safety of personnel and the surrounding community. The location of explosive magazines must be carefully considered, taking into account public risk, proximity to other magazines, and potential ignition sources. Access routes for vehicles must be designed to minimize risks during transportation, and activities within the site should be regulated to prevent unauthorized access. Weather protection is also essential to safeguard explosives from environmental conditions that could compromise their stability.

Magazine Fabrication Requirements: The design and construction of explosive magazines must adhere to strict safety standards. Portable magazines, relocatable magazines, and fixed or permanent magazines each have specific requirements regarding their construction materials, security features, and capacity. The exteriors of magazines should be robust to withstand external forces, while the interiors must be designed to minimize the risk of accidental detonation.

Security Measures: Security is a critical aspect of explosive storage. Remote security systems, locking devices, and perimeter fences are essential to prevent unauthorized access. Additionally, magazine doors should be equipped with secure hinges and locking mechanisms to ensure that only authorized personnel can enter. Segregation of different types of explosives is also important to prevent unintended reactions.

Utility Requirements: Proper utility installations, including water and electrical systems, are necessary for the safe operation of explosive storage facilities. Adequate lighting is essential for visibility during inspections and maintenance, while ventilation systems must be in place to prevent the accumulation of hazardous gases.

Environmental Conditions: The environmental conditions within explosive storage facilities must be carefully controlled. Adequate ventilation is necessary to disperse any potentially harmful gases, while shade roofs can protect explosives from direct sunlight and extreme temperatures.

Emergency Installations: Emergency installations, such as lightning protection systems and fire-fighting equipment, are critical for ensuring the safety of explosive storage facilities. Lightning protection systems help prevent accidental detonations caused by electrical discharges, while fire-fighting equipment must be readily available to address any fire hazards that may arise.

ME 02/M2: Basics for Rock Blast Design

Blasting Constraints

Rock blasting is a complex process that requires careful planning and design to achieve desired outcomes while minimizing risks. Blasting constraints include geological factors, environmental considerations, and safety regulations. Understanding the geological characteristics of the rock mass, such as its strength, density, and porosity, is essential for designing effective blasts.

Drilling and Blasting: The drilling process is a critical component of rock blasting, as it determines the placement and depth of explosive charges. A typical blast layout involves strategic placement of drill holes to optimize fragmentation and minimize ground vibrations. The blasting sequence, or the order in which charges are detonated, is also crucial for controlling the dynamics of the blast and reducing the risk of fly-rock and airblast.

Geological Influence on Blast Design

The geological characteristics of the rock mass significantly influence blast design. Factors such as weak zones, dynamic compressive strength, elastic moduli, and internal friction must be considered when planning a blast. The presence of water content and in situ static stress can also affect the performance of explosives. Bedding orientation, or the alignment of rock layers, can influence the direction of rock movement during blasting, making it essential to account for these factors in the design process.

ME 02/M3: Rock Blasting Environmental Impacts and Safety

Rock Blasting Environmental Impacts

Rock blasting can have significant environmental impacts that must be carefully managed. **Airblast** refers to the pressure wave generated by the detonation of explosives, which can cause noise pollution and disturb nearby communities. This pressure wave can also affect wildlife, particularly species sensitive to sound, leading to changes in behavior or habitat abandonment.

Fly-Rock is another critical concern during blasting operations. It refers to fragments of rock that are propelled away from the blast site at high velocities. Fly-rock can pose serious safety risks to personnel and equipment, as well as damage to nearby structures. Effective blast design and safety protocols are essential to minimize the risk of fly-rock incidents.

Disturbance of the Ground Profile occurs as a result of blasting, which can lead to changes in the landscape and potential erosion. The removal of rock can alter drainage patterns, affecting local hydrology and potentially leading to increased sedimentation in water bodies.

Dust and Fumes generated during blasting can have detrimental effects on air quality. Dust can travel significant distances, impacting vegetation and human health. Fumes from explosives can also release harmful gases, necessitating proper ventilation and monitoring to ensure the safety of workers and nearby communities.

Ground Vibrations are another environmental impact associated with rock blasting. Vibrations can cause structural damage to nearby buildings and infrastructure, as well as disturb local wildlife. Monitoring and controlling ground vibrations through careful blast design and timing are essential to mitigate these effects.

Geological Environmental Impacts

In addition to the immediate environmental impacts, rock blasting can lead to significant geological changes in the landscape. The alteration of the ground profile can result in the destabilization of slopes, increasing the risk of landslides and erosion. This is particularly concerning in areas with steep terrain or loose soil, where the removal of rock can compromise the integrity of surrounding geological formations.

The blasting process can also affect groundwater systems. Changes in the permeability of rock layers can alter the flow of groundwater, potentially leading to the contamination of aquifers or changes in water quality. The introduction of explosives into the subsurface can create fractures that may allow contaminants to migrate into groundwater supplies, posing risks to drinking water sources.

Furthermore, the disturbance of geological formations can impact local ecosystems. The removal of rock and alteration of habitats can lead to the displacement of flora and fauna, disrupting local biodiversity. Species that rely on specific geological features for habitat may be particularly vulnerable to the impacts of blasting.

Safety Tools and Materials

To ensure safety during rock blasting operations, various tools and materials must be utilized. **Strategic positioning** of personnel and equipment is critical to minimize exposure to hazards. The use of safety gear, such as helmets, goggles, and hearing protection, is essential for protecting workers from potential injuries.

A well-stocked **first aid kit** should be readily available on-site, containing essential items such as safety pins, sterile wound dressing material, disposable gloves, sterile triangular bandages, individual plasters, and eye pads. This preparedness ensures that immediate medical attention can be provided in the event of an accident.

Firefighting Equipment and Safety: Fire hazards are a significant concern in blasting operations due to the presence of explosives and flammable materials. Understanding the types of fires that may occur, such as those involving flammable liquids or electrical equipment, is crucial for effective fire prevention and response. Firefighting equipment, including extinguishers and fire blankets, should be readily accessible, and personnel should be trained in their use. Regular fire drills and adherence to fire safety tips can further enhance workplace safety.

Explosives Use and Regulations Enforcement

Purpose

This module course unit is designed to equip trainees with the knowledge and skills necessary for understanding safety measures for handling explosives and the regulations that govern their use. The module emphasizes the importance of enforcing environmental protection regulations, which are critical for the sustainable harnessing of mineral and rock deposits.

By the end of the course, trainees will appreciate the need for compliance with safety measures and regulations, recognizing their role in promoting safe practices in the mining and quarrying sectors.

Specific Objectives

By the end of this module course unit, trainees should be able to:

a) Explain safety measures for the use of explosives, including handling, storage, and transportation protocols.

b) Explain regulations governing the use of explosives in rock blasting, including compliance with local and international standards.

c) Demonstrate an understanding of the mining policy, legal framework, and basic governance functions for regulatory institutions concerning the enforcement of environmental protection in quarry and mining activities in South Sudan.

Specific Competences

By the end of this module course unit, participants will have acquired the abilities to:

1. Comply with the safety precautions that apply to the use of explosives, ensuring that all operations are conducted in a safe and responsible manner.

2. Comply with regulations that apply to the safe use of explosives in rock blasting activities, including adherence to licensing requirements and safety standards.

3. Comply with regulations that apply to safe mining, rehabilitation, and protection of mining and quarrying environments in accordance with the Mining Act 2016. This includes understanding the responsibilities of mining companies and operators in minimizing environmental impacts and promoting sustainable practices.

Conclusion

The modules on Explosives Storage, Rock Blasting Safety, and Environmental Impacts provide a comprehensive framework for understanding the complexities of using explosives in mining and quarrying operations. By equipping trainees with the necessary knowledge, skills, and competencies, these modules aim to foster a culture of safety and environmental responsibility within the industry.

Through effective training, participants will be better prepared to implement safety measures, comply with regulations, and respond to emergencies involving explosives. The emphasis on understanding geological environmental impacts highlights the importance of considering the broader implications of blasting activities on local ecosystems and geological stability.

Ultimately, these efforts will contribute to the sustainable development of mineral resources while protecting the health and safety of workers and the environment. By promoting best practices in explosive use and environmental management, the mining industry can work towards a more sustainable future that balances economic development with ecological preservation.

ME 04/M1: Quarries and Mines

Introduction

Quarries and mines are essential sources of raw materials used in various industries, including construction, manufacturing, and energy production. Understanding the types of materials extracted and their uses is crucial for effective resource management.

Types of Materials and Uses: Quarries typically extract materials such as aggregates, dimension stones, and sand, each serving specific purposes in construction and infrastructure development. Hard rock quarries focus on extracting durable materials like granite and limestone, while soft rock quarries may target sand and clay for use in concrete and ceramics.

Positive Impacts: While quarrying and mining can have negative environmental impacts, they also contribute to economic development, job creation, and infrastructure improvement. Responsible management practices can help maximize these benefits while minimizing adverse effects.

Types of Quarries

1. **Hard Rock Quarries**: These quarries extract durable materials such as granite, basalt, and limestone, which are essential for construction and road building. The extraction process often involves blasting, drilling, and crushing.

2. **Aggregate Quarries**: Aggregate quarries focus on producing materials like gravel and crushed stone, which are used in concrete production and as base materials for roads and foundations.

3. **Dimension Stone Quarries**: These quarries extract large blocks of stone for use in architectural applications, such as countertops, facades, and monuments.

4. **Artisanal Quarrying**: Artisanal quarrying involves small-scale operations that often rely on manual labor and traditional methods. While these operations can provide livelihoods for local communities, they may lack the regulatory oversight necessary to ensure environmental protection.

5. **Soft Rock Quarries**: Soft rock quarries extract materials like sand and clay, which are used in various applications, including construction, ceramics, and glass manufacturing.
6. **Aquatic and Terrestrial Deposits**: These quarries extract materials from both land and water bodies, such as river sand and gravel, which are essential for construction and landscaping.
7. **Residual Deposits**: Residual deposits, such as lateritic soils, are formed through weathering processes and can be mined for their mineral content.

Environmental Impacts

The extraction of materials from quarries and mines can lead to significant environmental impacts, including:

- **Infrastructure Impacts**: The construction of access roads and facilities can disrupt local ecosystems and lead to habitat fragmentation. The increased human activity associated with quarrying can also lead to disturbances in wildlife populations and migration patterns.
- **Erosion**: The removal of vegetation and soil can increase erosion rates, leading to sedimentation in nearby water bodies. This sedimentation can degrade water quality and disrupt aquatic habitats, affecting fish and other organisms.
- **Ecosystem Disruption**: Quarrying activities can lead to the destruction of habitats for flora and fauna, resulting in loss of biodiversity. The alteration of land can displace species that rely on specific habitats, leading to population declines or local extinctions.
- **Groundwater Depletion**: The extraction of materials can alter the natural flow of groundwater, potentially leading to depletion of aquifers. This can affect local water supplies for both human consumption and agricultural use.
- **Pollution**: Quarrying operations can generate dust, noise, and chemical runoff, which can contaminate soil and water resources. The use of heavy machinery and explosives can also contribute to air and noise pollution, impacting nearby communities.

Geological Environmental Impacts

In addition to the immediate environmental impacts, quarrying and mining can lead to significant geological changes in the landscape. These geological environmental impacts include:

- **Alteration of Landforms**: The extraction of rock and soil can dramatically change the topography of an area, creating steep slopes, pits, and depressions. These changes can affect drainage patterns and lead to increased runoff and erosion.

- **Soil Compaction and Degradation**: Heavy machinery used in quarrying can compact the soil, reducing its permeability and affecting its ability to support vegetation. This degradation can hinder the natural regeneration of ecosystems after mining activities cease.

- **Ground Instability**: The removal of rock can lead to ground instability, increasing the risk of landslides and subsidence. This is particularly concerning in areas with steep terrain or loose soil, where the removal of material can compromise the integrity of surrounding geological formations.

- **Hydrological Changes**: Quarrying can disrupt natural hydrological cycles by altering the flow of surface water and groundwater. Changes in water flow can lead to the drying up of springs and streams, affecting local ecosystems and water availability for communities.

- **Contamination of Geological Resources**: The introduction of chemicals and pollutants from quarrying activities can contaminate geological resources, including groundwater and soil. This contamination can have long-term effects on the health of ecosystems and the safety of water supplies.

- **Impact on Geological Features**: Quarrying can lead to the destruction of significant geological features, such as unique rock formations or fossil beds. The loss of these features can diminish the geological heritage of an area and reduce opportunities for scientific research and education.

ME 04/M2: Rehabilitation Processes

Benchmarking

Rehabilitation processes are essential for restoring quarries and mines to a state that is environmentally sustainable and beneficial for local communities. Effective rehabilitation requires comprehensive assessments to inform the design and implementation of restoration efforts.

Biodiversity Assessment: Understanding the existing flora and fauna is crucial for developing rehabilitation strategies that promote biodiversity. This assessment helps identify native species that can be reintroduced to the area and informs the selection of plants that will thrive in the rehabilitated environment.

Geology and Hydrology Assessments: Evaluating the geological and hydrological characteristics of the site is essential for designing effective rehabilitation measures. This includes understanding soil types, drainage patterns, and groundwater levels, which can influence plant growth and ecosystem recovery.

Social Assessment: Engaging with local communities and stakeholders is vital for understanding their needs and expectations regarding rehabilitation. Social assessments help ensure that rehabilitation efforts align with community interests and contribute to local economic development.

From Perception to Conception

The transition from perception to conception involves defining clear objectives for rehabilitation and orienting the design to meet these goals. Key considerations include:

- **Season of Intervention**: Timing is critical for successful rehabilitation. Understanding seasonal variations in weather and plant growth can inform the timing of planting and other restoration activities.
- **Ecological Continuities**: Maintaining ecological continuity involves preserving existing habitats and ensuring that rehabilitation efforts support the natural processes of ecosystem recovery.
- **Integrated Energy Design**: Incorporating sustainable energy practices into rehabilitation efforts can enhance the long-term viability of restored

sites. This may include using renewable energy sources for site operations.

- **Landscape Context**: The design of rehabilitation efforts should consider the surrounding landscape and how restored areas will integrate with existing ecosystems.
- **Site Features**: Understanding the unique features of the site, such as topography and soil composition, is essential for developing effective rehabilitation strategies.
- **Water Availability Onsite**: Assessing water availability is crucial for selecting appropriate plant species and ensuring their successful establishment.
- **Choice of Plants**: Selecting native and drought-resistant plant species can enhance the success of rehabilitation efforts and promote biodiversity.
- **Creating Masses of Plants**: Designing plantings in clusters or masses can improve habitat quality and enhance the aesthetic value of rehabilitated areas.
- **Rocks and Natural Dynamic Zones**: Incorporating natural features, such as rocks and dynamic zones, can enhance habitat diversity and support wildlife.
- **Technical Tools**: Utilizing appropriate tools and technologies can improve the efficiency and effectiveness of rehabilitation efforts.
- **Project Timeline, Budget, and Investment**: Developing a clear timeline and budget is essential for successful project implementation and securing necessary funding.

Implementation and Supervision

The implementation phase involves several key steps to ensure successful rehabilitation:

- **Securing the Working Area**: Establishing boundaries and securing the site is essential for safety and to prevent unauthorized access during rehabilitation activities.

- **Kick-off Meeting**: Conducting a kick-off meeting with all stakeholders ensures that everyone is aligned on project goals and responsibilities.
- **From Bare-Land to Structured-Land**: Transforming the site from bare land to a structured landscape involves constructing features such as stone walls, gabion walls, and substrate structures to stabilize the area and promote vegetation growth.
- **Open Excavated Ponds and Compacted Soil**: Creating open excavated ponds can enhance biodiversity and provide habitats for aquatic species. Compacted soil may be amended to improve its suitability for plant growth.

From Nursery to Site: Planting and Sowing

The successful establishment of vegetation is a critical component of rehabilitation:

- **Plant Establishment**: Collecting and storing seeds, preparing seedlings, and planting them in the rehabilitated area are essential steps in the restoration process.
- **Supervising and Documenting**: Ongoing supervision and documentation of rehabilitation activities are crucial for assessing progress and making necessary adjustments.
- **Coping with Reality and Adjusting the Design**: Flexibility in design and implementation allows for adjustments based on real-world conditions and challenges encountered during rehabilitation.
- **Maintenance**: Regular maintenance, including watering, weeding, and monitoring plant health, is essential for ensuring the long-term success of rehabilitation efforts.

Monitoring the Rehabilitation

Monitoring is a vital component of the rehabilitation process, allowing for the assessment of success and identification of areas for improvement:

- **Fauna Indicators**: Monitoring wildlife populations can provide insights into the ecological health of rehabilitated areas.

- **Flora Indicators**: Assessing plant diversity and health can indicate the success of vegetation establishment and ecosystem recovery.
- **Hydrology Indicators**: Monitoring water quality and availability can help assess the effectiveness of rehabilitation efforts in restoring natural hydrological processes.

ME 05: Active and Rehabilitated Quarry and Mine Visits

Purpose

This module course unit is intended to provide trainees with practical experience by applying theoretical knowledge learned in the classroom to real-world situations. Field visits to active mining or quarry sites and rehabilitated areas will allow trainees to observe and understand the complexities of quarrying and rehabilitation processes.

Specific Objectives

By the end of the module course unit, trainees should be able to:

a) Demonstrate an understanding of activities being undertaken to address environmental protection at active quarry or mining sites visited.

b) Demonstrate an understanding of the programs that were implemented to rehabilitate a quarry or mine site that ceased to be economically viable.

Specific Competences

1. Identify ongoing activities for the mine or quarry visited, including extraction methods, environmental management practices, and safety measures.
2. Identify mitigation measures that were put in place to address environmental impacts, such as erosion control, dust suppression, and habitat restoration.
3. Identify rehabilitation features that are responsible for new ecosystems in rehabilitated areas, including plantings, water features, and soil amendments.

MQ 05/M1: Active Quarry or Mine Pits

Site Location and Topography

During visits to active quarry or mine pits, trainees will assess the site location and topography, gaining insights into how these factors influence extraction activities and environmental management. Understanding the geotechnical setting is crucial for evaluating the stability of the site and the potential impacts of mining operations.

Extraction Activities

Trainees will observe extraction activities in real-time, gaining firsthand experience of the processes involved in quarrying and mining. This includes understanding the methods used for rock blasting, material handling, and transportation.

Environmental Impacts and Mitigation Measures

Trainees will learn about the environmental impacts associated with active quarrying, including erosion, habitat destruction, and pollution. They will also examine the mitigation measures implemented to address these impacts, such as sediment control practices, re-vegetation efforts, and water management strategies.

MQ 05/M2: Rehabilitated Quarry or Mine Pit

Site Location and Geotechnical Setting

Visits to rehabilitated quarry or mine pits will provide trainees with an opportunity to assess the effectiveness of rehabilitation efforts. Understanding the site location and geotechnical setting is essential for evaluating the success of restoration activities.

Ecosystems Generated

Trainees will observe the ecosystems that have developed in rehabilitated areas, assessing the diversity of flora and fauna and the overall health of the restored environment. This assessment will provide insights into the effectiveness of rehabilitation techniques and the potential for long-term ecological recovery.

Environmental Impacts Solved and Mitigation Measures Applied

Trainees will learn about the environmental impacts that have been addressed through rehabilitation efforts, including soil stabilization, water quality improvement, and habitat restoration. They will also examine the specific mitigation measures that were applied to achieve these outcomes, gaining practical knowledge that can be applied in future rehabilitation projects.

Conclusion

The modules on Enforcement Regulations, Quarry, Mine and Rehabilitation Processes, and Active and Rehabilitated Quarry and Mine Visits provide a comprehensive framework for understanding the complexities of quarrying and mining operations.

By equipping trainees with the necessary knowledge, skills, and competencies, these modules aim to foster a culture of safety, environmental responsibility, and sustainable practices within the mining and quarrying industry.

The **Enforcement Regulations** module emphasizes the importance of adhering to the Mining Act 2016 of Kenya, which provides a robust legal framework for managing mineral resources. Understanding the regulatory environment is crucial for ensuring that quarrying activities are conducted responsibly, with due consideration for health, safety, and environmental protection. The focus on charging operations and the proper handling of explosives further underscores the need for rigorous safety protocols to prevent accidents and mitigate risks associated with blasting activities.

The **Quarry, Mine and Rehabilitation Processes** module highlights the critical importance of rehabilitation in restoring ecosystems and ensuring the long-term viability of mined areas. By understanding the various types of quarries and the materials extracted, trainees can appreciate the broader implications of quarrying on the environment and local communities. The emphasis on assessment practices, design principles, and implementation strategies equips trainees with the tools necessary to develop effective rehabilitation plans that promote biodiversity and restore ecological balance.

The **Active and Rehabilitated Quarry and Mine Visits** module provides invaluable hands-on experience, allowing trainees to observe real-world applications of the concepts learned in the classroom. By visiting active sites, trainees can witness the challenges and opportunities associated with quarrying operations, while visits to rehabilitated areas offer insights into successful restoration practices. This experiential learning reinforces the importance of integrating theory with practice, preparing trainees to address the complexities of environmental management in the mining sector.

In conclusion, these modules collectively contribute to the development of a skilled workforce capable of navigating the challenges of the mining and quarrying industry. By fostering a deep understanding of regulatory frameworks, environmental impacts, and rehabilitation processes, trainees are better equipped to promote sustainable practices that benefit both the economy and the environment. As the industry continues to evolve, the knowledge and skills gained through these modules will be essential for ensuring that mineral resources are harnessed responsibly and sustainably, ultimately contributing to the well-being of communities and the preservation of natural ecosystems.

Concluding Remarks on the Importance of Developing Artisanal Small-Scale Mining (ASM) Appraisal Programs for Environmental Impact Assessment (EIA) in Kenya

The development of Artisanal Small-Scale Mining (ASM) appraisal programs focused on Environmental Impact Assessment (EIA) is of paramount importance in Kenya, particularly in light of the increasing prevalence of small-scale mining activities across the country. As defined by Kenya's National Environment Management Authority (NEMA), EIA is a systematic and thorough examination of the potential effects that proposed projects may have on the environment. This process is essential for ensuring that mining activities are conducted in a manner that minimizes adverse environmental impacts while promoting sustainable development.

Importance of EIA in ASM

The significance of EIA in the context of ASM cannot be overstated. Small-scale mining operations often occur in sensitive ecological areas, where the potential for environmental degradation is high. By implementing robust EIA processes, stakeholders can identify and assess the environmental impacts of mining activities, including habitat destruction, soil erosion, water pollution, and biodiversity loss. This proactive approach allows for the prediction of likely changes to the environment as a result of mining, enabling informed decision-making that prioritizes environmental protection.

Moreover, EIA serves as a critical tool for evaluating the impacts of various alternatives to proposed mining projects. By considering different approaches and methodologies, stakeholders can identify the most sustainable options that balance economic benefits with environmental stewardship. This is particularly important in ASM, where operators may lack the resources and technical expertise to conduct comprehensive environmental assessments independently.

Geological Environmental Impacts of Mining Sites

In addition to the broader environmental impacts associated with mining, it is crucial to recognize the specific geological environmental impacts that arise from mining activities, particularly in the context of ASM. Understanding these impacts is essential for developing effective EIA processes and mitigation strategies.

Mining activities can lead to significant alterations in landforms, creating open pits and waste dumps that disrupt natural drainage patterns and increase erosion. The heavy machinery used in mining can cause soil compaction and degradation, reducing the soil's ability to support vegetation and leading to biodiversity loss. Furthermore, mining can create ground instability, increasing the risk of landslides and subsidence, which pose hazards to both the environment and local communities.

Hydrological changes are another significant concern, as mining can disrupt groundwater flow and surface water drainage, affecting water availability for ecosystems and communities. The contamination of geological resources through the use of harmful chemicals poses serious health risks and can have long-term effects on the environment. Additionally, mining can lead to the destruction of significant geological features, diminishing the geological heritage of an area and reducing opportunities for scientific research and education.

Environmental Impacts and Remedial Measures of Rock Blasting

The environmental impacts of rock blasting in quarrying operations further illustrate the need for comprehensive EIA processes. Rock blasting, while essential for extracting materials, presents significant environmental challenges, including:

1. **Air Blast Noise**: High levels of air blast noise generated during rock blasting can lead to discomfort and health issues for nearby residents and wildlife. Understanding the factors contributing to air blast noise, such as explosive type and blast design, is essential for effective mitigation.

2. **Flyrock Hazards**: Flyrock poses significant risks, including property damage and injuries. Identifying the causes of flyrock incidents, such as incorrect blast design and geological conditions, is crucial for enhancing safety in quarrying operations.

3. **Dust Emissions**: Rock blasting produces substantial dust, adversely affecting air quality and posing respiratory health risks. Implementing effective dust control measures, such as water sprays and dust suppressants, is necessary to minimize these impacts.

4. **Toxic Fumes**: Blasting activities release toxic fumes, including nitrogen oxides and carbon monoxide, which pose health hazards. Identifying and implementing mitigation measures, such as using low-fume explosives and ensuring proper ventilation, are essential for protecting health.

5. **Ground Vibrations**: Ground vibrations from blasting can cause structural damage to nearby buildings and disturb local fauna. Optimizing blast design and drillhole geometry is crucial to minimize these vibrations and maintain stability.

6. **Geological Factors**: The effectiveness of blast design is influenced by geological characteristics, such as rock hardness and fracture patterns. Analyzing these factors is essential for optimizing blasting operations tailored to specific quarry conditions.

7. **Safety Measures**: Stringent safety measures, including proper training, warning signs, and emergency preparedness, are vital to prevent accidents and ensure the well-being of workers and nearby communities.

By addressing these environmental impacts through careful planning, monitoring, and the adoption of best practices, the negative consequences of rock blasting can be minimized, ensuring safer and more sustainable quarrying operations.

Mitigation Measures and Training Needs Assessments

The EIA process also plays a crucial role in proposing mitigation measures for any significant negative impacts identified. By outlining specific actions that can be taken to reduce environmental harm, EIA provides a framework for responsible mining practices. This is especially relevant for ASM, where operators may not be fully aware of the environmental consequences of their activities. Training Needs Assessments (TNAs) can be integrated into the EIA process to identify the knowledge gaps and training requirements of ASM operators. By equipping them with the necessary skills and knowledge, TNAs can enhance the capacity of small-scale miners to implement effective environmental management practices.

Furthermore, the involvement of local communities in the EIA process fosters a sense of ownership and responsibility towards environmental conservation. Engaging stakeholders in discussions about potential impacts and mitigation strategies ensures that the voices of those most affected by mining activities are heard. This participatory approach not only enhances the quality of the EIA but also builds trust between mining operators, regulatory agencies, and local communities.

Alignment with National Environmental Goals

Developing ASM appraisal programs that incorporate EIA aligns with Kenya's broader environmental goals and commitments. As the country strives to achieve sustainable development and protect its rich biodiversity, the integration of EIA into small-scale mining practices is essential. By ensuring that environmental considerations are at the forefront of mining operations, Kenya can work towards a mining sector that contributes positively to the economy while safeguarding its natural resources for future generations.

Enhancing Regulatory Frameworks

The establishment of ASM appraisal programs also necessitates the enhancement of regulatory frameworks governing small-scale mining. By creating clear guidelines and standards for conducting EIAs in ASM, NEMA can ensure that all mining activities adhere to best practices in environmental management. This includes establishing criteria for assessing the significance of environmental impacts, as well as protocols for monitoring compliance with mitigation measures. Strengthening regulatory frameworks will not only improve the quality of EIAs but also promote accountability among ASM operators.

Economic and Social Benefits

In addition to environmental protection, developing ASM appraisal programs focused on EIA can yield significant economic and social benefits. By promoting sustainable mining practices, these programs can help ensure the long-term viability of small-scale mining operations, which are often a vital source of income for local communities. Sustainable ASM can create jobs, stimulate local economies, and contribute to poverty alleviation. Furthermore, by fostering responsible mining practices, communities can benefit from improved health outcomes, enhanced access to clean water, and the preservation of natural resources that are essential for their livelihoods.

Capacity Building and Knowledge Transfer

The implementation of ASM appraisal programs also presents an opportunity for capacity building and knowledge transfer among stakeholders. Training programs can be developed to educate ASM operators about the importance of EIA, the potential environmental impacts of their activities, and the best practices for mitigating these impacts. By fostering a culture of environmental stewardship within the ASM sector, Kenya can empower small-scale miners to take an active role in protecting their environment while pursuing their economic interests.

Monitoring and Continuous Improvement

The establishment of ASM appraisal programs should include mechanisms for ongoing monitoring and evaluation of EIA processes and their outcomes. Continuous improvement is essential for adapting to changing environmental conditions, regulatory requirements, and community needs. Developing clear performance indicators can help assess the effectiveness of EIA processes and the implementation of mitigation measures. Regular monitoring of these indicators allows stakeholders to identify areas for improvement and make necessary adjustments to their practices.

Conclusion

In conclusion, the establishment of Artisanal Small-Scale Mining appraisal programs focused on Environmental Impact Assessment is vital for promoting sustainable mining practices in Kenya. By systematically evaluating the potential environmental impacts of mining activities, including geological impacts and the specific challenges associated with rock blasting, proposing effective mitigation measures, and addressing training needs, these programs can significantly reduce the adverse effects of ASM on the environment.

The integration of stakeholder engagement, technological innovation, and partnerships into ASM appraisal programs will enhance their effectiveness and ensure that they are responsive to the needs of local communities and the environment. As Kenya continues to develop its mining sector, prioritizing EIA will be crucial for ensuring that economic growth does not come at the expense of environmental integrity.

Through collaborative efforts among government agencies, mining operators, local communities, and NGOs, Kenya can pave the way for a responsible and sustainable mining industry that benefits all stakeholders involved. By investing in ASM appraisal programs that emphasize EIA, Kenya can protect its rich natural heritage while fostering a mining sector that is resilient, equitable, and aligned with the principles of sustainable development.

Ultimately, this comprehensive approach will contribute to the well-being of communities, the preservation of ecosystems, and the sustainable management of mineral resources for generations to come. By prioritizing environmental impact assessments in ASM, Kenya can set a precedent for responsible mining practices that serve as a model for other nations facing similar challenges in balancing economic development with environmental protection.

As we move forward, the commitment to effective EIA in ASM will be a cornerstone of Kenya's journey toward a sustainable future, ensuring that the benefits of mining are realized without compromising the health of the environment or the rights of local communities. The way ahead is one of collaboration, innovation, and resilience, where the lessons learned from EIA processes will guide us in creating a more sustainable and equitable world for all.

Final Thoughts

In summary, the integration of EIA into ASM practices is not just a regulatory obligation; it is a moral imperative that reflects our collective responsibility to protect the environment and promote social equity. By embracing a holistic approach to mining that prioritizes environmental sustainability, community engagement, and safety, Kenya can lead the way in establishing a mining sector that is not only economically viable but also environmentally sound and socially responsible. The future of mining in Kenya depends on our ability to learn from past experiences, adapt to new challenges, and work together towards a common goal of sustainable development.

Stakeholder Engagement and Community Involvement

A critical component of successful ASM appraisal programs is the active engagement of all stakeholders, including local communities, government agencies, non-governmental organizations (NGOs), and industry representatives. By fostering a collaborative environment, stakeholders can share knowledge, resources, and best practices that enhance the effectiveness of EIA processes.

Community Involvement: Local communities are often the most affected by mining activities, and their involvement in the EIA process is essential for ensuring that their concerns and needs are addressed. Community engagement can take various forms, including public consultations, workshops, and participatory assessments. By involving community members in the decision-making process, ASM operators can build trust and foster a sense of shared responsibility for environmental stewardship.

Feedback Mechanisms: Establishing feedback mechanisms allows communities to voice their concerns and provide input on mining activities and their impacts. This two-way communication can help identify potential issues early on and facilitate timely interventions. Moreover, it empowers communities to hold mining operators accountable for their environmental commitments.

Technological Innovation and Best Practices

The integration of technological innovations into ASM appraisal programs can significantly enhance the effectiveness of EIA processes. Advances in technology can facilitate more accurate data collection, analysis, and monitoring of environmental impacts.

Remote Sensing and GIS: The use of remote sensing and Geographic Information Systems (GIS) can provide valuable insights into land use changes, habitat loss, and environmental degradation associated with mining activities. These tools enable stakeholders to visualize and analyze spatial data, making it easier to assess potential impacts and develop targeted mitigation strategies.

Sustainable Mining Technologies: Encouraging the adoption of sustainable mining technologies can help minimize the environmental footprint of ASM operations. For example, the use of eco-friendly extraction methods, such as gravity separation or bioleaching, can reduce the reliance on harmful chemicals and lower the risk of pollution. Training programs can be developed to educate ASM operators about these technologies and their benefits.

The Role of Partnerships

Building effective partnerships among various stakeholders is crucial for the success of ASM appraisal programs. Collaborative efforts can leverage resources, expertise, and networks to enhance the capacity of ASM operators and improve environmental management practices.

Public-Private Partnerships (PPPs): Engaging in public-private partnerships can facilitate knowledge sharing and resource mobilization. Government agencies can work with private sector actors to develop training programs, provide technical assistance, and promote sustainable practices within the ASM sector. These partnerships can also help secure funding for EIA initiatives and rehabilitation projects.

Collaboration with NGOs: Non-governmental organizations often play a vital role in advocating for environmental protection and community rights. Collaborating with NGOs can enhance the credibility of ASM appraisal programs and ensure that they align with broader environmental and social goals. NGOs can also provide valuable training and capacity-building support to ASM operators and local communities.

Monitoring and Continuous Improvement

The establishment of ASM appraisal programs should include mechanisms for ongoing monitoring and evaluation of EIA processes and their outcomes. Continuous improvement is essential for adapting to changing environmental conditions, regulatory requirements, and community needs.

Performance Indicators: Developing clear performance indicators can help assess the effectiveness of EIA processes and the implementation of mitigation measures. Regular monitoring of these indicators allows stakeholders to identify areas for improvement and make necessary adjustments to their practices.

Adaptive Management: Embracing an adaptive management approach enables stakeholders to respond to new information and changing circumstances. By remaining flexible and open to change, ASM operators can enhance their resilience and ensure that their practices remain aligned with environmental sustainability goals.

Conclusion

In summary, the establishment of Artisanal Small-Scale Mining appraisal programs focused on Environmental Impact Assessment is not only vital for promoting sustainable mining practices in Kenya but also for fostering a holistic approach to environmental management. By systematically evaluating the potential environmental impacts of mining activities, proposing effective mitigation measures, and addressing training needs, these programs can significantly reduce the adverse effects of ASM on the environment.

The integration of stakeholder engagement, technological innovation, and partnerships into ASM appraisal programs will enhance their effectiveness and ensure that they are responsive to the needs of local communities and the environment. As Kenya continues to develop its mining sector, prioritizing EIA will be crucial for ensuring that economic growth does not come at the expense of environmental integrity.

Through collaborative efforts among government agencies, mining operators, local communities, and NGOs, Kenya can pave the way for a responsible and sustainable mining industry that benefits all stakeholders involved. By investing in ASM appraisal programs that emphasize EIA, Kenya can protect its rich natural heritage while fostering a mining sector that is resilient, equitable, and aligned with the principles of sustainable development.

Ultimately, this comprehensive approach will contribute to the well-being of communities, the preservation of ecosystems, and the sustainable management of mineral resources for generations to come. By prioritizing environmental impact assessments in ASM, Kenya can set a precedent for responsible mining practices that serve as a model for other nations facing similar challenges in balancing economic development with environmental protection.

5.9 Summary of Chapter 5

In this chapter, we explored the principles and processes of Environmental Impact Assessment (EIA) as a critical tool for evaluating the environmental implications of proposed projects. We discussed the importance of EIA in decision-making, public participation, and sustainable development. The EIA process was outlined, including key steps such as screening, scoping, impact assessment, and reporting. We examined various methodologies used in EIA and addressed the challenges faced in the process. Through case studies, we illustrated real-world applications of EIA and the lessons learned. Finally, we highlighted future directions for EIA, emphasizing the need for technological integration, climate change considerations, and enhanced public engagement.

Conclusion

Importance of Environmental Impact Assessment (EIA)

Environmental Impact Assessment (EIA) is an essential tool for promoting sustainable development and protecting the environment. By systematically evaluating the potential environmental effects of proposed projects, EIA enables informed decision-making that balances economic growth with ecological integrity. The importance of EIA cannot be overstated, as it serves to identify potential impacts early in the planning process, allowing for the consideration of alternatives and the implementation of mitigation measures.

EIA Process and Methodologies

The EIA process, with its structured steps of screening, scoping, impact assessment, reporting, review, and monitoring, provides a comprehensive framework for assessing environmental implications. Each step plays a crucial role in ensuring that environmental considerations are integrated into project planning and execution. The methodologies employed in EIA, whether qualitative or quantitative, enhance the robustness of the assessments and ensure that diverse perspectives are integrated into the decision-making process. However, challenges such as data limitations, political influences, and public engagement barriers must be addressed to improve the effectiveness of EIAs.

Case Studies and Lessons Learned

The case studies presented in this chapter illustrate the practical applications of EIA in various contexts, highlighting both successes and lessons learned.

These examples underscore the importance of stakeholder engagement and the need for adaptive management strategies that respond to evolving environmental challenges. The lessons learned from these case studies can inform best practices and guide future EIA efforts, ensuring that they are more effective and inclusive.

Future Directions for EIA

Looking ahead, the future of EIA will be shaped by advancements in technology, the need to incorporate climate change considerations, and the imperative to strengthen public participation. The integration of innovative technologies, such as remote sensing, Geographic Information Systems (GIS), and data analytics, can enhance the accuracy and efficiency of EIA processes. Additionally, incorporating climate change considerations into EIA frameworks will ensure that projects are resilient to future environmental changes and contribute to climate adaptation efforts.

Artisanal Small-Scale Mining (ASM) in Kenya

In this context, the development of Artisanal Small-Scale Mining (ASM) appraisal programs focused on EIA is particularly vital for Kenya. As small-scale mining activities increase, the need for robust EIA processes becomes even more critical. These programs can help identify and assess the specific geological environmental impacts of mining, such as alterations in landforms, soil compaction, ground instability, and hydrological changes. By systematically evaluating these impacts, ASM appraisal programs can propose effective mitigation measures that protect both the environment and local communities.

Capacity Building and Community Engagement

Furthermore, integrating Training Needs Assessments (TNAs) into the EIA process will enhance the capacity of ASM operators to implement sustainable practices. Training programs can educate miners about the importance of environmental stewardship, the potential impacts of their activities, and the best practices for minimizing harm. This capacity-building approach not only empowers ASM operators but also fosters a culture of responsibility and accountability within the sector.

Engaging local communities in the EIA process fosters a sense of ownership and responsibility, ensuring that their voices are heard and their concerns addressed. This participatory approach not only improves the quality of EIAs but also builds trust among stakeholders. By involving communities in discussions about potential impacts and mitigation strategies, ASM operators can create a collaborative environment that promotes sustainable practices and enhances community well-being.

Regulatory Frameworks and Accountability

The establishment of clear regulatory frameworks governing ASM and EIA processes is also essential. By creating guidelines and standards for conducting EIAs in ASM, the National Environment Management Authority (NEMA) can ensure that all mining activities adhere to best practices in environmental management. This includes establishing criteria for assessing the significance of environmental impacts, as well as protocols for monitoring compliance with mitigation measures. Strengthening regulatory frameworks will not only improve the quality of EIAs but also promote accountability among ASM operators.

Commitment to Sustainable Future

Ultimately, the commitment to effective Environmental Impact Assessment is crucial for fostering a sustainable future. By prioritizing environmental considerations in project planning and implementation, particularly in the ASM sector, we can protect our natural resources, enhance community well-being, and contribute to the resilience of ecosystems. As we continue to face pressing environmental issues, the role of EIA as a decision-making tool will remain vital in guiding us toward a more sustainable and equitable world.

Summary and Call to Action

In summary, by investing in ASM appraisal programs that emphasize EIA, Kenya can set a precedent for responsible mining practices that serve as a model for other nations. This comprehensive approach will contribute to the well-being of communities, the preservation of ecosystems, and the sustainable management of mineral resources for generations to come.

The Way Forward

As Kenya navigates the complexities of its mining sector, it is essential to recognize that the integration of EIA into ASM practices is not merely a regulatory requirement but a fundamental aspect of sustainable development. The way forward involves several key actions:

1. **Strengthening Institutional Capacity**: Government agencies, particularly NEMA, must enhance their capacity to oversee and enforce EIA processes effectively. This includes training personnel, improving data collection and analysis capabilities, and fostering inter-agency collaboration to ensure a holistic approach to environmental management.

2. **Promoting Sustainable Practices**: ASM operators should be encouraged to adopt sustainable mining practices that minimize environmental impacts. This can be achieved through incentives, such as access to funding for environmentally friendly technologies, and support for cooperative models that promote shared resources and knowledge among miners.

3. **Enhancing Public Awareness**: Raising awareness about the importance of EIA and sustainable mining practices among local communities and ASM operators is crucial. Educational campaigns can empower stakeholders to engage actively in the EIA process and advocate for their rights and interests.

4. **Fostering Multi-Stakeholder Collaboration**: Building partnerships among government, industry, NGOs, and local communities is essential for creating a collaborative framework for EIA in ASM. These partnerships can facilitate knowledge sharing, resource mobilization, and the development of best practices that benefit all stakeholders.

5. **Implementing Adaptive Management**: The dynamic nature of environmental challenges necessitates an adaptive management approach in EIA processes. Continuous monitoring and evaluation of mining activities, along with feedback mechanisms, will allow for timely adjustments to practices and policies, ensuring that they remain effective in addressing emerging issues.

6. **Incorporating Indigenous Knowledge**: Recognizing and integrating indigenous knowledge and practices into EIA processes can enhance the effectiveness of assessments. Local communities often possess valuable insights into their environments and can contribute to identifying potential impacts and mitigation strategies.

Conclusion

In conclusion, the establishment of Artisanal Small-Scale Mining appraisal programs focused on Environmental Impact Assessment is not only vital for promoting sustainable mining practices in Kenya but also for fostering a holistic approach to environmental management. By systematically evaluating the potential environmental impacts of mining activities, including geological impacts, proposing effective mitigation measures, and addressing training needs, these programs can significantly reduce the adverse effects of ASM on the environment.

The integration of stakeholder engagement, technological innovation, and partnerships into ASM appraisal programs will enhance their effectiveness and ensure that they are responsive to the needs of local communities and the environment. As Kenya continues to develop its mining sector, prioritizing EIA will be crucial for ensuring that economic growth does not come at the expense of environmental integrity.

Through collaborative efforts among government agencies, mining operators, local communities, and NGOs, Kenya can pave the way for a responsible and sustainable mining industry that benefits all stakeholders involved. By investing in ASM appraisal programs that emphasize EIA, Kenya can protect its rich natural heritage while fostering a mining sector that is resilient, equitable, and aligned with the principles of sustainable development.

Ultimately, this comprehensive approach will contribute to the well-being of communities, the preservation of ecosystems, and the sustainable management of mineral resources for generations to come. By prioritizing environmental impact assessments in ASM, Kenya can set a precedent for responsible mining practices that serve as a model for other nations facing similar challenges in balancing economic development with environmental protection.

As we move forward, the commitment to effective EIA in ASM will be a cornerstone of Kenya's journey toward a sustainable future, ensuring that the benefits of mining are realized without compromising the health of the environment or the rights of local communities. The journey ahead is one of collaboration, innovation, and resilience, where the lessons learned from EIA processes will guide us in creating a more sustainable and equitable world for all.

.

Call to Action

To achieve these goals, it is imperative that all stakeholders—government agencies, mining operators, local communities, and civil society—come together to champion the cause of sustainable mining. This collective effort will not only enhance the effectiveness of EIA processes but also ensure that the mining sector contributes positively to the socio-economic fabric of Kenya.

By fostering a culture of environmental stewardship and social responsibility, we can pave the way for a mining industry that respects both the environment and the communities it impacts. Together, we can build a future where mining serves as a catalyst for sustainable development, benefiting current and future generations.

CHAPTER 6

GEOLOGY AND CLIMATE CHANGE

6.0 Introduction: Understanding Climate Change

Climate change encompasses long-term alterations in temperature and weather patterns, which can be either natural or anthropogenic (human-induced). Since the 1800s, human activities, particularly the burning of fossil fuels, have been the primary drivers of climate change. This shift has led to significant changes in the Earth's climate system, with far-reaching implications for both natural ecosystems and human societies. The urgency of addressing climate change is underscored by the increasing frequency and intensity of climate-related events, which pose risks to global stability and sustainability.

The combustion of fossil fuels releases greenhouse gases (GHGs) such as carbon dioxide (CO2) and methane (CH4), which trap heat in the atmosphere, leading to global warming. These gases create a greenhouse effect, where heat is retained in the atmosphere, resulting in rising temperatures. The current trends indicate that the Earth's average surface temperature has risen approximately 1.2°C since the late 1800s, with the last decade being the warmest on record. This warming is unprecedented in the context of the last 100,000 years, highlighting the urgent need for action to mitigate its effects.

Impacts of Climate Change

Environmental Effects

The environmental impacts of climate change are profound and multifaceted. One of the most alarming consequences is the rising sea levels, which have increased by about 20 centimeters in the last century, with the rate accelerating in recent decades. This rise threatens coastal communities and ecosystems, leading to habitat loss and increased flooding. Additionally, the melting of ice sheets in Greenland and Antarctica contributes significantly to sea-level rise, further exacerbating the risks faced by low-lying areas.

Extreme weather events have also become more frequent and severe, with hurricanes, droughts, and floods increasingly linked to climate change. These events not only disrupt ecosystems but also pose significant challenges to human safety and infrastructure. Furthermore, biodiversity loss is a critical concern, as many species face extinction due to changing habitats and climate conditions. The delicate balance of ecosystems is being disrupted, leading to cascading effects on food webs and overall ecological health.

Human health is another area significantly impacted by climate change. The rise in temperatures exacerbates health issues, including heat-related illnesses and respiratory problems stemming from poor air quality. Additionally, the spread of vector-borne diseases, such as malaria and dengue fever, is influenced by changing climate patterns, posing new public health challenges.

Socioeconomic Consequences

The socioeconomic consequences of climate change are equally concerning. Food security is at risk as changes in climate affect agricultural productivity, leading to food shortages and increased prices, particularly in vulnerable regions. The impacts on crop yields can have devastating effects on communities that rely on agriculture for their livelihoods, exacerbating poverty and inequality.

The economic impact of climate change is projected to be substantial, with costs associated with disaster recovery and adaptation measures potentially reaching trillions of dollars globally. These financial burdens can strain national economies and divert resources from other critical areas, such as education and healthcare. Additionally, communities, especially in low-lying coastal areas, are increasingly at risk of displacement due to rising sea levels and extreme weather events. This displacement can lead to social unrest and further exacerbate existing inequalities.

Mitigation and Adaptation Strategies

To address the challenges posed by climate change, a combination of mitigation and adaptation strategies is essential. Reducing emissions is a critical component of any effective response. Transitioning to renewable energy sources, such as solar and wind, is crucial for reducing GHG emissions. Policies aimed at cutting emissions by at least 30% by 2030 are essential to limit global warming and prevent catastrophic climate impacts.

Adaptation measures are equally important in preparing for the inevitable changes that climate change will bring. Implementing strategies to adapt to climate impacts, such as building resilient infrastructure and developing early warning systems for natural disasters, can help mitigate risks and protect vulnerable communities. These measures can enhance resilience and ensure that societies are better equipped to cope with the challenges ahead.

International cooperation is vital in the fight against climate change. Global frameworks like the Paris Agreement aim to unite countries in their efforts to combat climate change, emphasizing the need for collective action and financial support for developing nations. By working together, countries can share knowledge, resources, and technologies to address the global nature of this crisis.

Conclusion

Addressing climate change requires a multifaceted approach that combines scientific research, policy-making, and community engagement. The urgency of the situation calls for immediate action to mitigate its effects and adapt to the changes already underway. By recognizing the interconnectedness of environmental, social, and economic factors, we can develop comprehensive strategies that promote sustainability and resilience in the face of climate change.

6.1 Navigating the Environmental Crisis: The Urgent Need for Action in the Face of Human-Induced Climate Change

Introduction

As humanity grapples with the consequences of its actions, the impact of climate change becomes increasingly evident. The rise in global temperatures has led to a cascade of environmental changes, including more frequent and severe weather events, shifts in ecosystems, and disruptions to food and water supplies. These changes not only threaten biodiversity but also challenge the very foundations of human society. The urgency of addressing these issues cannot be overstated, as the consequences of inaction will be felt for generations to come.

The Dual Role of Technology

In this context, technology plays a dual role. On one hand, it has contributed to the environmental crisis through industrialization, deforestation, and pollution. The rapid advancement of technology has often prioritized short-term gains over long-term sustainability, leading to significant environmental degradation. On the other hand, technology holds the potential to provide solutions. Innovations in renewable energy, sustainable agriculture, and waste management can help mitigate the effects of climate change. The development of smart technologies, such as artificial intelligence and the Internet of Things, can optimize resource use and reduce emissions, paving the way for a more sustainable future.

The Need for Robust Policy Frameworks

Addressing climate change requires robust policy frameworks and international cooperation. Governments must prioritize environmental sustainability in their agendas, implementing regulations that limit emissions and promote green technologies. The Paris Agreement serves as a critical framework for global collaboration, urging nations to commit to reducing their carbon footprints and supporting vulnerable communities in adapting to climate impacts. Effective governance is essential for ensuring that climate action is prioritized and that resources are allocated effectively to address the challenges ahead.

Grassroots Movements and Community Engagement

Beyond governmental action, grassroots movements and community engagement are vital in the fight against climate change. Educating the public about sustainable practices and the importance of conservation can empower individuals to make informed choices. Local initiatives, such as community gardens, recycling programs, and conservation efforts, can foster a sense of responsibility and collective action. By engaging communities in the fight against climate change, we can create a more informed and proactive society that is better equipped to address environmental challenges.

Ethical Implications and Social Justice

The ethical implications of climate change cannot be overlooked. The burden of environmental degradation disproportionately affects marginalized communities, who often lack the resources to adapt or recover from climate-related disasters. Addressing climate change thus requires a commitment to social justice, ensuring that all voices are heard and that vulnerable populations are supported in the transition to a sustainable future. By prioritizing equity and inclusivity in climate action, we can create a more just and resilient society.

Summary: Navigating the Environmental Crisis

The impact of human-induced climate change is increasingly evident, manifesting in rising global temperatures, severe weather events, and disruptions to ecosystems and food and water supplies. These changes threaten biodiversity and challenge the foundations of human society, highlighting the urgent need for action.

Technology plays a dual role in this crisis. While it has contributed to environmental degradation through industrialization and pollution, it also offers solutions through innovations in renewable energy, sustainable agriculture, and smart technologies that optimize resource use.

Addressing climate change requires robust policy frameworks and international cooperation. Governments must prioritize environmental sustainability, as exemplified by the Paris Agreement, which encourages nations to reduce carbon footprints and support vulnerable communities.

Grassroots movements and community engagement are essential in combating climate change. Educating the public and fostering local initiatives can empower individuals and create a proactive society.

The ethical implications of climate change must also be considered, as marginalized communities often bear the brunt of environmental degradation. A commitment to social justice is necessary to ensure that all voices are heard and that vulnerable populations receive support in transitioning to a sustainable future.

In conclusion, the urgent need for action in the face of human-induced climate change calls for a multifaceted approach that combines technological innovation, effective policies, community engagement, and social justice. By addressing these interconnected challenges, we can work towards a sustainable future that protects both our planet and its people.

6.2 Introduction to Geology and Climate Change

Climate change refers to significant and lasting changes in the Earth's climate, primarily driven by human activities such as the burning of fossil fuels, deforestation, and industrial processes. These activities increase the concentration of greenhouse gases in the atmosphere, leading to global warming and associated climatic shifts. Understanding the interplay between geology and climate change is crucial, as geological processes can influence climate over geological timescales, while climate change can have profound effects on geological systems. This chapter explores these interactions, the implications for geological systems, and strategies for adaptation and mitigation.

The relationship between geology and climate is dynamic and multifaceted. Geological processes such as plate tectonics, volcanism, and erosion can shape the Earth's climate by altering landforms, ocean currents, and atmospheric composition. Conversely, climate change can impact geological systems, leading to phenomena such as permafrost thawing, glacial retreat, and increased erosion. By examining these interactions, we can better understand the complex feedback mechanisms that exist between the Earth's geological and climatic systems.

6.3 Geological Processes Influencing Climate

Plate Tectonics

The movement of tectonic plates plays a significant role in shaping the Earth's climate over millions of years. The formation of mountain ranges can influence atmospheric circulation patterns, while the creation of ocean basins affects ocean currents. For example, the uplift of the Himalayas has contributed to changes in monsoon patterns in South Asia, impacting regional climates. The interaction between tectonic activity and climate is evident in the long-term geological record, where shifts in plate movements have been linked to significant climatic events, such as ice ages.

Volcanism

Volcanic eruptions can have both short-term and long-term effects on climate. During eruptions, ash and gases such as sulfur dioxide are released into the atmosphere, which can lead to temporary cooling by reflecting sunlight away from the Earth. Historical eruptions, such as the 1991 eruption of Mount Pinatubo, demonstrated this cooling effect, which can last for several years. The impact of volcanism on climate is not limited to cooling; it can also contribute to warming over geological timescales through the release of greenhouse gases during prolonged volcanic activity.

Weathering and Erosion

Weathering processes, particularly chemical weathering, can sequester carbon dioxide from the atmosphere. As rocks break down, they react with CO_2 to form carbonates, effectively removing greenhouse gases from the atmosphere. This process plays a crucial role in regulating the Earth's climate over geological timescales. Erosion, on the other hand, can transport carbon-rich sediments to the ocean, where they may be buried and sequestered for millions of years. Understanding these processes is essential for assessing the long-term carbon cycle and its implications for climate regulation.

Sedimentation

Sedimentation processes can influence carbon cycling and long-term climate regulation. For instance, the burial of organic matter in sediments can sequester carbon, while changes in sedimentation rates can affect nutrient cycling and ecosystem dynamics. The interplay between sedimentation and climate is complex, as variations in climate can alter sediment transport and deposition patterns, further influencing the geological record and the carbon cycle.

6.4 Impacts of Climate Change on Geological Systems

Permafrost Thawing

Thawing permafrost due to rising temperatures poses significant risks to geological stability. As permafrost thaws, it can lead to ground subsidence, increased greenhouse gas emissions (such as methane), and changes in hydrology. The destabilization of permafrost can also impact infrastructure in Arctic regions, leading to costly repairs and safety hazards. The release of methane, a potent greenhouse gas, further exacerbates climate change, creating a feedback loop that accelerates warming. Environmental impacts include the loss of habitat for species adapted to cold environments and the alteration of local hydrology, which can affect freshwater availability.

Glacial Retreat

The retreat of glaciers is one of the most visible impacts of climate change. Glacial melt contributes to rising sea levels and alters freshwater availability in regions dependent on glacial runoff. The loss of glaciers also affects local ecosystems and can lead to increased sediment transport in rivers. As glaciers recede, they expose underlying geology, which can influence erosion rates and sediment dynamics in adjacent areas. The implications of glacial retreat extend beyond immediate environmental changes, affecting water resources, biodiversity, and human communities. Ecosystems that rely on glacial meltwater for survival may face significant stress, leading to shifts in species composition and habitat loss.

Increased Erosion and Landslides

Changing precipitation patterns and extreme weather events, such as heavy rainfall and storms, can lead to increased erosion and landslide risks. Areas previously stable may become vulnerable to geological hazards, threatening infrastructure and ecosystems. The increased frequency and intensity of extreme weather events, driven by climate change, can overwhelm natural systems, leading to rapid erosion and destabilization of slopes. This poses significant challenges for land management and disaster preparedness. Environmental impacts include the loss of soil fertility, degradation of habitats, and increased sedimentation in waterways, which can harm aquatic ecosystems.

Coastal Erosion and Sea Level Rise

Rising sea levels due to climate change have significant implications for coastal geology. Coastal erosion can lead to habitat loss, changes in sediment dynamics, and increased salinity in freshwater systems. Communities along coastlines face heightened risks from flooding and erosion, which can threaten infrastructure, ecosystems, and livelihoods. The interaction between sea level rise and coastal geology is complex, as changes in sediment supply and wave dynamics can further exacerbate erosion and alter coastal landscapes. Environmental impacts include the loss of coastal habitats such as wetlands and mangroves, which serve as critical buffers against storms and provide essential ecosystem services.

6.5 Geological Carbon Sequestration

Concept of Carbon Sequestration

Geological carbon sequestration involves capturing carbon dioxide emissions from industrial sources and storing them underground in geological formations. This process is essential for mitigating climate change by reducing the amount of CO_2 in the atmosphere. By sequestering carbon, we can help stabilize atmospheric concentrations and limit global warming. Geological formations, such as depleted oil and gas reservoirs, saline aquifers, and unmineable coal seams, provide potential storage sites for captured CO_2.

Methods of Carbon Sequestration

Various methods of carbon sequestration include:

- **Enhanced Oil Recovery (EOR):** Injecting CO_2 into oil reservoirs to increase oil recovery while simultaneously storing carbon. This method not only helps to extract additional fossil fuels but also reduces greenhouse gas emissions.
- **Mineral Carbonation:** Reacting CO_2 with naturally occurring minerals to form stable carbonates. This process can permanently sequester carbon and has the potential to be applied at a large scale.
- **Depleted Oil and Gas Reservoirs:** Utilizing empty oil and gas reservoirs for long-term CO_2 storage. These reservoirs have been proven to contain hydrocarbons for millions of years, suggesting they can securely store CO_2.

Challenges and Opportunities

While geological carbon sequestration presents significant opportunities for climate change mitigation, challenges remain. These include the need for extensive monitoring to ensure the integrity of storage sites, public acceptance of sequestration projects, and the development of regulatory frameworks to govern their implementation. Addressing these challenges is crucial for the successful deployment of carbon sequestration technologies and for gaining public trust in their safety and efficacy.

6.6 Adaptation Strategies in Geology

Resilient Infrastructure

Designing infrastructure that can withstand the impacts of climate change is crucial. This includes building structures that can resist extreme weather events, such as floods and landslides, and incorporating adaptive designs that account for changing geological conditions. Resilient infrastructure not only protects human lives and property but also minimizes disruptions to essential services and ecosystems.

Land Use Planning

Effective land use planning can mitigate the impacts of climate change on geological systems. Zoning regulations that consider geological hazards, such as floodplains and landslide-prone areas, can help protect communities and ecosystems. By integrating geological assessments into land use planning, decision-makers can promote sustainable development while minimizing risks associated with climate change.

Ecosystem Restoration

Restoring natural ecosystems, such as wetlands and forests, can enhance resilience to climate change impacts. These ecosystems provide critical services, including erosion control, carbon storage, and habitat for biodiversity. Restoration efforts can also improve water quality and enhance the ability of landscapes to adapt to changing climatic conditions. By investing in ecosystem restoration, we can create more resilient landscapes that support both human and ecological communities.

6.7 Case Studies in Geology and Climate Change

Case Study 1: The Greenland Ice Sheet

The Greenland Ice Sheet is experiencing accelerated melting due to climate change, contributing significantly to global sea level rise. This case study examines the implications of ice sheet loss for global climate patterns, freshwater availability, and coastal communities. The melting of the Greenland Ice Sheet not only raises sea levels but also disrupts ocean circulation patterns, which can have far-reaching effects on global climate systems.

Case Study 2: Coastal Erosion in Louisiana

Louisiana's coastal regions face severe erosion due to a combination of sea level rise, sediment loss, and human activities. This case study explores the geological and climatic factors contributing to coastal erosion and the strategies being implemented to address these challenges. Efforts to restore wetlands and implement sustainable land use practices are critical for mitigating the impacts of erosion and protecting vulnerable communities.

Case Study 3: Coastal Erosion in Kenya's Recreational Facilities

Assuming a worst-case scenario of a one-meter rise in sea level due to the greenhouse effect, more than 200 square kilometers of Kenya's coastal area could be submerged by the year 2100. This alarming projection highlights the vulnerability of coastal regions, particularly in areas like Malindi, Lamu, Mombasa, Kilifi, and Kwale, where approximately 7 million people currently reside. Among these, about 2.5 million people live in areas susceptible to submergence. The potential inundation of these regions poses significant risks not only to human settlements but also to unique ecosystems, including mangroves and wetlands, which are critical for biodiversity and coastal protection.

The east coast of Africa is also increasingly threatened by tropical cyclones, which could inundate approximately 5% of its area, leading to the destruction of vital ecosystems. The loss of mangroves and wetlands would have cascading effects on coastal resilience, as these ecosystems provide essential services such as storm protection, carbon sequestration, and habitat for marine life. The degradation of these environments would further exacerbate the impacts of climate change, making it imperative to implement effective management strategies.

To address these challenges, the Coastal Development Authority (CDA) must implement Coastal Zone Management (CZM) strategies that identify appropriate and adaptive options for mitigating the impacts of sea level rise and climate change along the Kenyan coastline. CZM would provide essential information and recommendations to national bodies, such as the National Environment Management Authority (NEMA), and international policy centers regarding integrated coastal zone management strategies for the next twenty years. This includes long-term policies aimed at adapting to the impacts of global climate change.

Integrated Coastal Zone Management (CZM) and Recommendations for Adaptation

The integrated Coastal Zone Management (CZM) approach should encourage coastal communities to proactively adapt to sea level rise, not merely as a response to an impending catastrophe but as a means to seize opportunities for sustainable development and resilience (Figure 13). By acting now, communities can mitigate adverse impacts and enhance their adaptive capacity. This proactive stance is essential for ensuring that local populations can thrive despite the challenges posed by climate change.

Figure 13: Vulnerable Coastal structures (recreational facilities) along Mombasa Beaches

Recommendations for Better Management of Coastal Ecosystems

Coastal Zone Management Plans

Coastal areas should implement comprehensive Coastal Zone Management (CZM) plans in accordance with the National Environment Management Authority's (NEMA) Environmental Management and Coordination Act

(EMCA) and the Coastal Development Authority (CDA) Acts. These plans must address the unique challenges posed by climate change and rising sea levels. Additionally, stakeholder engagement processes should be included to ensure that local communities, businesses, and environmental groups are involved in decision-making. This participatory approach will help build trust and ensure that the plans reflect the needs and priorities of those most affected by coastal changes.

Identification of Hazard-Prone Areas

Coastal hazard-prone areas at risk should be identified and mapped to inform planning and development decisions. This will help prioritize areas that require immediate attention and resources for adaptation. Utilizing geographic information systems (GIS) and remote sensing technologies can enhance the accuracy of hazard mapping and provide valuable data for future planning.

Enforcement of Socio-Economic Development Activities

The Coastal Development Authority (CDA) should ensure that coastal socio-economic development activities do not increase the vulnerability of coastal structures to sea level rise. This includes regulating new developments and ensuring that existing structures are resilient to climate impacts. Development projects should undergo rigorous environmental impact assessments (EIAs) to evaluate their potential effects on coastal ecosystems and communities, with a transparent process that includes public consultations.

Emergency Preparedness and Response Mechanisms

Emergency preparedness and coastal zone response mechanisms need to be continuously reviewed and strengthened. This will enhance the resilience of coastal resources, both renewable and non-renewable, and ensure that communities are equipped to respond to climate-related disasters. Training programs for local authorities and community members on disaster response and recovery should be established to improve readiness for extreme weather events.

Proposed CRZ Allowances

The proposed CRZ should also allow for certain activities essential for coastal management and development, including:

Defensive Construction

Construction activities related to defense requirements that necessitate foreshore facilities, excluding classified operational components of defense projects.

Port and Harbor Operations

Operational constructions for ports, harbors, and lighthouses requiring water frontage to support trade and transportation.

Thermal Power Plants

Facilities for thermal power plants that include foreshore facilities for the transport of raw materials and cooling water intake and discharge.

Investment Activities

Activities with investments exceeding 10 million shillings to encourage significant investments in sustainable coastal development.

Discharge of Treated Effluents

Setting up and expanding facilities for discharging treated effluents into watercourses, with approval under the Water Act 2002.

Land Reclamation for Erosion Control

Land reclamation, bunding, or disturbing the natural course of seawater for coastal erosion control and maintenance of waterways.

Mining Rare Minerals

Mining rare minerals not available outside CRZ areas to support economic development while protecting sensitive ecosystems.

Groundwater Harvesting

Harvesting groundwater in the 200 to 500-meter zones, conducted through ordinary wells for drinking, horticulture, agriculture, and fisheries. This measure ensures sustainable water use, balancing the needs of local communities with environmental protection.

Alteration for Recreational Purposes

Dressing or altering sand dunes and natural features for beautification, recreational, and other purposes permissible under NEMA's Environmental Management and Coordination Act (EMCA) of 1999. This allowance promotes community engagement and recreational opportunities, enhancing the quality of life for residents and visitors alike.

Conclusion

The recommendations and proposed regulations outlined above aim to create a balanced approach to managing coastal ecosystems in the face of climate change. By implementing comprehensive Coastal Zone Management plans, identifying hazard-prone areas, enforcing socio-economic development activities, and enhancing emergency preparedness, we can protect coastal resources and communities. Additionally, the proposed prohibitions and allowances within the Coastal Regulation Zone are designed to safeguard sensitive ecosystems while allowing for necessary development and resource use. Through these measures, we can work towards a sustainable future for our coastal environments, ensuring their resilience and health for generations to come.

The rigorous implementation of NEMA's 1999 Act (EMCA), following the approval of Environmental Impact Assessments (EIAs), is crucial for conserving and preserving the fragile ecological coastal belt. This approach will not only protect the environment but also enhance the fishing industry and contribute to economic growth. The collaboration between NEMA and the CRZ Management Plan, in liaison with the CDA, should facilitate the approval of construction projects that adhere to sustainable practices. For instance, the construction of residential houses and hotels up to a height of 10 meters would allow for two- to four-storey structures that are less vulnerable to flooding.

However, it is essential to recognize that the interests of the fishing community must be protected. Unregulated development along the coast can lead to the displacement of fishing communities, forcing them further inland and jeopardizing their livelihoods. Therefore, the implementation of the CRZ Act must balance development with environmental conservation, ensuring that the needs of local communities are prioritized while also addressing the challenges posed by climate change.

In summary, the proposed Coastal Regulation Zone (CRZ) framework aims to create a balanced approach to coastal management that protects ecosystems while allowing for sustainable development. By implementing these recommendations, Kenya can enhance its resilience to climate change, safeguard its unique coastal environments, and support the livelihoods of millions of people who depend on these vital ecosystems. The proactive measures outlined in this case study are essential for ensuring a sustainable future for Kenya's coastal regions in the face of rising sea levels and increasing climate-related challenges.

Case Study 4: Carbon Sequestration Projects

The Sleipner Project in Norway is one of the world's first commercial-scale carbon sequestration projects. This case study analyzes the project's success in capturing and storing CO_2 in a depleted gas reservoir, highlighting the potential for similar initiatives worldwide. The Sleipner Project demonstrates the feasibility of geological carbon sequestration and provides valuable insights into monitoring and verification processes.

6.8 Future Directions in Geology and Climate Change

Research and Monitoring

Ongoing research and monitoring of geological processes in the context of climate change are essential for understanding and predicting future impacts. This includes studying the interactions between climate and geological systems and developing models to assess potential scenarios. Enhanced monitoring networks can provide critical data for assessing changes in geological stability, erosion rates, and carbon cycling.

Interdisciplinary Approaches

Addressing the challenges of climate change requires interdisciplinary collaboration among geologists, climatologists, policymakers, and communities. Integrating knowledge from various fields can lead to more effective solutions and strategies. Collaborative research initiatives can foster innovation and promote the development of comprehensive approaches to climate adaptation and mitigation.

Public Awareness and Education

Raising public awareness and promoting education about the geological aspects of climate change are crucial for fostering community engagement in mitigation efforts. Informing the public about the importance of geological processes in climate regulation can encourage support for sustainable practices. Educational programs that highlight the connections between geology and climate change can empower individuals and communities to take action.

6.9 Ecological and Geological Implications of Global Warming

Ecological Implications

Dependence on Natural Resources

Developing countries are often heavily reliant on natural resources for their economies and livelihoods, encompassing sectors such as agriculture, forestry, and fisheries. According to the United Nations Environment Programme (UNEP), approximately 70% of the population in developing countries depends on agriculture, which is particularly sensitive to climate variability (UNEP, 2016). Changes in temperature and precipitation patterns can directly affect crop yields, leading to food insecurity and economic instability. For instance, altered rainfall patterns can disrupt planting and harvesting schedules, while increased temperatures can exacerbate drought conditions, further threatening agricultural productivity. This dependence on climate-sensitive resources underscores the vulnerability of these nations to the adverse effects of global warming, necessitating urgent adaptation strategies to safeguard food security and economic resilience.

Limited Resources for Adaptation

Many low-income countries face significant barriers to adaptation due to existing socio-economic challenges. These nations often lack the financial and technical resources necessary to implement effective adaptation strategies. The Intergovernmental Panel on Climate Change (IPCC) has highlighted that developing countries are disproportionately affected by climate change due to their limited adaptive capacity (IPCC, 2014). The costs associated with infrastructure improvements, agricultural adjustments, and disaster preparedness can be prohibitive, leaving these nations vulnerable to climate impacts. For example, the implementation of irrigation systems, drought-resistant crop varieties, and early warning systems for extreme weather events requires substantial investment, which many developing countries cannot afford. Consequently, the lack of resources for adaptation exacerbates their vulnerability to climate change, perpetuating cycles of poverty and instability.

Vulnerability to Natural Disasters

Developing countries are particularly susceptible to natural disasters exacerbated by climate change, such as floods, droughts, and extreme weather events. The World Bank (2010) reports that climate change is expected to increase the frequency and intensity of these events, leading to greater economic losses and displacement of populations. For instance, rising sea levels threaten coastal communities with inundation, while increased rainfall can lead to flooding and landslides. The compounded effects of these disasters can overwhelm local response capacities, leading to humanitarian crises and long-term socio-economic challenges. The increased frequency of such events necessitates the development of robust disaster risk reduction strategies and infrastructure improvements to enhance resilience in vulnerable communities.

Food Security Concerns

Climate disruption poses a significant threat to food security, particularly in regions that serve as traditional breadbaskets. As climate conditions change, agricultural productivity may decline, leading to food shortages and increased prices. The Food and Agriculture Organization (FAO) has warned that climate change could reduce crop yields by up to 25% in some regions by 2050 (FAO, 2016). This decline in food availability can exacerbate poverty and malnutrition, particularly in vulnerable populations. The interdependence of food systems and climate stability highlights the urgent need for adaptive agricultural practices, such as crop diversification and sustainable land management, to mitigate the impacts of climate change on food security.

6.10 Geological Implications

Hydrological Cycle and CO2 Recycling

The hydrological cycle is a critical component of the Earth's climate system and plays a vital role in the recycling of carbon dioxide (CO2). According to Patwardhan (1999), the hydrological cycle influences both past and future climate changes. Increased temperatures can disrupt this cycle, leading to altered precipitation patterns and changes in water availability. The potential rise in average global temperatures by 3°C could have significant impacts on climatic zones and sea levels. Disruptions in the hydrological cycle can also affect freshwater resources, leading to increased competition for water among agricultural, industrial, and domestic users.

Impact of Temperature Rise

Projections indicate that global temperatures could rise between 0.5°C and 5.5°C by the middle of the next century. While a lower range of temperature increase may not exceed natural variability, a higher range could lead to significant melting of glaciers and polar ice, affecting freshwater availability and causing flooding. The melting of ice sheets contributes to rising sea levels, which can inundate coastal areas and displace populations. The geological implications of such temperature increases extend beyond immediate impacts, as they can alter geological processes, including erosion and sediment transport, further reshaping landscapes.

Pressure on Cultivated Land

The anticipated rise in temperature may impose tremendous pressure on land currently used for agriculture. Increased temperatures can lead to longer, hotter summers and shorter winters, reducing the time available for groundwater recharge. This can exacerbate drought conditions and reduce agricultural productivity. The IPCC (2014) emphasizes that changes in temperature and precipitation will significantly affect crop yields, particularly in regions already vulnerable to food insecurity. The resulting pressure on cultivated land necessitates innovative agricultural practices and land-use planning to ensure sustainable food production in the face of climate change.

Ocean Temperature and Water Distribution

The oceans, which store approximately 99% of the Earth's water, are expected to warm more slowly than the atmosphere. However, increased evaporation from warmer oceans can lead to cooling effects on surface temperatures while simultaneously decreasing the oceans' capacity to absorb atmospheric CO_2. This disruption can affect marine ecosystems, as the upwelling currents that support marine life may be altered. Changes in ocean temperature and circulation patterns can have cascading effects on global weather systems, further complicating climate predictions and adaptation strategies.

Carbon Dioxide and Marine Life

The interaction between the hydrosphere and atmosphere is essential for recycling CO_2 and other greenhouse gases. Disruption of this balance can affect marine ecosystems, as the upwelling currents that support marine life may be altered. Increased ocean temperatures can lead to coral bleaching and the decline of marine biodiversity, which are critical for maintaining healthy ocean ecosystems (Hughes et al., 2017). The loss of biodiversity in marine environments can have profound implications for global fisheries and food security, highlighting the interconnectedness of terrestrial and marine ecosystems in the context of climate change.

Plant Adaptation Challenges

While increased CO_2 levels may initially enhance photosynthesis in some plant species, not all plants can adapt quickly enough to rising temperatures and changing moisture levels. The IPCC (2014) notes that many species may face increased stress due to higher temperatures, leading to declines in biodiversity and disruptions in ecosystems. The loss of plant species can have cascading effects on food webs and ecosystem services, further complicating efforts to achieve sustainable resource management in a changing climate.

Historical Context of Climate Change

The Earth has experienced significant climatic changes throughout its geological history, including glaciation events. However, the current rate of change due to human activities, particularly fossil fuel combustion, is unprecedented. Geological records indicate that the Earth has undergone periods of warming and cooling, but the rapid pace of current changes poses unique challenges for adaptation. The historical context emphasizes the need for immediate action to mitigate the impacts of climate change, as the lessons learned from past climatic shifts can inform contemporary strategies for resilience and adaptation.

Need for Vigilance and Action

Although there is still uncertainty regarding the precise degree of global warming, it is imperative to take proactive measures to protect the planet's ecosystems. Large-scale afforestation, reduction of fossil fuel combustion, and sustainable land management practices can help mitigate temperature rise and its associated impacts. The IPCC (2014) emphasizes the importance of immediate action to limit global warming to below 2°C to avoid the most severe consequences of climate change. This call to action underscores the urgency of collaborative efforts among governments, industries, and communities to implement effective climate policies and practices that promote sustainability and resilience in the face of global warming.

Collaborative Approaches to Mitigation and Adaptation

To effectively address the ecological and geological implications of global warming, a collaborative approach is essential. Governments, non-governmental organizations, and local communities must work together to develop and implement strategies that promote both mitigation and adaptation. This includes investing in renewable energy sources, enhancing energy efficiency, and promoting sustainable agricultural practices. For instance, transitioning to solar, wind, and geothermal energy can significantly reduce greenhouse gas emissions while providing reliable energy sources for developing nations. Additionally, sustainable agricultural practices, such as agroforestry and permaculture, can enhance soil health, increase biodiversity, and improve resilience to climate impacts.

Education and Capacity Building

Education and capacity building are critical components of effective climate action. Raising awareness about the impacts of climate change and the importance of sustainable practices can empower communities to take proactive measures. Training programs that focus on climate resilience, sustainable resource management, and disaster preparedness can equip individuals and communities with the knowledge and skills needed to adapt to changing conditions. Furthermore, integrating climate education into school curricula can foster a generation of environmentally conscious citizens who are better prepared to tackle the challenges posed by global warming.

Policy Frameworks and Governance

Robust policy frameworks and governance structures are vital for facilitating effective climate action. Policymakers must prioritize climate change in national and local agendas, ensuring that environmental considerations are integrated into economic planning and development strategies. International cooperation is also essential, as climate change is a global challenge that transcends national borders. Agreements such as the Paris Agreement provide a platform for countries to commit to reducing emissions and enhancing resilience. However, for these agreements to be effective, they must be supported by transparent monitoring and accountability mechanisms that ensure compliance and foster trust among nations.

Research and Innovation

Investing in research and innovation is crucial for developing new technologies and practices that can mitigate the impacts of climate change. Scientific research can provide valuable insights into the mechanisms of climate change, enabling the development of more accurate models and predictions. Innovations in carbon capture and storage, sustainable agriculture, and renewable energy technologies can play a significant role in reducing emissions and enhancing resilience. Collaborative research initiatives that involve academia, industry, and government can accelerate the development and deployment of these solutions, fostering a more sustainable future.

Conclusion: A Call to Action

In conclusion, the ecological and geological implications of global warming present significant challenges that require immediate and concerted action. The interconnectedness of climate change impacts on natural resources, ecosystems, and human livelihoods underscores the urgency of addressing this global crisis. By fostering collaboration, investing in education and capacity building, implementing robust policy frameworks, and promoting research and innovation, we can work towards a sustainable and resilient future. The time for action is now; the choices we make today will determine the health of our planet and the well-being of future generations. It is imperative that we embrace a holistic approach to climate action, recognizing that the health of our ecosystems and the stability of our societies are inextricably linked. Through collective efforts, we can mitigate the impacts of global warming and pave the way for a more sustainable and equitable world.

Integrating Indigenous Knowledge and Practices

Valuing Traditional Ecological Knowledge

Indigenous communities possess a wealth of traditional ecological knowledge (TEK) that has been developed over generations through close interactions with their environments. This knowledge can provide valuable insights into sustainable land management practices, biodiversity conservation, and climate adaptation strategies. Integrating TEK with scientific research can enhance the effectiveness of climate action initiatives. For example, indigenous practices such as rotational grazing, controlled burns, and agroecology can improve soil health and resilience to climate impacts. Recognizing and respecting the rights of indigenous peoples in decision-making processes is essential for fostering inclusive and effective climate strategies.

Urban Resilience and Climate Adaptation

Building Climate-Resilient Cities

As urbanization continues to accelerate, cities are increasingly vulnerable to the impacts of climate change, including heatwaves, flooding, and sea-level rise. Developing climate-resilient urban infrastructure is critical for mitigating these risks. This includes investing in green infrastructure, such as parks and green roofs, which can help manage stormwater, reduce urban heat, and improve air quality. Additionally, enhancing public transportation systems and promoting sustainable urban planning can reduce greenhouse gas emissions and improve the quality of life for urban residents. Engaging communities in the planning process ensures that urban resilience strategies are tailored to local needs and conditions.

Economic Opportunities in the Green Transition

Harnessing the Green Economy

The transition to a low-carbon economy presents significant economic opportunities, particularly in developing countries. Investments in renewable energy, energy efficiency, and sustainable agriculture can create jobs, stimulate economic growth, and enhance energy security. For instance, the renewable energy sector is projected to create millions of jobs globally, from manufacturing and installation to maintenance and operation. By prioritizing green technologies and sustainable practices, countries can position themselves as leaders in the emerging green economy, attracting investment and fostering innovation.

Global Cooperation and Climate Justice

Addressing Inequities in Climate Action

Climate change disproportionately affects vulnerable populations, particularly in developing countries that have contributed the least to greenhouse gas emissions. Addressing these inequities is essential for achieving climate justice. Wealthier nations, which have historically been the largest emitters, have a moral obligation to support developing countries in their climate adaptation and mitigation efforts. This can be achieved through financial assistance, technology transfer, and capacity-building initiatives. International cooperation is crucial for ensuring that all countries have the resources and support needed to address the impacts of climate change effectively.

Monitoring and Evaluation of Climate Initiatives

Establishing Accountability Mechanisms

To ensure the effectiveness of climate action initiatives, it is essential to establish robust monitoring and evaluation frameworks. These frameworks should include clear indicators for assessing progress toward climate goals, as well as mechanisms for accountability and transparency. Regular reporting and assessments can help identify successful strategies and areas for improvement, enabling adaptive management of climate initiatives. Engaging stakeholders in the evaluation process fosters a sense of ownership and commitment to achieving climate objectives.

Public Awareness and Advocacy

Empowering Communities through Education

Raising public awareness about the impacts of climate change and the importance of sustainable practices is vital for fostering a culture of environmental stewardship. Educational campaigns, community workshops, and advocacy initiatives can empower individuals and communities to take action in their own lives. By promoting sustainable behaviors, such as reducing waste, conserving energy, and supporting local food systems, communities can contribute to broader climate goals. Grassroots movements and advocacy organizations play a crucial role in amplifying voices and driving policy changes at local, national, and international levels.

Conclusion: A Collective Responsibility

In summary, the ecological and geological implications of global warming necessitate a multifaceted and inclusive approach to climate action. By integrating indigenous knowledge, building climate-resilient urban environments, harnessing economic opportunities in the green transition, and addressing inequities in climate action, we can create a more sustainable and equitable future. The challenges posed by climate change are significant, but they also present an opportunity for innovation, collaboration, and transformation. It is a collective responsibility to act decisively and compassionately, ensuring that we protect our planet and its inhabitants for generations to come. The time for comprehensive and inclusive climate action is now; our future depends on it.

6.11 Understanding the Complex Impacts of Global Warming: Vulnerabilities, Geological Implications, and the Urgency for Action

Vulnerabilities of Developing Countries

A number of factors contribute to the heightened vulnerability of developing countries in the face of climate change. Firstly, these nations often rely heavily on natural resources and ecosystems, such as crop and grazing lands, forests, and fisheries, making them particularly susceptible to shifts in climate. Secondly, many developing countries are already grappling with various socio-economic challenges, which limits their financial and technical capacity to implement the necessary adaptations to climate change. Additionally, specific vulnerabilities, such as exposure to rising sea levels and increased frequency of natural disasters—like floods, droughts, and severe storms—further exacerbate their risk.

Geological Implications of Global Warming

The geological implications of global warming are equally concerning. The hydrological cycle, particularly the recycling of carbon dioxide (CO2), plays a central role in understanding both past and future climatic changes. The alarming rise in global temperatures, projected to increase by as much as 5.5°C by the middle of the next century, poses significant risks to the planet's ecosystems and human societies. A rise of just 3°C could lead to a corresponding increase in sea levels, with potentially catastrophic consequences for coastal regions.

Human Contributions to Climate Change

Human activities, particularly the combustion of fossil fuels, have significantly contributed to the concentration of greenhouse gases in the atmosphere. This has resulted in a greenhouse effect that enhances surface temperatures, leading to a range of environmental challenges. The contributions to global warming from various sectors—59% from fossil fuel combustion, 14% from agriculture,

9% from deforestation, and 20% from chlorofluorocarbons (CFCs)—highlight the urgent need for comprehensive mitigation strategies.

Historical Geological Changes

Geologically, the Earth has experienced significant changes over its 3.5 billion-year history, including periods of glaciation and fluctuating sea levels. The current warming trend threatens to disrupt the delicate balance of the planet's ecosystems, leading to increased desertification, altered precipitation patterns, and diminished freshwater resources. As the Earth's surface temperature rises, the implications for agriculture, water supply, and biodiversity become increasingly severe.

Conclusion

In conclusion, while the exact degree of global warming remains a subject of ongoing research, the potential consequences are too significant to ignore. The ecological and geological implications of global warming are complex and interconnected, affecting ecosystems, agriculture, and human societies in profound ways. As global temperatures continue to rise, we are witnessing shifts in weather patterns, increased frequency of extreme weather events, and disruptions to natural systems that sustain life on Earth.

Developing countries, in particular, face significant challenges due to their reliance on natural resources and limited capacity for adaptation. These nations often bear the brunt of climate impacts despite contributing the least to greenhouse gas emissions. Vulnerable populations in these regions are at heightened risk of food insecurity, water scarcity, and displacement, making it imperative that we prioritize their needs in climate action efforts.

As global temperatures rise, the impacts will become increasingly pronounced, necessitating a proactive and collaborative approach to address the challenges posed by climate change. It is essential to recognize the urgency of the situation and take collective action to mitigate climate change, protect vulnerable populations, and ensure a sustainable future for generations to come.

To achieve this, we must invest in sustainable practices that promote environmental stewardship and resilience. This includes transitioning to renewable energy sources, implementing energy-efficient technologies, and adopting sustainable agricultural practices that enhance food security while minimizing environmental degradation. By enhancing community resilience through education, infrastructure improvements, and disaster preparedness, we can empower individuals and communities to adapt to the changing climate.

Reducing greenhouse gas emissions is a critical component of our response to climate change. This requires not only national and international commitments to emissions reduction targets but also grassroots movements that advocate for sustainable practices at the local level. Understanding the intricate relationships between climate, ecosystems, and human activities is crucial for developing effective solutions that promote resilience and sustainability in the face of a changing climate.

It is imperative that we act now to protect our planet and ensure the survival of future generations, as the window for effective action is rapidly closing. The choices we make today will have lasting implications for the health of our planet and the well-being of all its inhabitants. By fostering collaboration among governments, businesses, and communities, we can create a unified response to the climate crisis that is equitable and effective.

In summary, the challenge of global warming is immense, but it is not insurmountable. Together, we have the power to shape a sustainable future that prioritizes the health of our planet and the welfare of its people. Let us rise to the occasion, embrace the urgency of this moment, and commit to meaningful action that will safeguard our environment and ensure a livable world for generations to come.

6.12 Summary of Chapter 6

The Relationship Between Geology and Climate Change

In this chapter, we explored the intricate relationship between geology and climate change, emphasizing how geological processes influence climate and how climate change impacts geological systems. We discussed key topics such as permafrost thawing, glacial retreat, and coastal erosion, as well as the role of geological carbon sequestration in mitigating climate change. Each of these elements plays a crucial role in understanding the broader implications of climate change on both natural environments and human societies.

Coastal Erosion and Recreational Facilities in Kenya

A significant focus of this chapter was the case study on coastal erosion in Kenya, particularly concerning recreational facilities along the coastline. The Kenyan coast, which includes popular tourist destinations such as Mombasa, Malindi, and Lamu, is facing severe threats from climate change. Projections indicate that a one-meter rise in sea level by the year 2100 could submerge over 200 square kilometers of coastal land, directly impacting both human settlements and natural ecosystems.

The coastal recreational facilities, which are vital for tourism and local economies, are particularly vulnerable to these changes. These facilities not only provide leisure opportunities for residents and tourists but also support local businesses and employment. The potential inundation of these areas due to rising sea levels and increased storm intensity poses a significant risk to the livelihoods of millions of people who depend on coastal resources. The loss of beaches, recreational parks, and marine environments would not only diminish the aesthetic and recreational value of these areas but also lead to economic downturns in tourism-dependent communities.

The Need for Coastal Zone Management (CZM)

To address the challenges posed by climate change, effective Coastal Zone Management (CZM) strategies are essential. The Coastal Development Authority (CDA) in Kenya must implement comprehensive CZM plans that consider the unique challenges of rising sea levels and climate change impacts. These plans should be designed to protect coastal ecosystems, including mangroves and wetlands, which play a critical role in buffering against storm surges and providing habitat for diverse marine life.

CZM strategies should also involve stakeholder engagement processes to ensure that local communities, businesses, and environmental groups are actively involved in decision-making. This participatory approach is vital for building trust and ensuring that the plans reflect the needs and priorities of those most affected by coastal changes. By integrating local knowledge and perspectives, CZM can foster a sense of ownership and responsibility among community members, enhancing the effectiveness of adaptation measures.

Recommendations for Better Management of Coastal Ecosystems

Based on the analysis of current challenges and opportunities, several policy-relevant recommendations have emerged to enhance the management of coastal ecosystems in Kenya. First and foremost, comprehensive Coastal Zone Management (CZM) plans should be developed in accordance with the National Environment Management Authority's (NEMA) Environmental Management and Coordination Act (EMCA) and the Coastal Development Authority (CDA) Acts. These plans must specifically address the unique challenges posed by climate change and rising sea levels, ensuring that they are adaptable to future conditions. By creating a robust framework for managing coastal resources, these plans can help mitigate the impacts of climate change while promoting sustainable development.

Another critical recommendation is the identification and mapping of hazard-prone areas along the coast. This process is essential for informing planning and development decisions. Utilizing geographic information systems (GIS) and remote sensing technologies can significantly enhance the accuracy of hazard mapping, providing valuable data for future planning. By prioritizing areas that require immediate attention and resources for adaptation, stakeholders can make informed decisions that protect both human and ecological communities.

The enforcement of socio-economic development activities is also vital. The CDA should ensure that any socio-economic development does not increase the vulnerability of coastal structures to sea level rise. This includes regulating new developments and ensuring that existing structures are resilient to climate impacts. Development projects should undergo rigorous environmental impact assessments (EIAs) to evaluate their potential effects on coastal ecosystems and communities. This proactive approach will help safeguard the environment while allowing for necessary economic growth.

Additionally, continuous review and strengthening of emergency preparedness and response mechanisms are necessary to enhance the resilience of coastal resources. Training programs for local authorities and community members on disaster response and recovery should be established to improve readiness for extreme weather events. By equipping communities with the knowledge and skills to respond effectively to climate-related disasters, we can reduce vulnerability and enhance adaptive capacity.

Proposed Coastal Regulation Zone (CRZ) Prohibitions

To protect coastal ecosystems and mitigate the impacts of climate change, the proposed Coastal Regulation Zone (CRZ) should prohibit several activities that could exacerbate environmental degradation. First, new industrial developments and the expansion of existing ones should be prohibited, except for those directly related to waterfront activities. This restriction is essential to prevent industrial pollution and habitat destruction in sensitive coastal areas.

Furthermore, the manufacturing, handling, storing, or disposing of hazardous materials should be strictly prohibited to protect coastal ecosystems from contamination and safeguard public health. Establishing and expanding fish processing units should also be limited, as these activities can lead to pollution and overfishing, threatening marine biodiversity. Additionally, expanding units for the disposal of waste and effluents must be prohibited to prevent the degradation of water quality in coastal areas, which is crucial for both human health and marine life.

Discharging untreated wastes and effluents from industries, cities, or towns must be prohibited to maintain the health of marine ecosystems and protect human health. Moreover, dumping city or town waste for landfilling or other purposes can lead to soil and water contamination, adversely affecting coastal habitats and the organisms that depend on them. Land reclamation activities that disturb the natural course of seawater should also be restricted, as they can disrupt natural coastal processes and lead to habitat loss, further exacerbating the impacts of climate change.

Proposed CRZ Allowances

While certain activities should be prohibited, the proposed CRZ should also allow for essential activities that support coastal management and development. For instance, construction activities related to defense requirements that necessitate foreshore facilities should be permitted, excluding classified operational components of defense projects. This allowance is important for maintaining coastal security and infrastructure.

Operational constructions for ports, harbors, and lighthouses requiring water frontage are vital for trade and transportation, supporting the local economy while ensuring that they are built with environmental considerations in mind. Facilities for thermal power plants that include foreshore facilities for the transport of raw materials and cooling water intake and discharge should also be allowed, provided that environmental safeguards are in place to minimize impacts on coastal ecosystems.

Encouraging activities with investments exceeding 10 million shillings can promote significant investments in sustainable coastal development. Additionally, setting up and expanding facilities for discharging treated effluents into watercourses, with approval under the Water Act 2002, ensures that wastewater is managed responsibly, protecting water quality in coastal areas. Finally, land reclamation for coastal erosion control and maintenance of waterways should be permitted to protect coastal infrastructure.

Conclusion

The ecological and geological implications of global warming, coupled with the pressing need for sustainable development, underscore the importance of a comprehensive and integrated approach to environmental management. The rigorous implementation of NEMA's 1999 Act (EMCA), alongside the approval of Environmental Impact Assessments (EIAs), is essential for conserving the fragile ecological coastal belt. This strategy not only aims to protect the environment but also seeks to enhance the fishing industry and contribute to economic growth. Collaboration between NEMA, the CRZ Management Plan, and the CDA is vital to ensure that construction projects adhere to sustainable practices, such as limiting the height of residential and commercial buildings to reduce vulnerability to flooding.

However, it is crucial to prioritize the interests of local communities, particularly fishing communities that may be displaced by unregulated coastal development. Balancing development with environmental conservation is imperative to safeguard livelihoods and maintain social equity. The challenges posed by climate change are significant, yet they also present opportunities for innovation, collaboration, and transformation. By integrating indigenous knowledge, fostering climate-resilient urban environments, and harnessing economic opportunities in the green transition, we can work towards a more sustainable and equitable future.

As we navigate the complexities of climate change, it is our collective responsibility to act decisively and compassionately. The time for comprehensive and inclusive climate action is now; our future depends on it. By ensuring that environmental regulations are enforced and that the needs of vulnerable populations are addressed, we can protect our planet and its inhabitants for generations to come. Together, we can pave the way for a resilient and thriving future, where economic development and environmental stewardship go hand in hand.

CHAPTER 7

GEOLOGY AND NATURAL RESOURCES

7.1 Introduction to Natural Resources

Natural resources are fundamental materials and components found in the environment that humans utilize for a variety of purposes, including energy production, manufacturing, and sustenance. These resources can be broadly categorized into two main types: renewable and non-renewable resources. Renewable resources, such as solar energy, wind, and biomass, are those that can be replenished naturally over time, making them sustainable options for long-term use. In contrast, non-renewable resources, including fossil fuels and minerals, are finite and cannot be replaced within a human timescale, leading to concerns about their depletion and the environmental impacts associated with their extraction and use.

The significance of natural resources cannot be overstated; they form the foundation of human civilization, driving economic development, technological advancement, and overall quality of life. From the energy that powers our homes and industries to the raw materials that are essential for manufacturing goods, natural resources play a critical role in shaping our societies. Understanding the geological processes that lead to the formation and distribution of these resources is essential for their sustainable management and utilization. This knowledge allows us to make informed decisions about resource extraction, conservation, and the development of alternative energy sources, ultimately ensuring that future generations can benefit from these vital assets.

7.2 Geological Formation of Natural Resources

The geological formation of natural resources is a complex process that involves various mechanisms through which minerals and fossil fuels are created. Mineral resources, for instance, are naturally occurring inorganic substances characterized by a defined chemical composition and crystalline structure. They can form through several geological processes. Igneous processes involve the crystallization of minerals from molten rock (magma) as it cools, leading to the formation of valuable minerals such as gold and copper. Sedimentary processes, on the other hand, result in the accumulation of sediments and organic materials over time, giving rise to minerals like limestone and gypsum. Additionally, metamorphic processes occur when existing rocks are altered under heat and pressure, producing valuable resources such as marble and slate.

Minerals are classified into several categories based on their properties and uses. Metallic minerals, such as iron and aluminum, are essential for construction and manufacturing, while non-metallic minerals, like quartz and talc, have applications in various industries, including electronics and cosmetics. Industrial minerals, such as potash and gypsum, are crucial for agriculture and construction. Each type of mineral has unique economic importance and applications, making them vital components of modern economies.

Fossil fuels, including coal, oil, and natural gas, are another critical category of natural resources. These fuels are formed from the remains of ancient organic matter, primarily plants and marine organisms. The formation process involves sedimentation, where organic material accumulates in sedimentary basins and is buried by layers of sediment. Over millions of years, the accumulated organic matter is subjected to heat and pressure, transforming it into fossil fuels through processes such as coalification for coal and kerogen maturation for oil and gas. The geological conditions necessary for fossil fuel formation include the presence of source rocks, reservoir rocks, and cap rocks, which trap hydrocarbons and allow for their extraction.

Renewable resources, such as geothermal energy, hydropower, and biomass, also have geological components. For example, geothermal energy is derived from the heat stored within the Earth, often found in volcanic regions or areas with high geothermal gradients. The availability of water resources for hydropower generation is influenced by geological features such as river systems and topography, highlighting the interconnectedness of geology and renewable energy sources.

7.3 Extraction and Utilization of Natural Resources

The extraction of natural resources is a critical process that involves various methods tailored to the type of resource being harvested. For mineral resources, mining and quarrying are the primary methods of extraction. Surface mining is one approach that involves removing overburden to access minerals located near the surface. Techniques such as open-pit mining and strip mining are commonly employed in this method. In contrast, underground mining is used for deeper mineral deposits, requiring the creation of tunnels or shafts to access the resources. Quarrying is another process specifically used for extracting building materials such as stone, gravel, and sand.

While mining and quarrying are essential for resource extraction, they can have significant environmental impacts. These activities often lead to habitat destruction, soil erosion, and water pollution. Sustainable mining practices aim to minimize these impacts through responsible land management and rehabilitation efforts, ensuring that ecosystems are preserved and restored after extraction activities are completed.

The extraction of oil and gas involves several processes that are critical for meeting global energy demands. Drilling is the first step, where wells are drilled into the Earth to access hydrocarbon reservoirs. Hydraulic fracturing, commonly known as fracking, is a technique used to release trapped oil and gas by injecting high-pressure fluid into rock formations. Once extracted, crude oil undergoes refining to produce various petroleum products, including gasoline, diesel, and petrochemicals. However, the environmental concerns associated with fossil fuel extraction are significant, including greenhouse gas emissions, potential spills, and habitat disruption. Transitioning to cleaner energy sources is essential for reducing these impacts and promoting environmental sustainability.

Sustainable resource management is a crucial aspect of ensuring the availability of natural resources for future generations while minimizing environmental impacts. Strategies such as recycling promote the reuse of materials, reducing the demand for new resource extraction and minimizing waste. Responsible sourcing practices ensure that resources are obtained sustainably and ethically, protecting ecosystems and communities. Additionally, implementing conservation measures, such as reducing consumption and improving efficiency, can help preserve natural resources and promote a more sustainable future.

7.4 Environmental Impacts of Resource Extraction

The environmental impacts of resource extraction are profound and multifaceted, affecting land, water, and air quality. One of the most significant consequences of resource extraction is land degradation. Activities such as mining and drilling can lead to deforestation, soil erosion, and loss of biodiversity. The removal of vegetation and soil disrupts ecosystems and can result in long-term changes in land use. The loss of habitat for wildlife and the alteration of natural landscapes can have cascading effects on local biodiversity, threatening species and disrupting ecological balance.

Water resources are also significantly impacted by resource extraction activities. Mining operations can lead to the contamination of surface and groundwater with heavy metals, chemicals, and sediments. For instance, acid mine drainage from coal mining can severely affect aquatic ecosystems and drinking water sources, posing risks to both human health and wildlife. Additionally, the extraction of oil and gas can result in spills and leaks, further threatening water resources and the organisms that depend on them.

Air quality is another critical concern associated with resource extraction. The processes involved in extracting and processing natural resources can contribute to air pollution, releasing particulate matter, volatile organic compounds (VOCs), and greenhouse gases into the atmosphere. These emissions can have serious health implications for nearby communities, leading to respiratory issues and other health problems. Moreover, the combustion of fossil fuels contributes to climate change, exacerbating environmental challenges and necessitating a transition to cleaner energy sources.

7.5 Case Studies in Natural Resource Management

To illustrate the complexities of natural resource management, several case studies provide valuable insights into the environmental and social impacts of resource extraction. The first case study focuses on mining in the Amazon Rainforest, a region renowned for its rich biodiversity and indigenous communities. This case examines the environmental and social consequences of mining activities, highlighting issues such as deforestation, habitat destruction, and the rights of indigenous peoples. The study underscores the urgent need for sustainable practices that protect vulnerable ecosystems and respect the rights of local communities.

The second case study explores oil extraction in the Niger Delta, Nigeria, one of the most oil-rich areas in the world. Despite its wealth in natural resources, the region has faced significant environmental degradation and social unrest due to oil extraction activities. This case study delves into the challenges faced by local communities, including oil spills, gas flaring, and loss of livelihoods. It emphasizes the necessity for responsible resource management and the involvement of local stakeholders in decision-making processes to ensure that the benefits of resource extraction are equitably shared.

The third case study highlights geothermal energy utilization in Iceland, a country that has successfully harnessed the heat from its volcanic landscape for electricity generation and heating. This case examines the geological conditions that make geothermal energy viable in Iceland and discusses the benefits of renewable energy in reducing greenhouse gas emissions. It serves as a model for other regions seeking to adopt similar sustainable energy practices, demonstrating the potential of renewable resources to meet energy needs while minimizing environmental impacts.

7.6 Future Directions in Natural Resource Management

As the world grapples with the challenges of climate change, there is a growing shift towards renewable energy sources, presenting both challenges and opportunities for geological resource management. The transition to renewable resources requires careful consideration of geological factors to ensure sustainability and minimize environmental impacts. For instance, the development of solar, wind, and geothermal energy technologies must take into account local geological conditions to optimize resource utilization and reduce ecological footprints.

Technological innovations are also transforming the way natural resources are extracted and managed. Advancements in automation, remote sensing, and data analytics can improve resource efficiency, reduce waste, and enhance environmental monitoring. These technologies can help identify new resource deposits and optimize extraction processes while minimizing ecological impacts. By leveraging technology, the resource management sector can become more sustainable and responsive to environmental challenges.

Effective policies and governance frameworks are essential for promoting sustainable resource management. Governments and regulatory bodies must establish clear guidelines for resource extraction, prioritize environmental protection, and engage stakeholders in decision-making processes. International cooperation and agreements can also play a crucial role in addressing transboundary resource management challenges, ensuring that resources are managed sustainably across borders.

7.7 Review of Geothermal Resources as Manifestations of Volcanism in Eastern Africa

Abstract—Igneous rocks are naturally occurring materials that evolve from a molten state (magma) through cooling processes. The solid materials that result from these processes are often crystalline compounds known as minerals. Magmatic chambers located at shallow depths frequently form high-temperature geothermal systems, deriving their heat from magmatic processes that culminate in volcanic activities when magma is extruded onto the ground surface. This extrusion occurs in the form of lava or volcanic ash as a result of violent volcanic activities. Volcanic activities are often related to geothermal processes under favorable hydrogeological conditions, leading to the development of geothermal reservoirs that can be encountered during exploration. These reservoirs are distributed globally, particularly along plate boundaries and continental rift systems, as observed in Eastern Africa.

Volcanic activities are closely associated with plate tectonics and continental drifting. For instance, the Great Rift Valley, which traverses the African continent from North to East and Central Africa, hosts numerous volcanoes, both active and dormant, located on the valley floor and its flanks. This chapter examines geothermal activities manifested at the ground surface in various forms, with a focus on geothermal features in Eastern Africa, particularly those that are prominent in the Kenya Rift System. The features discussed include hot springs, warm springs or geysers, steam jets, fumaroles, hot grounds, and surface deposits of sulfur.

Additionally, the chapter captures the major global distribution of volcanic centers and relates them to their plate tectonic settings. It also highlights the main operational geothermal investments in these volcanic centers. For instance, the Kenyan government continues to invest heavily in the exploration of geothermal resources as a preferred form of green energy to spur development across all sectors of the economy. Significant reserves are being exploited to contribute generated energy to the national grid.

Introduction

Geothermal energy, derived from the Earth's internal heat, is a sustainable and renewable resource that has gained prominence as a viable alternative to fossil fuels. In Eastern Africa, particularly within the Great Rift Valley, geothermal resources are abundant due to the region's unique geological setting characterized by active volcanism and tectonic activity. This review aims to explore the manifestations of geothermal resources in Eastern Africa, focusing on the Kenya Rift System, and to discuss the implications of these resources for energy production and economic development.

Geothermal Features in Eastern Africa

The geothermal features in Eastern Africa are diverse and include various manifestations of volcanic activity. Hot springs, warm springs, geysers, steam jets, fumaroles, and hot grounds are some of the prominent features observed in the region. These geothermal manifestations are indicative of the underlying magmatic processes and provide valuable insights into the geothermal potential of the area.

1. **Hot Springs and Geysers**: Hot springs are natural water sources heated by geothermal energy, often found in volcanic regions. Geysers, a specific type of hot spring, periodically erupt with steam and hot water, showcasing the dynamic nature of geothermal systems.

2. **Fumaroles**: Fumaroles are openings in the Earth's crust that emit steam and gases, often found in volcanic areas. They indicate the presence of high-temperature geothermal systems and can provide information about the geothermal gradient and subsurface conditions.

3. **Steam Jets and Hot Grounds**: Steam jets are localized areas where steam escapes from the ground, while hot grounds refer to surface areas with elevated temperatures due to geothermal activity. Both features are indicative of active geothermal systems and can be explored for potential energy extraction.

4. **Surface Deposits of Sulfur**: The presence of sulfur deposits around geothermal features is a common occurrence, resulting from the interaction of volcanic gases with the surrounding environment. These

deposits can be economically significant and are often associated with geothermal energy production.

Global Distribution of Volcanic Centers

The global distribution of volcanic centers is closely linked to plate tectonic settings. Regions along divergent and convergent plate boundaries are often characterized by high volcanic activity and geothermal potential. The Great Rift Valley in Eastern Africa serves as a prime example of a continental rift system where tectonic forces create conditions favorable for volcanic activity and geothermal resource development.

Geothermal Investments in Eastern Africa

The Kenyan government has recognized the potential of geothermal energy as a sustainable resource and has made significant investments in exploration and development. Geothermal energy, derived from the Earth's internal heat, is a clean and renewable source of power that can play a crucial role in addressing the energy needs of the country while reducing reliance on fossil fuels.

Abundant Geothermal Resources

Kenya is home to some of the largest geothermal reserves in Africa, primarily located in the East African Rift Valley, which runs through the country. The Olkaria Geothermal Plant, for instance, is one of the largest geothermal power stations in Africa and has been instrumental in demonstrating the viability of geothermal energy in the region. The country's geothermal potential is estimated to be around 10,000 megawatts, with only a fraction currently harnessed. This untapped potential presents a significant opportunity for both energy production and economic development.

Ongoing Projects and Developments

Ongoing projects aim to harness this energy to meet the growing demand for electricity, which has been driven by rapid urbanization, industrialization, and population growth. The Kenyan government, in collaboration with various stakeholders, including international investors and development partners, is actively pursuing the expansion of geothermal energy production. Recent initiatives include the drilling of new wells, the construction of additional power plants, and the enhancement of existing facilities to increase their capacity and efficiency.

One notable project is the Olkaria V Geothermal Power Station, which is expected to add an additional 158 megawatts to the national grid. This project, along with others in the pipeline, is part of Kenya's broader strategy to increase the share of renewable energy in its energy mix, aiming for at least 70% of the country's electricity to come from renewable sources by 2030.

Integration into the National Grid

By integrating geothermal energy into the national grid, Kenya is positioning itself as a leader in renewable energy production in the region. The government has implemented policies and frameworks to facilitate the integration of geothermal power into the existing energy infrastructure, ensuring that it can effectively meet the increasing demand for electricity. This integration not only enhances energy security but also stabilizes the national grid, reducing the frequency of power outages that have historically plagued the country.

Economic and Environmental Benefits

The shift towards geothermal energy also brings significant economic and environmental benefits. By investing in renewable energy, Kenya is reducing its carbon footprint and contributing to global efforts to combat climate change. Geothermal energy is known for its low greenhouse gas emissions compared to fossil fuels, making it a more environmentally friendly option for electricity generation.

Moreover, the development of geothermal resources creates job opportunities in various sectors, including construction, engineering, and maintenance. Local communities benefit from increased access to electricity, which can spur economic activities and improve living standards. The government is also working to ensure that the benefits of geothermal development are shared with local communities, promoting social equity and sustainable development.

Regional Leadership and Collaboration

Kenya's advancements in geothermal energy position it as a regional leader in renewable energy production, attracting interest from neighboring countries looking to develop their geothermal resources. The government has engaged in regional collaborations and partnerships to share knowledge, technology, and best practices in geothermal energy development. This collaborative approach not only enhances Kenya's energy security but also contributes to regional stability and economic growth.

In conclusion, the Kenyan government's commitment to geothermal energy development (Figs. 14 & 15) represents a significant step towards achieving sustainable energy goals. By harnessing its abundant geothermal resources, Kenya is not only addressing its energy needs but also setting an example for other countries in Eastern Africa. The ongoing investments in geothermal energy are paving the way for a cleaner, more sustainable energy future, with far-reaching benefits for the economy, the environment, and the well-being of its citizens.

Figure 14: Steam discharge from geothermal reservoir testing phase (A) and Volcanic profile in the background Olkaria field (B), Nakuru County

Figure 15: Above Olkaria Geothermal Steam Geysers (A) and Plant System (B)

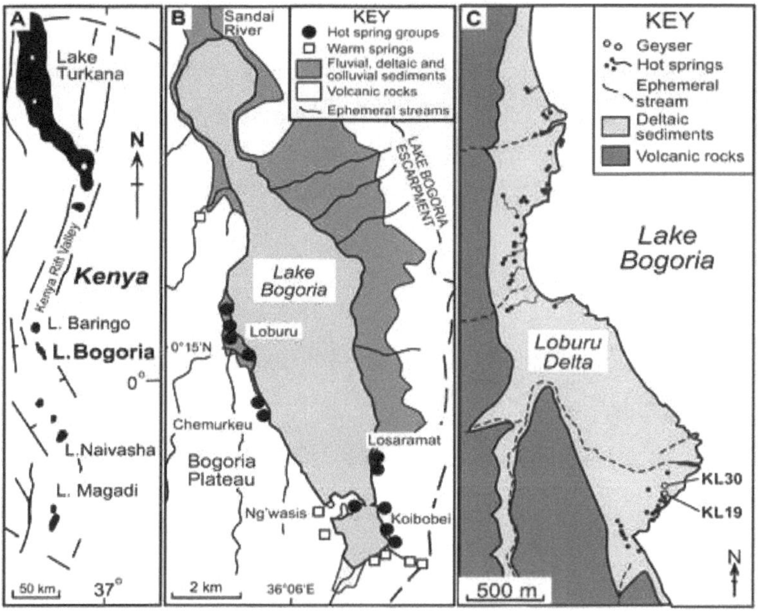

Figure 16: A part of the geothermal features of Lake Bogoria

Lake Bogoria, with a salinity approximately twice that of seawater, lies north of the equator in a narrow half-graben basin in the central Kenya Rift Valley. Typical active geysers and fumaroles of the lake are as shown in above and below figures 16 and 17.

Figure 17: Typical active geysers and fumarole in Lake Bogoria

Socio-Environmental Impacts of Geothermal Projects

While geothermal energy presents numerous benefits, including reduced greenhouse gas emissions and a reliable energy source, it also poses socio-environmental challenges that must be addressed.

1. **Land Use and Displacement**: The establishment of geothermal power plants often necessitates significant land use, which can lead to the displacement of local communities and disruption of traditional land use practices. Engaging with affected communities is essential to ensure their rights are respected and to seek equitable solutions.

2. **Water Resource Management**: Geothermal projects can impact local water resources, particularly in areas where water is already scarce. The extraction of geothermal fluids can lead to changes in groundwater levels and quality, potentially affecting agriculture and drinking water supplies. Sustainable water management practices must be implemented to mitigate these impacts.

3. **Environmental Degradation**: The construction and operation of geothermal facilities can lead to environmental degradation, including habitat destruction and pollution. Careful environmental assessments

and monitoring are necessary to minimize these impacts and ensure compliance with environmental regulations.

4. **Community Health and Safety**: The release of gases such as hydrogen sulfide from geothermal systems can pose health risks to nearby communities. Monitoring air quality and implementing safety measures are crucial to protect public health.
5. **Economic Opportunities and Challenges**: Geothermal projects can create jobs and stimulate local economies, but they can also lead to economic disparities if benefits are not equitably distributed. Ensuring that local communities benefit from geothermal development is essential for fostering social acceptance and long-term sustainability.

Conclusion

Geothermal Resources and Sustainable Development

In conclusion, geothermal resources in Eastern Africa, particularly within the Kenya Rift System, represent a significant opportunity for sustainable energy development. The diverse manifestations of geothermal activity, including hot springs, fumaroles, and geysers, underscore the region's volcanic heritage and its potential for harnessing geothermal energy. However, it is imperative to address the socio-environmental impacts associated with geothermal projects to ensure that development is both equitable and sustainable.

Economic Growth and Energy Security

As investments in geothermal exploration and development continue, Eastern Africa can leverage these resources to support economic growth and contribute to global efforts in transitioning to renewable energy sources. The integration of geothermal energy into the national grid not only enhances energy security but also promotes environmental sustainability, making it a crucial component of the region's energy strategy.

Holistic Approach to Resource Management

To maximize the benefits of geothermal energy while minimizing adverse effects, it is essential to adopt a holistic approach that includes stakeholder engagement, rigorous environmental assessments, and the implementation of best practices in resource management. By prioritizing responsible development practices, Eastern Africa can harness its geothermal potential while safeguarding the well-being of its communities and ecosystems.

Future Prospects and Collaboration

In summary, the future of geothermal energy in Eastern Africa is promising, but it requires a concerted effort from governments, industry stakeholders, and local communities to navigate the complexities of sustainable development. By fostering collaboration and ensuring that the voices of affected communities are heard, the region can achieve a balanced approach to energy production that respects both the environment and the rights of its people. This commitment to sustainability will not only enhance the resilience of local economies but also contribute to the broader goal of achieving a sustainable energy future for all.

7.8 Summary of Chapter 7

Geological Processes and Resource Formation

In this chapter, we delved into the intricate relationship between geology and natural resources, focusing on the geological processes that lead to the formation of essential materials such as minerals, fossil fuels, and renewable resources. Understanding these processes is crucial for comprehending how these resources are distributed across the Earth and how they can be sustainably managed. We examined the various geological mechanisms involved in the formation of mineral resources, including igneous, sedimentary, and metamorphic processes, which contribute to the diversity of minerals available for extraction and use. Additionally, we explored the formation of fossil fuels, which are derived from ancient organic matter subjected to heat and pressure over millions of years, highlighting the geological conditions necessary for their creation.

Environmental Impacts of Resource Extraction

The chapter also addressed the extraction and utilization of these natural resources, emphasizing the methods employed in mining, drilling, and harvesting. However, we underscored the significant environmental impacts associated with resource extraction, which can have far-reaching consequences for ecosystems and human health. Land degradation is one of the most pressing issues, as mining and drilling activities often lead to deforestation, soil erosion, and the destruction of habitats. The removal of vegetation and soil disrupts local ecosystems, resulting in the loss of biodiversity and altering natural landscapes. This degradation can have cascading effects, threatening wildlife populations and diminishing the ecological services that healthy ecosystems provide.

Water Quality and Air Pollution Concerns

As water scarcity becomes more prevalent, implementing water conservation techniques is essential for sustainable resource management. Strategies such as rainwater harvesting, greywater recycling, and water-efficient irrigation systems like drip irrigation can significantly enhance water use efficiency. Rainwater harvesting collects and stores rainwater from rooftops, reducing dependence on municipal supplies and alleviating pressure on local water systems during dry spells. Greywater recycling reuses water from sinks and showers for irrigation, further decreasing overall consumption. Additionally, drip irrigation allows for targeted watering at plant roots, minimizing evaporation and promoting healthier growth. Restoring wetlands enhances groundwater recharge and improves water availability while providing critical habitats for wildlife and supporting biodiversity. These practices not only improve water quality by reducing runoff but also contribute to healthier ecosystems and help mitigate climate change impacts.

CHAPTER 8

GEOHAZARDS AND RISK MANAGEMENT

8.1 Introduction to Geohazards

Geohazards are natural phenomena that pose significant risks to human life, property, and the environment. These hazards can be broadly categorized into three main types: geological hazards, hydrological hazards, and meteorological hazards. Geological hazards include events such as earthquakes, landslides, and volcanic eruptions, while hydrological hazards encompass floods and droughts. Meteorological hazards involve extreme weather events like hurricanes, tornadoes, and tsunamis. Understanding geohazards is crucial for public safety, environmental protection, and sustainable development, as these events can have devastating consequences for communities and ecosystems. The increasing frequency and intensity of geohazards, often exacerbated by climate change and human activities, underscore the urgent need for effective risk assessment and management strategies to mitigate their impacts.

8.2 Types of Geohazards

Earthquakes

Earthquakes are one of the most significant geological hazards, resulting from the sudden release of energy in the Earth's crust, typically due to tectonic plate movements. The impacts of earthquakes can be catastrophic, leading to the destruction of infrastructure, loss of life, and long-term economic challenges. Regions located along tectonic plate boundaries, such as the Pacific Ring of Fire, are particularly vulnerable to seismic activity. The intensity of an earthquake is measured using the Richter scale or the moment magnitude scale, which quantifies the energy released. Earthquakes can trigger secondary hazards, such as tsunamis and landslides, further complicating their impacts. Effective earthquake preparedness involves implementing robust building codes, conducting regular drills, and establishing early warning systems to alert communities of impending seismic events.

Landslides

Landslides occur when there is a sudden failure of a slope, often triggered by factors such as heavy rainfall, earthquakes, volcanic activity, or human activities like deforestation and construction. The consequences of landslides can be severe, resulting in property damage, loss of life, and environmental degradation. Landslides can also disrupt transportation networks and contaminate water sources. Understanding the geological and hydrological conditions that contribute to landslides is essential for effective risk management. This includes assessing slope stability, soil composition, and drainage patterns. Mitigation strategies may involve engineering solutions such as retaining walls, reforestation efforts to stabilize slopes, and land-use planning that restricts development in high-risk areas.

Volcanic Eruptions

Volcanic eruptions are another significant geohazard, caused by the movement of magma from beneath the Earth's surface. Eruptions can be explosive or effusive, with varying hazards associated with each type. Explosive eruptions can produce ash clouds, pyroclastic flows, and volcanic gases, while effusive eruptions typically result in lava flows. The impacts of volcanic eruptions can be devastating, affecting air quality, agriculture, and water resources. Communities near active volcanoes must be prepared for potential eruptions and their associated risks. This preparation includes establishing monitoring systems to detect signs of volcanic activity, developing evacuation plans, and conducting public education campaigns to inform residents about the hazards and appropriate responses.

Floods

Floods are hydrological hazards that can result from heavy rainfall, rapid snowmelt, dam failures, or storm surges. The impacts of floods can be devastating, affecting communities, agriculture, and ecosystems. Flooding can lead to loss of life, displacement of populations, and significant economic losses. Flood risk management involves understanding the causes of flooding and implementing measures to reduce vulnerability. This may include floodplain zoning, infrastructure improvements such as levees and flood barriers, and the restoration of natural floodplains to absorb excess water.

Additionally, early warning systems can provide timely alerts to communities at risk of flooding, allowing for proactive measures to be taken.

Tsunamis

Tsunamis are large ocean waves generated by underwater earthquakes, volcanic eruptions, or landslides. The potential impacts of tsunamis on coastal communities can be catastrophic, leading to loss of life, extensive property damage, and long-term environmental consequences. Tsunamis can inundate low-lying areas, erode coastlines, and disrupt marine ecosystems. Early warning systems and community preparedness are critical for minimizing tsunami risks. These systems rely on seismic monitoring and ocean buoys to detect changes in sea level, providing alerts to coastal populations. Community education and evacuation drills are essential components of tsunami preparedness, ensuring that residents know how to respond quickly in the event of a tsunami warning.

8.3 Risk Assessment and Management

Risk assessment is a systematic process that involves identifying hazards, analyzing vulnerabilities, and evaluating risks. It is essential for understanding the potential impacts of geohazards and informing risk management strategies. Key components of effective risk assessment include data collection, modeling, and historical analysis. Data collection involves gathering information on past geohazard events, geological conditions, and population demographics. Modeling techniques, such as probabilistic seismic hazard assessment, help quantify the likelihood and potential impacts of geohazards.

Historical analysis provides insights into patterns and trends, allowing for better predictions of future events.

Risk Management Strategies

Risk management strategies encompass a combination of approaches aimed at reducing the impacts of geohazards. Mitigation involves implementing measures to reduce the likelihood and severity of geohazards. This can include land-use planning that restricts development in high-risk areas, engineering solutions such as retaining walls for landslides, and ecosystem restoration to enhance natural defenses against flooding. Preparedness focuses on developing emergency response plans, conducting drills, and educating communities about geohazards to enhance readiness. Preparedness efforts may involve training emergency responders, establishing communication protocols, and ensuring that resources are available for rapid response.

Response involves coordinating emergency services and resources during and after a geohazard event to ensure effective response efforts. This includes mobilizing first responders, providing medical assistance, and coordinating evacuation efforts. Effective response requires clear communication and collaboration among various agencies and organizations. Finally, recovery entails planning for long-term recovery and rebuilding efforts after a geohazard event. This phase includes assessing damage, providing support to affected communities, and implementing strategies for rebuilding that incorporate lessons learned from the event. Community engagement is crucial during the recovery phase to ensure that the needs and priorities of affected populations are addressed.

8.4 Case Studies in Geohazards and Risk Management

Case Study 1: The 2010 Haiti Earthquake

The 2010 Haiti Earthquake serves as a poignant example of the devastating impacts of geological hazards. The earthquake struck Haiti on January 12, 2010, resulting in widespread destruction and loss of life, with estimates of over 230,000 fatalities and millions displaced. This case study analyzes the causes of the earthquake, which occurred along the Enriquillo-Plantain Garden fault zone, and its impacts on infrastructure and communities. The response and recovery efforts that followed highlighted the importance of improving earthquake preparedness in vulnerable regions. Lessons learned from this event emphasize the need for better building codes, community education, and international support for disaster response.

Case Study 2: Landslides in the Himalayas

Landslides in the Himalayas illustrate the complex interplay of geological conditions and human activities. The Himalayan region is prone to landslides due to its steep terrain, geological conditions, and heavy monsoon rains. This case study examines the factors contributing to landslides in the Himalayas, including climate change, which has led to increased rainfall and glacial melting, as well as human activities such as deforestation, urbanization, and road construction. These factors exacerbate the risk of landslides, leading to significant impacts on local communities, infrastructure, and ecosystems. Mitigation strategies discussed in this case study include community engagement in landslide risk reduction, sustainable land management practices, and the implementation of early warning systems. By involving local populations in monitoring and preparedness efforts, communities can enhance their resilience to landslides and reduce the potential for loss of life and property.

Case Study 3: Volcanic Eruptions in Indonesia

Volcanic eruptions in Indonesia present significant challenges for risk management due to the country's numerous active volcanoes, including Mount Merapi and Mount Sinabung. This case study explores the complexities of managing volcanic hazards in Indonesia, where millions of people live in close proximity to active volcanoes. The role of monitoring and early warning systems is critical in reducing risks associated with volcanic eruptions. Indonesia has established a network of volcano observatories that monitor seismic activity, gas emissions, and ground deformation to provide timely alerts to communities at risk. The importance of community preparedness and education is emphasized, as local populations must be informed about the hazards and trained in evacuation procedures. This case study highlights the need for continuous investment in monitoring technology, public education, and community engagement to effectively manage volcanic risks.

8.5 Future Directions in Geohazard Research and Management

Looking ahead, several future directions in geohazard research and management are emerging. Advancements in technology are transforming the field of geohazard monitoring and management. Remote sensing technologies, such as satellite imagery and LiDAR, provide valuable data for assessing geological conditions and detecting changes over time. Geographic Information Systems (GIS) enable the integration of various data sources to create detailed hazard maps and models, enhancing our ability to assess risks and respond effectively to geohazards. Early warning systems, which utilize real-time data from seismic sensors and ocean buoys, are becoming increasingly sophisticated, allowing for quicker and more accurate alerts to communities at risk.

Climate Change Considerations

Climate change considerations are becoming increasingly important in geohazard management. Climate change is expected to influence the frequency and intensity of geohazards, making it essential to incorporate climate change impacts into risk assessments and management strategies. For example, changing precipitation patterns may increase the risk of flooding and landslides, while rising sea levels can exacerbate coastal erosion and tsunami impacts. Understanding how these changing climate patterns affect geohazards will be critical for developing adaptive strategies that can mitigate risks associated with extreme weather events and other climate-related phenomena.

Community Involvement

Community involvement is vital for building resilience against geohazards. Engaging communities in geohazard preparedness and response efforts can enhance the effectiveness of risk management strategies. Local knowledge and participation can foster a sense of ownership among community members, leading to more effective implementation of preparedness plans and a stronger collective response during emergencies. Community-based approaches that incorporate local insights and practices can improve the relevance and acceptance of risk management strategies, ultimately leading to better outcomes in disaster response and recovery.

Interdisciplinary Collaboration

Future geohazard research and management will also benefit from interdisciplinary collaboration among scientists, engineers, urban planners, and policymakers. By integrating knowledge from various fields, stakeholders can develop comprehensive strategies that address the multifaceted nature of geohazards. For instance, collaboration between geologists and urban planners can lead to better land-use policies that minimize risks in vulnerable areas. Similarly, partnerships between environmental scientists and community organizations can enhance public education and engagement efforts, ensuring that communities are well-informed and prepared for potential hazards.

8.6 Summary of Chapter 8

Overview of Geohazards

In this chapter, we delved into the various types of geohazards, including earthquakes, landslides, volcanic eruptions, floods, and tsunamis. Each of these geohazards presents unique challenges and risks, necessitating tailored approaches to risk assessment and management. We emphasized the importance of developing comprehensive risk assessment and management strategies to mitigate the impacts of these hazards. This involves systematic processes that identify potential hazards, analyze vulnerabilities within communities and ecosystems, and evaluate the associated risks to inform effective preparedness and response measures.

Environmental Impacts of Geohazards

The environmental impacts of geohazards are profound and multifaceted, affecting not only human life and infrastructure but also the natural environment. For instance, earthquakes can lead to soil liquefaction, a phenomenon where saturated soil loses its strength and stiffness, resulting in significant damage to buildings and infrastructure. This process can also alter landscapes and disrupt ecosystems, leading to long-term ecological consequences.

Landslides, often triggered by heavy rainfall or seismic activity, can result in substantial soil erosion, loss of vegetation, and habitat destruction. These events can severely affect biodiversity, as the displacement of soil and vegetation disrupts the habitats of numerous species, leading to declines in local populations and altering ecosystem dynamics.

Volcanic eruptions pose additional environmental challenges, as they can release ash and gases that contaminate air and water resources. The fallout from eruptions can lead to respiratory issues in humans and animals, while the deposition of volcanic materials can alter soil chemistry and affect agricultural productivity. Lava flows can obliterate habitats and agricultural land, leading to significant economic losses and long-term changes in land use.

Floods, on the other hand, can inundate ecosystems, leading to altered water quality and the displacement of wildlife. The rapid influx of water can wash away habitats, disrupt breeding cycles, and lead to the introduction of pollutants into aquatic systems. Tsunamis, often triggered by underwater earthquakes or volcanic eruptions, can cause extensive coastal erosion and damage to marine habitats, further threatening biodiversity in coastal regions.

Case Studies in Geohazard Management

Through case studies, we illustrated the complexities of geohazard management and the lessons learned from past events. The 2010 Haiti Earthquake served as a stark reminder of the critical need for improved earthquake preparedness, highlighting the importance of not only protecting human life but also safeguarding the environment from the secondary effects of seismic activity.

The landslides in the Himalayas underscored the importance of sustainable land management and community engagement, as improper land use practices can exacerbate environmental degradation and increase vulnerability to landslides. The volcanic eruptions in Indonesia demonstrated the effectiveness of monitoring and early warning systems in reducing risks associated with volcanic hazards, emphasizing the need for proactive measures to protect both communities and ecosystems.

Future Directions in Geohazard Research and Management

Finally, we emphasized the need for advancements in technology, climate change considerations, and community involvement in future geohazard research and management efforts. Climate change is expected to influence the frequency and intensity of geohazards, making it essential to incorporate environmental considerations into risk assessments and management strategies.

By understanding the nature of geohazards and their environmental impacts, and by implementing effective strategies, we can better protect lives, infrastructure, and the environment from the devastating consequences of these natural phenomena.

Building Resilience and Sustainable Development

The integration of scientific research, technological innovation, and community participation will be essential for building resilience and ensuring sustainable development in the face of increasing geohazard risks. This holistic approach will not only enhance human safety but also promote the health of ecosystems, ensuring a balanced coexistence between communities and their natural environments.

By fostering collaboration among scientists, policymakers, and local communities, we can develop adaptive strategies that address the challenges posed by geohazards while also prioritizing environmental sustainability and community well-being.

Other Important Geohazard Technologies

1. Geohazard Monitoring Technologies

Geohazard monitoring technologies play a crucial role in understanding and mitigating the impacts of natural hazards. These technologies include:

- **Seismic Sensors**: These devices detect and record ground motion caused by earthquakes. Advanced seismic networks can provide real-time data, enabling early warning systems that alert communities before the shaking reaches them.
- **Satellite Imagery and Remote Sensing**: Satellites equipped with various sensors can monitor land deformation, vegetation changes, and water levels. This data is invaluable for assessing risks related to landslides, floods, and volcanic activity.
- **LiDAR (Light Detection and Ranging)**: This technology uses laser pulses to create high-resolution topographic maps. LiDAR can identify areas at risk of landslides and help in flood modeling by providing detailed terrain information.
- **Drones**: Unmanned aerial vehicles (UAVs) can be deployed for rapid assessment of disaster-affected areas, providing real-time imagery and data collection that can aid in response and recovery efforts.

2. Public Policy and Legislation

Public policy and legislation are essential for effective geohazard risk management. Key aspects include:

- **Building Codes**: Enforcing strict building codes can ensure that structures are designed to withstand specific hazards, such as earthquakes and floods. This can significantly reduce damage and loss of life during disasters.
- **Land-Use Planning**: Policies that restrict development in high-risk areas, such as floodplains or near fault lines, can minimize exposure to

geohazards. Zoning regulations can guide sustainable development and protect vulnerable communities.

- **Disaster Preparedness Plans**: Governments should develop and regularly update disaster preparedness plans that outline response strategies, resource allocation, and coordination among agencies.
- **Funding and Resources**: Legislation can facilitate funding for research, infrastructure improvements, and community education programs aimed at enhancing resilience to geohazards.

3. Economic Impacts of Geohazards

The economic consequences of geohazards can be profound and multifaceted:

- **Direct Costs**: These include immediate expenses related to emergency response, infrastructure repair, and recovery efforts. For example, the cost of rebuilding after a major earthquake can run into billions of dollars.
- **Indirect Costs**: Long-term economic impacts may include decreased productivity, loss of business, and reduced property values. Communities may face prolonged economic decline, especially if they rely heavily on industries vulnerable to geohazards.
- **Insurance and Risk Financing**: The role of insurance in managing economic risks is critical. However, high-risk areas may face challenges in obtaining affordable coverage, leading to financial vulnerability for homeowners and businesses.

4. Social Vulnerability and Equity

Social factors significantly influence vulnerability to geohazards:

- **Poverty and Inequality**: Low-income communities often lack the resources to prepare for and recover from disasters. They may live in high-risk areas and have limited access to information, infrastructure, and emergency services.
- **Education and Awareness**: Communities with lower levels of education may be less aware of risks and preparedness measures. Public education campaigns are essential to empower all community members to respond effectively to geohazards.

- **Cultural Factors**: Cultural beliefs and practices can influence how communities perceive and respond to risks. Engaging with local leaders and incorporating cultural perspectives into risk management strategies can enhance effectiveness.

5. Climate Change Adaptation Strategies

As climate change intensifies the frequency and severity of geohazards, adaptation strategies become increasingly important:

- **Infrastructure Resilience**: Upgrading infrastructure to withstand extreme weather events, such as floods and hurricanes, is essential. This may involve reinforcing levees, improving drainage systems, and retrofitting buildings.
- **Ecosystem Restoration**: Restoring natural ecosystems, such as wetlands and forests, can enhance resilience to flooding and landslides. These ecosystems act as natural buffers, absorbing excess water and stabilizing soil.
- **Community-Based Adaptation**: Involving communities in developing adaptation strategies ensures that local knowledge and needs are considered. This can lead to more effective and sustainable solutions.

6. Community-Based Disaster Risk Reduction (CBDRR)

CBDRR emphasizes the importance of local involvement in disaster preparedness and response:

- **Local Knowledge**: Communities possess valuable knowledge about their risks and vulnerabilities. Engaging them in risk assessments and planning can lead to more relevant and effective strategies.
- **Capacity Building**: Training community members in emergency response, first aid, and risk management can enhance local capacity to respond to disasters. This empowerment fosters resilience and self-sufficiency.
- **Participatory Approaches**: Involving community members in decision-making processes ensures that their voices are heard and that strategies are culturally appropriate and widely accepted.

7. International Collaboration and Aid

International collaboration is vital for effective geohazard risk management:

- **Knowledge Sharing**: Countries can share knowledge and best practices in geohazard risk management through international forums, conferences, and partnerships. This exchange of information can lead to improved strategies and technologies for disaster preparedness and response.
- **Technical Assistance**: Developed nations and international organizations can provide technical assistance to countries with limited resources. This may include training programs, funding for infrastructure projects, and the provision of advanced monitoring technologies.
- **Disaster Relief and Recovery**: In the aftermath of a geohazard event, international aid can be crucial for recovery efforts. Humanitarian organizations often mobilize resources to provide immediate assistance, such as food, shelter, and medical care, to affected populations.
- **Global Initiatives**: Programs like the United Nations' Sendai Framework for Disaster Risk Reduction promote global cooperation in reducing disaster risks and enhancing resilience. These initiatives encourage countries to adopt comprehensive risk management strategies.

8. Psychological and Social Impacts of Disasters

The psychological and social impacts of geohazards can be profound and long-lasting:

- **Mental Health Issues**: Survivors of geohazards may experience trauma, anxiety, depression, and post-traumatic stress disorder (PTSD). Addressing mental health needs is essential for recovery and rebuilding efforts.
- **Community Cohesion**: Disasters can either strengthen community bonds or exacerbate social tensions. Effective recovery efforts should focus on fostering social cohesion and rebuilding trust among community members.

- **Support Systems**: Establishing support networks, including counseling services and community groups, can help individuals cope with the emotional aftermath of disasters. Community engagement in recovery processes can also enhance resilience.

9. Ecosystem-Based Approaches to Risk Reduction

Ecosystem-based approaches leverage natural systems to reduce risks associated with geohazards:

- **Natural Buffers**: Wetlands, mangroves, and forests can act as natural barriers against floods, storm surges, and landslides. Protecting and restoring these ecosystems can enhance community resilience.

- **Biodiversity Conservation**: Healthy ecosystems support biodiversity, which is essential for maintaining ecological balance. Biodiversity can enhance ecosystem resilience, making them more capable of withstanding and recovering from geohazards.

- **Sustainable Land Management**: Implementing sustainable land management practices, such as agroforestry and permaculture, can reduce soil erosion and improve land stability, thereby mitigating landslide risks.

10. Education and Awareness Programs

Education and awareness are critical components of effective geohazard risk management:

- **Public Awareness Campaigns**: Informing communities about the risks associated with geohazards and the importance of preparedness can empower individuals to take proactive measures.

- **School Programs**: Integrating disaster risk education into school curricula can help instill a culture of preparedness among young people. Students can learn about hazards, safety measures, and emergency response.

- **Community Workshops**: Organizing workshops and training sessions can provide practical skills and knowledge to community members, enhancing their ability to respond to geohazards effectively.

11. Case Studies of Successful Risk Management

Examining successful case studies can provide valuable insights into effective geohazard risk management:

- **Japan's Earthquake Preparedness**: Japan has implemented rigorous building codes, early warning systems, and public education programs that have significantly reduced the impacts of earthquakes. The country's experience highlights the importance of preparedness and resilience.
- **Flood Management in the Netherlands**: The Netherlands has developed an extensive flood management system that includes dikes, storm surge barriers, and innovative water management practices. This case demonstrates the effectiveness of proactive planning and investment in infrastructure.
- **Community-Led Recovery in Nepal**: Following the 2015 earthquake, community-led recovery efforts in Nepal emphasized local participation and traditional knowledge. This approach not only facilitated faster recovery but also strengthened community resilience.

CHAPTER 9

GEOLOGY AND ENVIRONMENTAL POLICY

9.1 Introduction to Environmental Policy

Environmental policy refers to the comprehensive framework of regulations, laws, and practices designed to manage natural resources and protect the environment. This policy landscape encompasses a wide array of issues, including air and water quality, land use, waste management, and biodiversity conservation. The significance of environmental policy lies in its ability to guide decision-making processes that balance economic development with ecological sustainability. Effective environmental policies are essential for addressing the pressing challenges posed by climate change, habitat destruction, and resource depletion.

Understanding the role of geology in environmental policy is crucial, as geological knowledge informs decisions related to resource management, hazard mitigation, and land use planning. Geology provides insights into the physical characteristics of the Earth, including soil composition, rock formations, groundwater resources, and natural hazards such as earthquakes and landslides. By integrating geological insights into environmental policies, we can better protect ecosystems, manage natural resources sustainably, and reduce risks associated with natural hazards. This integration is vital for fostering resilience in communities and ensuring the long-term health of the environment.

9.2 Key Environmental Policies and Regulations

National Environmental Policy Act (NEPA): Enacted in 1969, NEPA is a cornerstone of environmental legislation in the United States. It requires federal agencies to assess the environmental impacts of their proposed actions before making decisions. This includes conducting Environmental Assessments (EAs) and Environmental Impact Statements (EISs) that consider geological factors such as soil stability, water resources, and potential geohazards. NEPA serves as a critical tool for integrating geological considerations into federal project planning, ensuring that potential impacts on the environment are thoroughly evaluated and addressed. The act promotes transparency and public participation, allowing stakeholders to voice their concerns and contribute to the decision-making process.

Clean Water Act (CWA): The CWA aims to restore and maintain the integrity of the nation's waters by regulating discharges of pollutants and setting water quality standards. Geological factors play a significant role in understanding watershed management, as the geology of an area influences groundwater recharge, surface water flow, and sediment transport. Effective implementation of the CWA requires collaboration between geologists and water resource managers to ensure that water quality is protected and that aquatic ecosystems are preserved. By considering geological characteristics, such as soil permeability and landforms, policymakers can develop strategies that enhance water quality and protect aquatic habitats.

Clean Air Act (CAA): The CAA establishes national air quality standards to protect public health and the environment from air pollution. Geological factors, such as topography and local geology, can influence air quality by affecting pollutant dispersion and accumulation. For example, valleys may trap pollutants, leading to poor air quality, while elevated areas may facilitate dispersion. Understanding these geological factors is essential for developing effective air quality management strategies that consider the unique characteristics of different regions. The CAA also emphasizes the importance of monitoring and assessing air quality to ensure compliance with established standards.

Resource Conservation and Recovery Act (RCRA): The RCRA governs the management of hazardous waste, ensuring that waste is handled in a manner that protects human health and the environment. Geological considerations are critical in determining suitable sites for waste disposal, as the geology of an area can influence groundwater contamination and the long-term stability of waste storage facilities. Proper geological assessments are necessary to minimize risks associated with hazardous waste management. The RCRA also promotes waste minimization and encourages recycling and recovery of materials, aligning with sustainable resource management practices.

9.3 Land Use Planning and Geology

Importance of Land Use Planning: Land use planning is essential for balancing development needs with environmental protection. Effective land use planning considers geological factors to minimize risks associated with natural hazards and ensure sustainable resource management. By integrating geological data into land use decisions, planners can identify areas that are suitable for development while protecting vulnerable ecosystems and communities. This proactive approach helps mitigate potential conflicts between development and environmental conservation.

Geological Considerations in Land Use: Geological factors, such as soil stability, groundwater resources, and natural hazards, significantly influence land use decisions. For example, areas prone to landslides or flooding may require special zoning regulations to mitigate risks. Geologists play a vital role in providing data and expertise to inform land use planning processes, ensuring that decisions are based on sound geological principles. Assessments of soil types, slope stability, and groundwater availability are critical for determining the suitability of land for various uses, including agriculture, residential development, and infrastructure projects.

Zoning Regulations: Zoning regulations can incorporate geological considerations to reduce risks associated with natural hazards. For instance, areas identified as high-risk for landslides may be designated as restricted zones for development. Successful zoning practices that integrate geological data can enhance community resilience and safety, protecting both human life and the environment from the impacts of natural hazards. Examples of effective zoning practices include establishing buffer zones around floodplains, implementing building codes that account for seismic risks, and designating areas for conservation to protect sensitive geological features.

9.4 The Role of Geologists in Policy-Making

Geologists as Advisors: Geologists serve as valuable advisors to policymakers, providing scientific expertise and data to inform decision-making. Their knowledge of geological processes and resources is essential for developing effective environmental policies that address complex challenges. By contributing their expertise, geologists can help ensure that policies are grounded in scientific understanding. This collaboration is particularly important in areas such as land use planning, resource management, and hazard mitigation, where geological insights can significantly influence outcomes.

Public Engagement and Advocacy: Geologists can play a crucial role in engaging the public and advocating for sound environmental policies. By communicating the importance of geological issues and their relevance to environmental challenges, geologists can help raise awareness and promote informed decision-making. Public engagement is essential for fostering support for policies that protect geological resources and manage hazards effectively. Geologists can participate in community outreach programs, educational initiatives, and public forums to share their knowledge and encourage dialogue on environmental issues.

Interdisciplinary Collaboration: Collaboration between geologists, environmental scientists, policymakers, and community stakeholders is essential for developing effective environmental policies. Interdisciplinary approaches can lead to more comprehensive solutions that address complex environmental issues, ensuring that all relevant factors are considered in decision-making processes. By working together, professionals from different fields can share insights, identify synergies, and develop integrated strategies that promote sustainability and resilience.

9.5 Challenges in Integrating Geology and Environmental Policy

Data Limitations: One of the primary challenges in integrating geology into environmental policy is the availability and quality of geological data. In many regions, geological data may be incomplete, outdated, or not readily accessible, hindering effective decision-making. For instance, a lack of detailed geological surveys can prevent policymakers from understanding the risks associated with natural hazards, such as landslides or groundwater contamination. Addressing these data limitations is crucial for ensuring that policies are based on accurate and reliable information. Investments in geological research, data collection, and the establishment of comprehensive geological databases can help overcome these challenges.

Political and Economic Influences: Political and economic factors can significantly impact the implementation of environmental policies and the consideration of geological factors. Policymakers may face pressure from various interest groups, including industry stakeholders, developers, and local communities, which can complicate the integration of scientific data into policy decisions. Economic considerations, such as the costs associated with implementing geological assessments or mitigation measures, can also influence the prioritization of environmental policies. Balancing these political and economic pressures with the need for sound geological practices is a critical challenge in effective environmental policy-making. Policymakers must navigate these complexities while striving to uphold environmental integrity and public safety.

Public Perception and Awareness: Public understanding of geological issues and their relevance to environmental policy is crucial for effective engagement and advocacy. A lack of awareness can lead to resistance against necessary policies or initiatives that aim to protect geological resources and manage hazards. For example, communities may oppose zoning regulations that restrict development in high-risk areas due to a lack of understanding of the geological risks involved. Educating the public about the importance of geology in environmental issues can foster greater support for sound policies. Initiatives that promote public awareness and understanding of geological processes, risks, and the benefits of sustainable practices are essential for building a more informed citizenry that can advocate for effective environmental policies.

9.6 Case Studies in Geology and Environmental Policy

Case Study 1: The California Environmental Quality Act (CEQA): The CEQA requires state and local agencies in California to assess the environmental impacts of their proposed projects. This case study examines how CEQA incorporates geological considerations in project assessments, such as evaluating seismic risks and soil stability. The act has led to increased awareness of geological hazards and has prompted developers to implement mitigation measures to protect communities. For example, projects in earthquake-prone areas must include detailed geological studies to assess potential risks and develop appropriate engineering solutions. CEQA has also facilitated public participation, allowing communities to voice their concerns about geological hazards and advocate for safer development practices.

Case Study 2: Groundwater Management in the Ogallala Aquifer: The Ogallala Aquifer, one of the largest aquifers in the United States, is critical for agricultural irrigation in the Great Plains. This case study explores the policies and regulations governing groundwater management in the region, focusing on the role of geology in sustainable water use. The challenges of over-extraction and the need for conservation measures highlight the importance of integrating geological knowledge into water management policies. For instance, understanding the geological characteristics of the aquifer, such as recharge rates and soil types, is essential for developing effective management strategies that ensure the long-term sustainability of this vital resource. Collaborative efforts among farmers, policymakers, and geologists have led to the implementation of best practices for groundwater conservation, including regulated pumping and the adoption of more efficient irrigation techniques.

Case Study 3: Coastal Zone Management in Florida: Florida faces significant challenges related to coastal erosion and sea level rise. This case study examines the policies and practices aimed at managing coastal resources and mitigating risks associated with these geological hazards. The integration of geological data into coastal zone management plans has been essential for protecting ecosystems and communities from the impacts of climate change. For example, Florida's coastal management policies incorporate geological assessments to identify vulnerable areas and prioritize conservation efforts. Strategies such as beach nourishment, habitat restoration, and the establishment of setback regulations for new developments are informed by geological studies that evaluate sediment dynamics and erosion patterns. These proactive measures help safeguard coastal communities and preserve critical habitats.

9.7 Future Directions in Geology and Environmental Policy

Integrating Geoscience into Policy Frameworks: There is a growing recognition of the need to incorporate geoscience into environmental policy frameworks. This integration can enhance decision-making processes and lead to more effective management of natural resources and hazards. Policymakers should prioritize collaboration with geoscientists to ensure that policies are informed by the best available scientific data. By fostering partnerships between geologists and policymakers, we can create more resilient and adaptive policies that address the complexities of environmental challenges. This collaboration can also facilitate the development of innovative solutions that leverage geological insights to promote sustainability.

Promoting Sustainable Practices: Environmental policies should promote sustainable practices in resource management and land use planning. This includes encouraging the use of renewable resources, protecting critical habitats, and implementing conservation measures that consider geological factors. Sustainable practices not only help preserve geological resources but also contribute to the overall health of ecosystems and communities. Policymakers can incentivize sustainable practices through grants, tax breaks, and educational programs that raise awareness about the importance of responsible resource management.

Advocating for Science-Based Policies: The development of science-based policies is essential for addressing environmental challenges effectively. Policymakers should prioritize evidence-based decision-making and engage with the scientific community to ensure that policies are grounded in sound geological and environmental research. By advocating for policies that are informed by scientific data, we can enhance the effectiveness of environmental management and promote long-term sustainability. This approach requires ongoing collaboration between scientists, policymakers, and stakeholders to ensure that policies remain relevant and responsive to emerging environmental issues.

Overview of Geology and Environmental Policy

In this chapter, we explored the intersection of geology and environmental policy, highlighting the importance of integrating geological knowledge into policy decisions. We discussed key environmental policies and regulations, including the National Environmental Policy Act (NEPA), Clean Water Act (CWA), Clean Air Act (CAA), and Resource Conservation and Recovery Act (RCRA), emphasizing how geological considerations are essential for effective implementation. Each of these policies plays a critical role in managing environmental impacts, such as pollution, habitat degradation, and resource depletion, which can have far-reaching consequences for ecosystems and human health.

Environmental Impacts of Inadequate Geological Considerations

The environmental impacts associated with inadequate geological considerations in policy decisions can be severe. For instance, pollution from industrial activities can contaminate soil and groundwater, leading to long-term health risks for communities and wildlife. Habitat degradation, often a result of unregulated land use and resource extraction, can lead to the loss of biodiversity and the disruption of ecological balance. Resource depletion, particularly of non-renewable resources, can result in environmental degradation that affects not only the immediate area but also broader ecosystems that depend on those resources.

Land Use Planning and Zoning Regulations

We examined the significance of land use planning and zoning regulations in mitigating risks associated with natural hazards, such as landslides, flooding, and earthquakes. These geological hazards can lead to significant environmental impacts, including soil erosion, water contamination, and loss of biodiversity. For example, landslides can displace large amounts of soil and vegetation, leading to increased sedimentation in waterways, which can harm aquatic habitats and degrade water quality. Flooding can inundate ecosystems, altering their structure and function, while earthquakes can cause ground shaking that disrupts habitats and leads to secondary effects such as soil liquefaction and landslides.

Incorporating Geological Data into Policy Decisions

By incorporating geological data into land use decisions, policymakers can better protect vulnerable areas and promote sustainable development practices that minimize environmental degradation. This proactive approach can help mitigate the adverse effects of natural hazards, ensuring that communities are better prepared for potential geological events and that ecosystems are preserved.

The Role of Geologists in Policy Making

The critical role of geologists in advising policymakers and engaging the public was also highlighted. Geologists provide essential expertise that informs decision-making processes, ensuring that policies are grounded in scientific understanding. Their insights can help identify areas at risk of geological hazards and inform strategies to mitigate these risks. Public engagement is vital for fostering awareness of geological issues and their relevance to environmental challenges, which can lead to more informed community support for sustainable practices. When communities understand the geological context of their environment, they are more likely to advocate for policies that protect both human health and ecological integrity.

Case Studies in Integrating Geology into Environmental Policy

Through case studies, we illustrated the complexities of integrating geology into environmental policy and the challenges faced in this process. The

California Environmental Quality Act (CEQA) demonstrated how geological considerations can enhance project assessments, leading to more resilient infrastructure and reduced environmental impacts. For instance, incorporating geological data into the planning process can help identify suitable locations for development that minimize risks from earthquakes and landslides, ultimately protecting both human life and the environment.

Groundwater management in the Ogallala Aquifer emphasized the need for sustainable practices informed by geological data, as over-extraction can lead to significant environmental consequences, including reduced water quality and habitat loss. Unsustainable groundwater extraction can result in the depletion of aquifers, leading to increased salinity and contamination of water supplies, which can have dire consequences for both agricultural productivity and ecosystem health.

Coastal zone management in Florida showcased the importance of integrating geological factors to address the impacts of climate change, such as sea level rise and coastal erosion, which threaten both ecosystems and human communities. The erosion of coastal habitats can lead to the loss of critical ecosystems such as mangroves and salt marshes, which provide essential services such as storm protection, carbon sequestration, and habitat for diverse species.

Future Directions for Sustainable Practices

Finally, we emphasized the need for future directions that promote sustainable practices, advocate for science-based policies, and enhance collaboration between geoscientists and policymakers. By recognizing the critical role of geology in environmental policy, we can work towards more effective management of natural resources and a healthier environment for future generations. The integration of geological insights into policy frameworks will not only enhance decision-making but also contribute to the resilience and sustainability of communities in the face of ongoing environmental challenges.

Conclusion: A Holistic Approach to Environmental Management

This holistic approach is essential for mitigating environmental impacts and ensuring that both human and ecological systems can thrive in a changing world. By fostering collaboration among scientists, policymakers, and communities, we can create a more sustainable future that respects the intricate connections between geology, the environment, and human well-being.

9.8 Summary of Chapter 9

Overview of Geology and Environmental Policy

In this chapter, we explored the intersection of geology and environmental policy, highlighting the importance of integrating geological knowledge into policy decisions. We discussed key environmental policies and regulations, including the National Environmental Policy Act (NEPA), Clean Water Act (CWA), Clean Air Act (CAA), and Resource Conservation and Recovery Act (RCRA), emphasizing how geological considerations are essential for effective implementation. Each of these policies plays a critical role in managing environmental impacts, such as pollution, habitat degradation, and resource depletion, which can have far-reaching consequences for ecosystems and human health.

Environmental Impacts of Inadequate Geological Considerations

The environmental impacts associated with inadequate geological considerations in policy decisions can be severe. For instance, pollution from industrial activities can contaminate soil and groundwater, leading to long-term health risks for communities and wildlife. Habitat degradation, often a result of unregulated land use and resource extraction, can lead to the loss of biodiversity and the disruption of ecological balance. Resource depletion, particularly of non-renewable resources, can result in environmental degradation that affects not only the immediate area but also broader ecosystems that depend on those resources.

Land Use Planning and Zoning Regulations

We examined the significance of land use planning and zoning regulations in mitigating risks associated with natural hazards, such as landslides, flooding, and earthquakes. These geological hazards can lead to significant environmental impacts, including soil erosion, water contamination, and loss of biodiversity. For example, landslides can displace large amounts of soil and vegetation, leading to increased sedimentation in waterways, which can harm aquatic habitats and degrade water quality. Flooding can inundate ecosystems, altering their structure and function, while earthquakes can cause ground shaking that disrupts habitats and leads to secondary effects such as soil liquefaction and landslides.

Incorporating Geological Data into Policy Decisions

By incorporating geological data into land use decisions, policymakers can better protect vulnerable areas and promote sustainable development practices that minimize environmental degradation. This proactive approach can help mitigate the adverse effects of natural hazards, ensuring that communities are better prepared for potential geological events and that ecosystems are preserved.

The Role of Geologists in Policy Making

The critical role of geologists in advising policymakers and engaging the public was also highlighted. Geologists provide essential expertise that informs decision-making processes, ensuring that policies are grounded in scientific understanding. Their insights can help identify areas at risk of geological hazards and inform strategies to mitigate these risks. Public engagement is vital for fostering awareness of geological issues and their relevance to environmental challenges, which can lead to more informed community support for sustainable practices. When communities understand the geological context of their environment, they are more likely to advocate for policies that protect both human health and ecological integrity.

Case Studies in Integrating Geology into Environmental Policy

Through case studies, we illustrated the complexities of integrating geology into environmental policy and the challenges faced in this process. The California Environmental Quality Act (CEQA) demonstrated how geological considerations can enhance project assessments, leading to more resilient infrastructure and reduced environmental impacts. For instance, incorporating geological data into the planning process can help identify suitable locations for development that minimize risks from earthquakes and landslides, ultimately protecting both human life and the environment. This proactive approach not only safeguards communities but also promotes sustainable land use practices that align with the natural geological landscape.

Groundwater management in the Ogallala Aquifer emphasized the need for sustainable practices informed by geological data, as over-extraction can lead to significant environmental consequences, including reduced water quality and habitat loss. Unsustainable groundwater extraction can result in the depletion of aquifers, leading to increased salinity and contamination of water supplies, which can have dire consequences for both agricultural productivity and ecosystem health. By integrating geological assessments into groundwater management strategies, stakeholders can develop more effective policies that balance water needs with ecological preservation, ensuring the long-term viability of this critical resource.

Coastal zone management in Florida showcased the importance of integrating geological factors to address the impacts of climate change, such as sea level rise and coastal erosion, which threaten both ecosystems and human communities. The erosion of coastal habitats can lead to the loss of critical ecosystems such as mangroves and salt marshes, which provide essential services such as storm protection, carbon sequestration, and habitat for diverse species. By incorporating geological insights into coastal management plans, policymakers can better anticipate and mitigate the effects of climate change, fostering resilience in both natural and human systems.

Future Directions for Sustainable Practices

Finally, we emphasized the need for future directions that promote sustainable practices, advocate for science-based policies, and enhance collaboration between geoscientists and policymakers. By recognizing the critical role of geology in environmental policy, we can work towards more effective management of natural resources and a healthier environment for future generations. The integration of geological insights into policy frameworks will not only enhance decision-making but also contribute to the resilience and sustainability of communities in the face of ongoing environmental challenges.

To achieve these goals, it is essential to invest in education and outreach initiatives that raise awareness about the importance of geology in environmental decision-making. Engaging local communities in the policy process can foster a sense of ownership and responsibility towards natural resources, leading to more sustainable practices at the grassroots level. Additionally, interdisciplinary collaboration among geoscientists, ecologists, urban planners, and policymakers can facilitate the development of comprehensive strategies that address complex environmental issues holistically.

Conclusion: A Holistic Approach to Environmental Management

This holistic approach is essential for mitigating environmental impacts and ensuring that both human and ecological systems can thrive in a changing world. By fostering collaboration among scientists, policymakers, and communities, we can create a more sustainable future that respects the intricate connections between geology, the environment, and human well-being. Emphasizing the integration of geological knowledge into environmental policy not only enhances our understanding of natural systems but also empowers us to make informed decisions that promote resilience and sustainability for generations to come.

CHAPTER 10

THE FUTURE OF GEOLOGY AND SUSTAINABILITY

10.1 Introduction to Sustainability in Geology

Sustainability refers to the ability to meet present needs without compromising the ability of future generations to meet their own needs. In the context of geology, sustainability emphasizes the balance between resource use and environmental protection. This balance is crucial as we navigate the complexities of resource extraction, land use, and environmental conservation. Geological knowledge is essential for achieving sustainable development goals, particularly in areas such as resource management, hazard mitigation, and environmental protection. As the world faces increasing challenges related to climate change, population growth, and resource depletion, the role of geology in promoting sustainability becomes ever more critical.

10.2 The Role of Geology in Sustainable Resource Management

Mineral Resources: Sustainable practices in mineral extraction are vital for minimizing environmental impacts and ensuring the long-term availability of resources. Responsible sourcing, recycling, and waste reduction are key components of sustainable mining practices. Geological assessments play a crucial role in identifying areas suitable for mining while minimizing ecological disruption. By integrating geological knowledge into mining operations, companies can adopt practices that reduce waste and enhance resource efficiency. For example, advanced geological mapping can help identify high-grade ore deposits, reducing the need for extensive excavation and minimizing habitat destruction.

Water Resources: Geology is fundamental to the sustainable management of water resources. Understanding groundwater recharge processes, aquifer characteristics, and watershed dynamics is essential for protecting water quality and availability. Geological studies inform water management strategies, helping to ensure that water resources are used sustainably and equitably. Protecting aquifers and promoting responsible land use practices are critical for maintaining the integrity of water resources. For instance, geological surveys can identify areas prone to contamination, allowing for targeted protection measures that safeguard drinking water supplies.

Energy Resources: The transition to renewable energy sources is a key aspect of sustainable development. Geology plays a significant role in harnessing geothermal, wind, and solar energy. For example, geothermal energy relies on geological formations that can provide heat for electricity generation and direct use. Additionally, carbon capture and storage (CCS) technologies, which aim to mitigate climate change by storing CO2 underground, depend on geological formations that can safely contain carbon emissions. Understanding the geological context is essential for the successful implementation of these technologies, ensuring that they are both effective and environmentally sound.

10.3 Emerging Trends in Geological Research

Geoinformatics and Big Data: The advent of geoinformatics and big data has transformed geological research. Technologies such as remote sensing, Geographic Information Systems (GIS), and data analytics enable geologists to analyze complex geological processes and patterns at unprecedented scales. These advancements enhance our understanding of geological phenomena and inform sustainable practices, from resource management to hazard assessment. For example, big data analytics can help identify trends in resource consumption and environmental impacts, guiding more sustainable decision-making.

Climate Change Research: Geology plays a vital role in climate change research by providing insights into past climate events through geological records, such as ice cores and sediment layers. Understanding historical climate patterns helps inform current climate adaptation and mitigation strategies. Geologists contribute to climate models and assessments, providing critical data that can guide policy decisions and resource management. By studying geological records, scientists can better predict future climate scenarios and develop strategies to mitigate their impacts.

Interdisciplinary Approaches: Addressing complex geological and environmental challenges requires interdisciplinary collaboration. Partnerships between geologists, environmental scientists, policymakers, and community stakeholders can lead to innovative solutions that integrate diverse perspectives and expertise. Interdisciplinary approaches foster a holistic understanding of sustainability challenges and promote effective decision-making. For instance, collaborative projects that combine geological research with social sciences can enhance community resilience to natural hazards.

10.4 Education and Public Engagement

Importance of Geoscience Education: Geoscience education is essential for promoting awareness of geological issues and their relevance to sustainability. Educational programs at all levels can help cultivate a deeper understanding of geological processes and their impact on the environment. Initiatives that engage students in hands-on geological research and fieldwork can inspire the next generation of geoscientists and environmental stewards. By fostering a strong foundation in geoscience, we can empower individuals to make informed decisions about resource use and environmental protection.

Community Involvement: Community engagement is crucial for promoting sustainable geological practices. Citizen science initiatives, where community members participate in data collection and monitoring, can enhance public understanding of geological issues and foster local stewardship. Involving communities in decision-making processes empowers them to take an active role in managing their natural resources and addressing environmental challenges. For example, local stewardship programs can encourage residents to monitor water quality and participate in conservation efforts, leading to more sustainable practices at the community level.

10.5 Challenges and Opportunities for the Future

Challenges in Sustainable Geology: Promoting sustainable practices in geology faces several challenges, including economic pressures, regulatory barriers, and public perception. The demand for natural resources often conflicts with environmental protection goals, making it essential to find innovative solutions that balance these competing interests. Additionally, regulatory frameworks may not always adequately address geological considerations, hindering effective resource management. Public perception of geology and its importance in sustainability can also be limited, necessitating efforts to raise awareness and understanding.

Opportunities for Advancing Sustainability: Despite these challenges, there are significant opportunities for advancing sustainability in geology. Technological innovations, such as improved monitoring and assessment tools, can enhance resource management and environmental protection. Policy reforms that prioritize sustainability and integrate geological knowledge can lead to more effective decision-making. Collaborative efforts among geologists, policymakers, and communities can drive progress toward a more sustainable future. For instance, partnerships that focus on sustainable land use planning can help mitigate the impacts of urbanization on geological resources.

10.6 The Future of AI and Robotics in Mining

The mining industry is undergoing a profound transformation due to the emergence of advanced technologies such as artificial intelligence (AI) and robotics. These innovations are revolutionising ore deposit exploration and extraction processes, enhancing operational efficiency, safety, and environmental sustainability. AI facilitates sophisticated data analysis and predictive modelling, enabling the identification of mineral deposits with greater accuracy. Concurrently, robotics enhances operational capabilities by automating drilling and processing tasks, thereby reducing human exposure to hazardous conditions.

Key Emerging Technologies in Ore Deposit Exploration and Extraction

Artificial Intelligence (AI) and Machine Learning (ML)

AI and machine learning are at the forefront of technological advancements in mining. These technologies enable the analysis of extensive geological datasets, allowing for the identification of potential mining sites and the optimisation of extraction processes. AI algorithms can process various types of geological data, including remote sensing images, geochemical assays, and historical exploration data, to enhance target identification and resource estimation. Machine learning models are particularly adept at predicting ore quality and quantity, thereby improving decision-making in exploration activities.

Robotics and Automation

The integration of robotics into mining operations is transforming the industry by increasing precision and safety. Autonomous drilling systems and robotic miners are being developed to perform tasks in hazardous environments, significantly reducing the risk to human workers. Additionally, automated haulage trucks and drones are employed for transporting materials and surveying sites, which enhances operational efficiency and reduces costs.

Remote Sensing Technologies

Remote sensing technologies, particularly drones equipped with advanced sensors and imaging capabilities, are revolutionising site monitoring and mapping. These drones provide real-time data that facilitate better planning and resource management. Furthermore, LiDAR (Light Detection and Ranging) technology creates detailed three-dimensional maps of mining areas, aiding in the identification of transport corridors and stockpiling locations.

Digital Twins

The concept of digital twins—virtual models of mining operations—allows for the simulation and optimisation of processes. By integrating real-time data, digital twins enable predictive maintenance and operational adjustments, enhancing strategic decision-making. This technology allows mining companies to model various scenarios and assess their potential impacts on operations, leading to more informed and effective management strategies.

Geophysical Exploration Techniques

Innovations in geophysical exploration techniques, such as 3D seismic imaging, have improved the ability to locate and assess ore deposits without extensive drilling. These advanced methods reduce costs and minimise environmental impacts associated with traditional exploration techniques.

IoT and Wearable Technologies

The Internet of Things (IoT) and wearable technologies are increasingly being utilised in mining to monitor equipment performance and worker safety. IoT devices provide valuable data that can be analysed to predict maintenance needs, thereby enhancing operational efficiency. Wearable technology for miners tracks health metrics, ensuring safety and well-being in challenging environments.

Demand for Critical Minerals and Exploration Strategies

The growing demand for critical minerals, particularly those essential for renewable energy technologies, is reshaping exploration strategies. The surge in demand for lithium and cobalt, driven by the rise of electric vehicles and renewable energy storage solutions, has prompted intensified exploration efforts for deposits rich in these minerals. Companies are investing in new technologies that facilitate the efficient extraction and processing of critical minerals, leading to the discovery of new ore deposits. Furthermore, sustainability considerations are increasingly influencing exploration practices, with a focus on minimising environmental impacts while meeting the demand for these essential resources.

Environmental Impacts of Mining Technologies

While the integration of AI and robotics in mining offers numerous benefits, it is essential to consider the environmental impacts associated with these technologies.

1) **Reduced Land Disturbance**: Advanced exploration techniques, such as remote sensing and geophysical methods, can significantly reduce the need for extensive drilling and land clearing. This minimises habitat destruction and preserves biodiversity in mining areas.

2) **Lower Carbon Footprint**: Automation and robotics can lead to more efficient operations, reducing energy consumption and greenhouse gas emissions. For instance, automated haulage systems can optimise routes and reduce fuel usage, contributing to a lower carbon footprint for mining operations.

3) **Waste Management**: AI can enhance waste management practices by predicting the quantity and type of waste generated during mining operations. This allows for better planning and implementation of waste reduction strategies, minimising the environmental impact of mining activities.

4) **Water Management**: The use of IoT devices can improve water management in mining operations by monitoring water usage and quality in real-time. This helps to prevent contamination of local water sources and ensures compliance with environmental regulations.

5) **Rehabilitation and Restoration**: Emerging technologies can aid in the rehabilitation of mining sites post-extraction. For example, drones can be used to monitor the success of re-vegetation efforts and assess the recovery of ecosystems, ensuring that mining companies fulfil their environmental responsibilities.

Role of Ore Deposits in Renewable Energy Transition

Ore deposits play a crucial role in the transition to renewable energy by supplying essential materials for technologies such as solar panels, wind turbines, and batteries. The mining of these deposits not only supports economic growth but also creates job opportunities in regions transitioning to green technologies. Understanding ore deposits is vital for developing sustainableresource management practices that align with global energy transition goals.

Interdisciplinary Approaches to Understanding Ore Deposits

A comprehensive understanding of ore deposits necessitates interdisciplinary collaboration. Integrating geology, environmental science, economics, and engineering enhances the exploration and extraction of these resources. Data sharing and analysis among interdisciplinary teams can leverage diverse data sources and analytical techniques, leading to improved exploration strategies. A holistic approach that considers social, environmental, and economic factors is essential for fostering sustainable mining practices and making informed decisions regarding ore deposits.

1. **Collaboration Across Disciplines**: By bringing together experts from various fields, mining companies can develop more effective exploration and extraction strategies. For instance, geologists can work alongside data scientists to analyse geological data using AI, while environmental scientists can assess the potential impacts of mining activities on local ecosystems.

2. **Data Sharing and Analysis**: Interdisciplinary teams can leverage diverse data sources, such as geological surveys, environmental assessments, and economic analyses, to create comprehensive models of ore deposits. This collaborative approach enhances the accuracy of predictions regarding mineral locations and extraction feasibility.

3. **Holistic Resource Management**: A comprehensive approach that considers social, environmental, and economic factors leads to more sustainable mining practices. By understanding the broader implications of mining activities, companies can make better decisions that align with community needs and environmental conservation efforts.

Conclusion

The future of mining is being shaped by the integration of emerging technologies such as AI, robotics, and advanced geophysical methods. These innovations not only enhance the efficiency and safety of mining operations but also contribute to more environmentally responsible practices. As the demand for critical minerals continues to rise, the mining industry must adapt by embracing these technologies and adopting interdisciplinary approaches to ensure sustainable resource management.

The ongoing evolution of mining practices will play a pivotal role in supporting the global transition to renewable energy and addressing the challenges of resource scarcity. By prioritising environmental sustainability and fostering collaboration across disciplines, the mining sector can navigate the complexities of modern resource extraction while minimising its ecological footprint.

Questions and Answers on Emerging Technologies in Ore Exploration and Extraction

1. AI and Machine Learning

Q: How are AI and machine learning algorithms specifically applied in the exploration of ore deposits?

A: AI and machine learning algorithms are applied to analyse vast datasets, including geological, geochemical, and geophysical information, to identify patterns that indicate the presence of mineral deposits. These technologies enhance predictive accuracy and improve decision-making in exploration activities.

Q: What types of geological data are most effectively analysed using AI technologies?

A: AI technologies are particularly effective at analysing remote sensing data, historical exploration results, and geochemical assays. These data types provide valuable insights that can lead to more accurate predictions of mineral locations.

2. Robotics and Automation

Q: What are the current capabilities of robotic systems in mining operations, and how do they enhance safety and efficiency?

A: Robotic systems in mining can automate hazardous tasks such as drilling, blasting, and material handling, significantly reducing human exposure to risks. They can operate continuously without fatigue, provide real-time monitoring, and improve precision, leading to enhanced safety and operational efficiency.

Q: How do autonomous drilling systems compare to traditional drilling methods in terms of cost and productivity?

A: Autonomous drilling systems reduce labor costs by minimising the need for human operators and can operate continuously, leading to higher productivity. In contrast, traditional drilling methods are limited by human work hours and can incur higher labor costs due to fatigue and potential errors.

3. Remote Sensing Technologies

Q: What advancements in drone technology are being utilised for mineral exploration and site monitoring?

A: Recent advancements in drone technology include automated data capture, high-resolution imaging, and the ability to create 3D models of mining sites. These innovations enhance exploration efficiency and allow for rapid and accurate mapping of mineral deposits.

Q: How does LiDAR technology improve the accuracy of geological mapping compared to traditional methods?

A: LiDAR technology provides centimeter-level positioning accuracy and can penetrate vegetation to capture detailed terrain data. This capability allows for more precise geological mapping and better identification of geological features compared to traditional surveying techniques.

4. Digital Twins

Q: What are the key benefits of using digital twins in mining operations, and how do they contribute to decision-making?

A: Digital twins allow for the simulation and optimisation of mining processes, enabling companies to analyse various scenarios and make informed decisions based on real-time data. This leads to improved resource management and operational efficiency.

Q: How can digital twin technology be integrated with real-time data from mining operations?

A: Digital twin technology can be integrated with real-time data through IoT devices and sensors that continuously monitor operational parameters. This integration allows for predictive maintenance and operational adjustments based on current conditions.

5. Geophysical Exploration Techniques

Q: What are the advantages of using 3D seismic imaging over conventional drilling methods in ore deposit assessment?

A: 3D seismic imaging provides a clearer understanding of subsurface structures, reducing the need for extensive drilling and minimising environmental impact. It allows for better identification of ore deposits and enhances exploration strategies.

Q: How do geophysical methods complement traditional geological surveys in identifying potential mining sites?

A: Geophysical methods provide additional data that can enhance the accuracy of traditional geological surveys. They help identify subsurface features and anomalies that may indicate the presence of mineral deposits, leading to more informed exploration decisions.

6. IoT and Wearable Technologies

Q: How can IoT devices improve equipment maintenance and operational efficiency in mining?

A: IoT devices can monitor equipment performance in real-time, predicting maintenance needs and optimising operational efficiency. This proactive approach reduces downtime and enhances overall productivity in mining operations.

Q: What role do wearable technologies play in enhancing worker safety in hazardous mining environments?

A: Wearable technologies can track health metrics and provide real-time data on environmental conditions, ensuring that workers are protected in challenging conditions. They enhance safety protocols and help prevent accidents in hazardous areas.

7. Environmental Impact

Q: How do emerging technologies contribute to reducing the environmental impact of mining operations?

A: Emerging technologies, such as drones and automation, can lead to more efficient resource management, reduced land disturbance, and lower carbon emissions. These advancements help mining companies align their operations with environmental sustainability goals.

Q: In what ways can technology help in the rehabilitation of mining sites post-extraction?

A: Technology can facilitate effective rehabilitation by monitoring the success of re-vegetation efforts and assessing ecosystem recovery. Drones and remote sensing technologies can provide valuable data to ensure compliance with environmental regulations.

8. Cost-Benefit Analysis

Q: What are the economic implications of adopting advanced technologies in ore deposit exploration and extraction?

A: The adoption of advanced technologies can lead to significant cost savings through improved efficiency, reduced labor costs, and enhanced productivity. However, companies must also consider the initial investment required for implementation.

Q: How do the initial costs of implementing new technologies compare to the long-term savings and efficiencies gained?

A: While the initial costs of implementing new technologies can be high, the long-term savings from increased productivity, reduced operational costs, and improved resource management often outweigh these initial investments, making them economically viable.

9. Future Trends

Q: What future technological advancements are anticipated to further revolutionise ore deposit exploration and extraction?

A: Future advancements may include enhanced AI algorithms, more sophisticated robotics, and improved data analytics capabilities. These innovations are expected to further optimise exploration processes and increase the efficiency of resource extraction.

Q: How might the integration of artificial intelligence and robotics evolve in the mining industry over the next decade?

A: The integration of AI and robotics is likely to become more seamless, with AI-driven systems managing robotic operations autonomously. This evolution could lead to fully automated mining operations that enhance safety, efficiency, and sustainability.

10. Interdisciplinary Collaboration

Q: How can collaboration between geologists, engineers, and data scientists enhance the effectiveness of emerging technologies in mining?
A: Interdisciplinary collaboration allows for the sharing of expertise and insights, leading to more effective exploration and extraction strategies. By combining geological knowledge with engineering and data analysis, mining companies can develop innovative solutions to complex challenges.

Q: What role does interdisciplinary research play in the development and implementation of new mining technologies?

A: Interdisciplinary research fosters innovation by bringing together diverse perspectives and expertise. This collaborative approach can drive the development of new technologies and ensure that they are effectively implemented in mining operations, addressing both technical and environmental challenges.

These questions can facilitate discussions and research on the impact of emerging technologies in the mining sector, particularly in the context of ore deposit exploration and extraction. By addressing these inquiries, stakeholders can better understand the potential benefits and challenges associated with the integration of advanced technologies in mining, ultimately leading to more sustainable and responsible practices in the industry.

10.7 Remote Sensing Technologies

Advancements in Drone Technology for Mineral Exploration and Site Monitoring

Recent advancements in drone technology have significantly enhanced mineral exploration and site monitoring capabilities. Innovations such as improved aerial navigation, automated data capture, and hyperspectral imaging collectively improve efficiency, reduce costs, and enhance safety. Drones now allow for rapid and accurate mapping and analysis of mineral deposits, transforming traditional exploration methods.

One of the key advancements is **automated data capture**. Modern drones are equipped with advanced autonomous navigation systems that facilitate automated data collection. This capability significantly reduces the time required for surveys, enabling the generation of detailed three-dimensional environment maps that enhance situational awareness and decision-making. Additionally, the integration of **high-resolution sensors**, including thermal and multispectral imaging, allows drones to capture detailed imagery and data for mineral identification and analysis. This technology enables the detection of minerals that may not be visible to the naked eye, thereby improving exploration accuracy.

Drones also excel in **3D modeling and mapping**, creating precise three-dimensional models of mining sites and stockpiles. This capability facilitates accurate volume calculations and resource assessments, which are particularly useful for monitoring changes in stockpile sizes over time, allowing for better resource management. Furthermore, by utilizing drones for site monitoring, mining companies can minimize the need for personnel to enter hazardous areas, thereby enhancing safety and reducing the risk of accidents. This is particularly important in environments where traditional surveying methods pose significant risks to human workers.

Another significant advantage of drone technology is **real-time data processing**. Drones equipped with onboard processing capabilities can analyze data in real-time, providing immediate insights that can inform decision-making during exploration and monitoring activities. This rapid feedback loop enhances operational efficiency and responsiveness. Moreover, new drone designs are capable of operating in challenging conditions, such as high electromagnetic interference (EMI) environments, ensuring reliable data collection even in difficult terrains. This adaptability expands the range of environments in which drones can be effectively deployed.

Improvements in Geological Mapping with LiDAR Technology

LiDAR (Light Detection and Ranging) technology has revolutionized geological mapping by providing significant improvements over traditional methods. One of the primary benefits of LiDAR is its **increased accuracy**, offering centimeter-level positioning accuracy that greatly enhances the precision of geological mapping. This high level of detail allows for better identification of geological features and structures, leading to more informed exploration decisions.

LiDAR systems can also collect vast amounts of data quickly, significantly reducing the time required for mapping projects. For instance, drones equipped with LiDAR can gather data that would take traditional surveying methods much longer to achieve, thereby accelerating project timelines. Additionally, one of the key advantages of LiDAR is its ability to penetrate vegetation and capture the underlying terrain. This capability is particularly beneficial in densely forested areas where traditional methods may struggle, allowing for a clearer view of the ground surface and geological features.

The data collected through LiDAR can be used to create detailed three-dimensional models of geological formations, enhancing the understanding of spatial relationships and geological structures. This provides valuable insights for exploration and resource management. Furthermore, LiDAR technology offers a **cost-effective** solution for geological mapping compared to conventional surveying techniques by reducing the need for extensive ground control points and minimizing fieldwork. This cost efficiency is particularly advantageous for large-scale mining operations.

Finally, LiDAR data can be easily integrated with other geological and geospatial data, allowing for comprehensive analysis and improved resource modeling. This integration enhances the overall understanding of mineral deposits and their potential for extraction.

Robotics and Automation

Integration of Robotic Systems in Mining Operations

Robotic systems are increasingly being integrated into mining operations, enhancing safety and efficiency by automating hazardous tasks, improving precision, and enabling real-time monitoring. Their integration leads to reduced human exposure to risks, increased productivity, and optimized resource management, ultimately transforming mining practices for the better.

Automation of Hazardous Tasks

One of the primary applications of robotic systems is the automation of hazardous tasks. These systems are employed to perform high-risk activities such as drilling, blasting, and material handling, significantly reducing the exposure of human workers to dangerous environments and thereby enhancing overall safety. Additionally, robotic machines can operate continuously, functioning 24/7 without the need for breaks, fatigue, or shift changes. This capability leads to increased operational hours and productivity, allowing mining operations to maximize output.

Advanced Monitoring Capabilities

Equipped with sophisticated sensors, robotic systems can also provide advanced monitoring of environmental conditions in real-time. This critical data helps prevent accidents and enhances safety protocols, ensuring a safer working environment. Moreover, automated drilling systems achieve higher accuracy in borehole placement, which improves the overall effectiveness of mining operations and resource extraction. This precision reduces the likelihood of resource wastage and enhances operational efficiency.

Environmental Sustainability

In terms of environmental sustainability, robotics can optimize fuel efficiency and reduce emissions, aligning mining practices with environmental regulations and sustainability goals. This focus on sustainability is increasingly important in the context of global environmental challenges, as mining operations seek to minimize their ecological footprint. Furthermore, many robotic systems can be operated from remote locations, allowing for safer management of mining operations without the need for personnel in hazardous areas. This capability enhances safety and operational flexibility, particularly in environments that pose significant risks to human workers.

Enhancing Safety and Productivity

The integration of robotics and automation in mining not only enhances safety and efficiency but also contributes to the overall sustainability of mining operations. By reducing the need for human presence in hazardous areas, companies can protect their workforce while maintaining productivity. Additionally, the ability to operate continuously without fatigue allows for more consistent output and resource management, which is crucial in meeting the demands of the industry.

Future of Mining with Advanced Technologies

As the mining sector continues to evolve, the adoption of advanced technologies such as drones, LiDAR, and robotics will play a pivotal role in shaping the future of mineral exploration and extraction. These innovations not only improve operational efficiency but also provide valuable data that can inform better decision-making and resource management practices. By leveraging these technologies, mining companies can enhance their ability to navigate the complexities of modern mining challenges while prioritizing safety, environmental sustainability, and economic viability.

Conclusion: Transforming the Mining Landscape

In conclusion, the advancements in remote sensing technologies, particularly through drone and LiDAR applications, alongside the integration of robotics and automation, are transforming the landscape of mineral exploration and mining operations. These technologies enable more accurate data collection, enhance safety protocols, and promote sustainable practices, ultimately leading to a more efficient and responsible mining industry. As these technologies continue to develop, they will undoubtedly play a crucial role in addressing the challenges faced by the mining sector and ensuring its long-term viability in a rapidly changing world.

Comparison of Autonomous Drilling Systems to Traditional Drilling Methods

The comparison between autonomous drilling systems and traditional drilling methods reveals significant advantages for the former in terms of cost, productivity, safety, and precision.

1. Cost Efficiency

Autonomous Drilling: One of the most significant advantages of autonomous drilling systems is their ability to reduce labor costs. By minimising the need for human operators, these systems can operate independently, which not only cuts down on direct labor expenses but also lowers operational costs through optimised fuel consumption and maintenance. Autonomous drilling rigs are designed to operate efficiently, utilising advanced technologies that monitor and adjust their performance in real-time. This capability allows for the identification of maintenance needs before they become critical, thereby reducing downtime and repair costs. While the initial investment in autonomous systems may be substantial, the long-term savings in operational expenses often justify this expenditure. Over time, the return on investment can be significant, as companies benefit from lower labor costs and enhanced operational efficiency.

Traditional Drilling: In contrast, traditional drilling methods typically involve higher labor costs due to the reliance on human operators. This approach can lead to inefficiencies stemming from human factors such as fatigue, error, and variability in performance. Workers may require breaks, and their productivity can fluctuate based on physical and mental fatigue, which can result in inconsistent output. Additionally, the costs associated with training personnel, implementing safety measures, and managing potential accidents can further inflate operational expenses. The need for skilled labor in traditional drilling also means that companies must invest in ongoing training and development, which can strain budgets and resources.

2. Productivity

Autonomous Drilling: Autonomous drilling systems are capable of continuous operation, which significantly enhances productivity in mining operations. These systems can work around the clock without the need for breaks, shift changes, or downtime associated with human fatigue. This capability allows for faster completion of drilling tasks, leading to higher output and more efficient resource extraction. The ability to operate continuously means that mining companies can maximise their operational hours, ultimately increasing the overall productivity of their operations. Furthermore, autonomous systems can be programmed to optimise drilling parameters based on real-time data, ensuring that they perform at peak efficiency throughout their operational lifespan.

Traditional Drilling: Traditional drilling methods, on the other hand, are inherently limited by human work hours. Workers need to take breaks, and shifts must be changed, which can lead to delays and interruptions in the drilling process. These factors contribute to lower overall productivity compared to autonomous systems. Additionally, human factors can introduce variability in performance, leading to inconsistencies in output. For example, a worker's performance may vary from day to day based on their physical condition or mental focus, which can affect the quality and speed of drilling. This variability can result in inefficiencies and increased costs, as companies may need to allocate additional resources to compensate for these fluctuations.

3. Safety

Autonomous Drilling: One of the most compelling benefits of autonomous drilling systems is their ability to enhance safety in mining operations. By removing human operators from dangerous environments, these systems significantly reduce the risk of accidents and injuries. This is particularly important in high-risk areas where traditional drilling methods expose workers to potential hazards, such as falling rocks, equipment malfunctions, or exposure to harmful gases. Autonomous systems can operate in these challenging conditions without putting human lives at risk, thereby improving overall safety standards in the mining industry. Additionally, the use of advanced sensors and monitoring technologies in autonomous drilling systems allows for real-time assessment of environmental conditions, further enhancing safety protocols.

Traditional Drilling: In contrast, traditional drilling methods require human presence in hazardous conditions, which increases the likelihood of accidents and injuries. The need for personnel to operate drilling equipment in dangerous environments can lead to higher insurance costs and liability issues for mining companies. Furthermore, the potential for accidents can create a culture of fear among workers, impacting morale and productivity. Companies must invest in extensive safety training and protocols to mitigate these risks, which can further strain resources and budgets. The reliance on human operators in hazardous environments underscores the importance of transitioning to safer, more efficient autonomous systems.

4. Precision

Autonomous Drilling: Autonomous drilling systems utilise advanced algorithms and sensors to achieve a high level of precision in drilling operations. This technological sophistication improves the quality of the work and resource extraction, as these systems can accurately position drill bits and adjust drilling parameters in real-time. The precision offered by autonomous systems reduces the likelihood of drilling errors, which can lead to costly resource wastage and inefficiencies. For example, precise drilling can ensure that boreholes are optimally placed to maximise resource recovery while minimising environmental impact. This level of accuracy not only enhances operational efficiency but also contributes to more sustainable mining practices.

Traditional Drilling: Traditional drilling methods may suffer from human error, which can result in less accurate drilling and potential resource wastage. Variability in human performance can lead to inconsistencies in drilling quality, as factors such as fatigue, distraction, or lack of experience can affect a worker's ability to perform tasks accurately. Inaccurate drilling can result in misaligned boreholes, which may necessitate additional drilling efforts or lead to the loss of valuable resources. The reliance on human operators for precision tasks underscores the need for advanced technologies that can enhance accuracy and efficiency in mining operations. Transitioning to autonomous drilling systems can help mitigate these issues and improve overall resource management.

In summary, the shift from traditional drilling methods to autonomous systems presents numerous advantages in terms of cost efficiency, productivity, safety, and precision. As the mining industry continues to evolve, embracing these advanced technologies will be crucial for enhancing operational performance and ensuring sustainable practices.

5. Environmental Impact

Autonomous Drilling: The environmental impact of mining operations is a growing concern, and autonomous drilling systems can play a significant role in minimising this impact. By utilising advanced technologies, these systems can optimise drilling processes to reduce land disturbance and resource wastage. For instance, precise drilling techniques can ensure that boreholes are strategically placed, minimising the need for extensive land clearing and reducing the ecological footprint of mining activities. Additionally, autonomous systems can operate with greater fuel efficiency, leading to lower greenhouse gas emissions compared to traditional drilling methods. The ability to monitor environmental conditions in real-time also allows for immediate adjustments to operations, further mitigating potential negative impacts on local ecosystems.

Traditional Drilling: Traditional drilling methods often result in greater environmental disruption due to the reliance on human operators and less precise drilling techniques. The need for extensive land clearing and the potential for drilling errors can lead to habitat destruction and resource wastage. Furthermore, the emissions associated with traditional drilling equipment can contribute to air and noise pollution, negatively affecting surrounding communities and wildlife. The environmental regulations governing mining operations are becoming increasingly stringent, and companies that continue to rely on traditional methods may face higher compliance costs and reputational risks. Transitioning to more sustainable practices, such as autonomous drilling, can help mitigate these challenges and align mining operations with environmental stewardship goals.

6. Technological Integration

Autonomous Drilling: The integration of autonomous drilling systems with other advanced technologies, such as artificial intelligence (AI), machine learning, and data analytics, can further enhance their effectiveness. By leveraging AI algorithms, these systems can analyse vast amounts of geological data to optimise drilling parameters and improve decision-making. For example, machine learning models can predict the best drilling locations based on historical data, geological surveys, and real-time environmental conditions. This integration not only increases the efficiency of drilling operations but also enhances the overall effectiveness of resource extraction strategies. As technology continues to advance, the potential for autonomous systems to work in conjunction with other innovations will likely lead to even greater improvements in mining operations.

Traditional Drilling: Traditional drilling methods often lack the technological integration that characterises modern autonomous systems. While some companies may utilise basic data collection and analysis tools, the reliance on human operators limits the ability to fully leverage advanced technologies. This lack of integration can result in missed opportunities for optimisation and efficiency gains. As the mining industry evolves, companies that fail to adopt new technologies may find themselves at a competitive disadvantage. Embracing technological advancements and integrating them into drilling operations will be essential for maintaining productivity and sustainability in the face of increasing industry challenges.

7. Future Outlook

Autonomous Drilling: The future of mining is likely to be heavily influenced by the continued development and adoption of autonomous drilling systems. As technology advances, these systems will become more sophisticated, incorporating enhanced AI capabilities, improved sensors, and better data analytics tools. This evolution will enable even greater precision, efficiency, and safety in drilling operations. Additionally, as the demand for minerals and resources continues to rise, the need for more sustainable and efficient mining practices will drive the adoption of autonomous systems. Companies that invest in these technologies will be better positioned to meet regulatory requirements, address environmental concerns, and respond to market demands.

Traditional Drilling: The future of traditional drilling methods appears increasingly uncertain as the mining industry shifts towards automation and advanced technologies. While traditional methods may still have a place in certain contexts, their limitations in terms of cost, productivity, safety, and environmental impact will likely lead to a decline in their use. Companies that continue to rely on traditional drilling may face challenges in attracting skilled labor, meeting regulatory standards, and maintaining competitiveness in a rapidly evolving industry. As the mining sector embraces innovation, the transition to more advanced drilling techniques will be essential for ensuring long-term viability and success.

Conclusion

In conclusion, the transition from traditional drilling methods to autonomous drilling systems presents numerous advantages across various dimensions, including cost efficiency, productivity, safety, precision, environmental impact, technological integration, and future outlook. As the mining industry continues to evolve, embracing these advanced technologies will be crucial for enhancing operational performance, ensuring sustainable practices, and meeting the growing demand for resources. Companies that invest in autonomous drilling systems and other innovative technologies will not only improve their bottom line but also contribute to a more sustainable and responsible mining industry. The future of mining lies in the ability to adapt to changing technologies and environmental expectations, and autonomous drilling represents a significant step forward in this journey.

Conclusion

The integration of remote sensing technologies, robotics, and automation in mining operations is transforming the industry, enhancing efficiency, safety, and environmental sustainability. As these technologies continue to advance, they offer significant potential for minimising the environmental impact of mining activities while meeting the growing demand for critical minerals. By embracing these innovations and prioritising sustainable practices, the mining sector can navigate the challenges of resource extraction in a responsible manner, contributing to a more sustainable future.

Questions and Answers on Emerging Technologies in Ore Exploration and Extraction

1. Remote Sensing Technologies

Q: What advancements in drone technology are being utilised for mineral exploration and site monitoring?

A: Recent advancements in drone technology for mineral exploration and site monitoring include the integration of high-resolution cameras, multispectral and hyperspectral sensors, and advanced autonomous navigation systems. These drones can capture detailed aerial imagery and collect data over large areas quickly and efficiently. The use of automated flight paths allows for consistent data collection, while real-time data processing capabilities enable immediate analysis of geological features. This technology significantly enhances the speed and accuracy of mineral exploration, allowing companies to identify potential sites more effectively.

Q: How does LiDAR technology improve the accuracy of geological mapping compared to traditional methods?

A: LiDAR (Light Detection and Ranging) technology improves the accuracy of geological mapping by providing high-resolution, three-dimensional data of the terrain. Unlike traditional methods that may rely on ground surveys, LiDAR can penetrate vegetation and capture detailed topographical information, allowing for a clearer understanding of geological features. This technology offers centimeter-level accuracy, which is crucial for identifying subtle geological structures and formations. As a result, LiDAR enhances the precision of geological assessments and reduces the likelihood of errors associated with conventional mapping techniques.

2. Robotics and Automation

Q: What are the current capabilities of robotic systems in mining operations, and how do they enhance safety and efficiency?

A: Robotic systems in mining operations are capable of performing a variety of tasks, including drilling, blasting, material handling, and site monitoring. These systems enhance safety by automating hazardous tasks, thereby reducing the need for human presence in dangerous environments. For example, robotic drilling rigs can operate in high-risk areas without exposing workers to potential hazards. Additionally, these systems improve efficiency by operating continuously without fatigue, allowing for increased productivity and reduced operational downtime. The integration of sensors and real-time monitoring also enables proactive maintenance, further enhancing operational efficiency.

Q: How do autonomous drilling systems compare to traditional drilling methods in terms of cost and productivity?

A: Autonomous drilling systems generally offer significant advantages over traditional drilling methods in both cost and productivity. By minimising the need for human operators, autonomous systems reduce labor costs and can operate continuously, leading to higher output and faster completion of drilling tasks. In contrast, traditional drilling methods are limited by human work hours and can experience delays due to fatigue and shift changes. The initial investment in autonomous systems may be higher, but the long-term savings in operational expenses and increased productivity often justify this expenditure, making them a more economically viable option in the long run.

3. Environmental Impact

Q: How do emerging technologies contribute to reducing the environmental impact of mining operations?

A: Emerging technologies contribute to reducing the environmental impact of mining operations in several ways. For instance, advanced remote sensing technologies, such as drones and LiDAR, enable more precise mapping and monitoring of mining sites, which can help minimise land disturbance and habitat destruction. Additionally, autonomous systems can optimise resource extraction processes, reducing waste and energy consumption. The use of data analytics and IoT devices allows for better monitoring of environmental conditions, enabling companies to respond quickly to potential issues and comply with environmental regulations. Overall, these technologies promote more sustainable mining practices and help mitigate the industry's ecological footprint.

Q: In what ways can technology help in the rehabilitation of mining sites post-extraction?

A: Technology plays a crucial role in the rehabilitation of mining sites post-extraction by providing tools for effective monitoring and assessment of environmental recovery. For example, drones equipped with multispectral sensors can monitor vegetation regrowth and soil health, allowing companies to evaluate the success of rehabilitation efforts. Additionally, data analytics can be used to model and predict the ecological recovery of disturbed areas, guiding restoration strategies. Technologies such as remote sensing and GIS (Geographic Information Systems) can also assist in planning and implementing rehabilitation projects, ensuring that they are effective and compliant with environmental standards.

4. Future Trends

Q: What future technological advancements are anticipated to further revolutionise ore deposit exploration and extraction?

A: Future technological advancements expected to revolutionise ore deposit exploration and extraction include the continued development of artificial intelligence (AI) and machine learning algorithms that can analyse geological data more effectively. Innovations in robotics, such as more sophisticated autonomous drilling systems and automated material handling, will further enhance operational efficiency. Additionally, advancements in sensor technology and data analytics will enable more precise monitoring of mining operations and environmental conditions. The integration of blockchain technology for supply chain transparency and sustainability tracking is also anticipated to gain traction in the mining sector.

Q: How might the integration of artificial intelligence and robotics evolve in the mining industry over the next decade?

A: The integration of artificial intelligence and robotics in the mining industry is expected to evolve significantly over the next decade, leading to fully automated mining operations. AI will play a crucial role in optimising drilling and extraction processes by analysing real-time data and making autonomous decisions based on predictive models. Robotics will become increasingly sophisticated, capable of performing complex tasks with minimal human intervention. This evolution will enhance safety, reduce operational costs, and improve overall productivity. As these technologies continue to advance, the mining industry will likely see a shift towards more sustainable and efficient practices, aligning with global demands for responsible resource extraction.

These questions and answers provide a comprehensive overview of the impact of emerging technologies in ore exploration and extraction, highlighting their potential benefits and challenges in the mining industry. By addressing these inquiries, stakeholders can better understand the potential benefits and challenges associated with the integration of advanced technologies in mining, ultimately leading to more sustainable and responsible practices in the industry.

10.7 Call to Action

As we look to the future, it is imperative that all stakeholders—geologists, policymakers, educators, and the public—commit to a collaborative approach in addressing the challenges of sustainability. To facilitate this, several actionable steps can be taken. First, advocating for science-based policies is crucial. Policymakers should be encouraged to base their decisions on scientific evidence and geological data. This can be achieved through active engagement with scientific communities and participation in public forums, ensuring that informed decisions are made for the benefit of both society and the environment.

Investing in education and outreach is another vital step. Supporting educational initiatives that promote geoscience literacy among students and the general public can foster a culture of sustainability. Programs that emphasize the importance of geology in understanding environmental issues can empower individuals to make informed choices and advocate for responsible resource management. By enhancing geoscience education, we can cultivate a generation that values and understands the significance of sustainable practices.

Promoting sustainable practices within industries and communities is essential as well. Encouraging the adoption of responsible sourcing, waste reduction, and the implementation of technologies that minimize environmental impacts can lead to more sustainable resource management. By highlighting successful examples of sustainable practices, we can inspire others to follow suit and contribute to a more sustainable future.

Engaging in community-based initiatives is also critical. Local efforts to monitor and protect geological resources can significantly enhance public awareness and stewardship. Community involvement in conservation projects and citizen science initiatives allows individuals to take an active role in protecting their environment, fostering a sense of ownership and responsibility toward local resources.

Finally, fostering interdisciplinary collaboration is key to addressing complex sustainability challenges. Supporting partnerships between geologists and other disciplines, such as environmental science, urban planning, and public policy, can lead to innovative solutions that benefit both society and the environment. By working together, we can leverage diverse perspectives and expertise to tackle the multifaceted issues we face.

Final Thoughts

The future of geology is not just about understanding the Earth's processes; it is about applying that knowledge to create a sustainable world. As we navigate the complexities of modern challenges—climate change, resource depletion, and environmental degradation—geology offers valuable insights and solutions. By embracing sustainability as a guiding principle, we can ensure that geological resources are managed responsibly and that the environment is protected for generations to come.

In conclusion, the integration of geology into sustainability efforts is essential for achieving a balanced relationship between human activities and the natural world. As we move forward, let us commit to leveraging geological knowledge to foster a sustainable future where both people and the planet can thrive.

10.8 Overall Conclusion of Chapter 10: The Future of Mining and Geology

The future of mining is being shaped by the integration of emerging technologies such as artificial intelligence (AI), robotics, and advanced geophysical methods. These innovations not only enhance the efficiency and safety of mining operations but also contribute to more environmentally responsible practices. As the demand for critical minerals continues to rise, the mining industry must adapt by embracing these technologies and adopting interdisciplinary approaches to ensure sustainable resource management. The ongoing evolution of mining practices will play a pivotal role in supporting the global transition to renewable energy and addressing the challenges of resource scarcity. By prioritising environmental sustainability and fostering collaboration across disciplines, the mining sector can navigate the complexities of modern resource extraction while minimising its ecological footprint.

Transformative Technologies in Mining

The integration of remote sensing technologies, robotics, and automation in mining operations is transforming the industry, enhancing efficiency, safety, and environmental sustainability. As these technologies continue to advance, they offer significant potential for minimising the environmental impact of mining activities while meeting the growing demand for critical minerals. By embracing these innovations and prioritising sustainable practices, the mining sector can navigate the challenges of resource extraction in a responsible manner, contributing to a more sustainable future.

The Role of Geology in Sustainability

Moreover, the future of geology is intrinsically linked to sustainability. As we face pressing global challenges, the role of geology in promoting sustainable practices and informed decision-making becomes increasingly important. By integrating geological knowledge into resource management, hazard mitigation, and environmental policy, we can work towards a more sustainable future. It is essential for geologists, policymakers, and the public to collaborate and advocate for science-based solutions that address the complex challenges of our time. Together, we can harness the power of geology to create resilient communities, protect our natural resources, and ensure a sustainable environment for future generations.

Socio-Environmental Considerations

The socio-environmental aspects of mining are also critical to consider as we adopt these emerging technologies. For instance, the use of drones and remote sensing can facilitate better engagement with local communities by providing transparent data on mining activities and their potential impacts. This transparency fosters trust and allows for more informed public participation in decision-making processes. Additionally, technologies that enhance environmental monitoring can help mitigate negative impacts on local ecosystems and communities, ensuring that mining operations do not compromise the health and well-being of surrounding populations.

Collaborative Approaches to Sustainability

As we look to the future, it is imperative that all stakeholders—geologists, policymakers, educators, and the public—commit to a collaborative approach in addressing the challenges of sustainability. Actionable steps include advocating for science-based policies, investing in education and outreach, promoting sustainable practices, engaging in community-based initiatives, and fostering interdisciplinary collaboration. These efforts can enhance public awareness and stewardship, leading to innovative solutions that address complex sustainability challenges.

Final Thoughts on Geology and Sustainability

Final thoughts emphasize that the future of geology is not just about understanding the Earth's processes; it is about applying that knowledge to create a sustainable world. As we navigate the complexities of modern challenges—climate change, resource depletion, and environmental degradation—geology offers valuable insights and solutions. By embracing sustainability as a guiding principle, we can ensure that geological resources are managed responsibly and that the environment is protected for generations to come.

Commitment to a Balanced Relationship

In conclusion, the integration of geology into sustainability efforts is essential for achieving a balanced relationship between human activities and the natural world. As we move forward, let us commit to leveraging geological knowledge to foster a sustainable future where both people and the planet can thrive. The mining industry, equipped with advanced technologies and a commitment to socio-environmental responsibility, can play a pivotal role in this endeavor. By prioritizing collaboration, transparency, and sustainability, we can ensure that the benefits of mining extend beyond economic gains to include social equity and environmental stewardship.

A Call to Action for All Stakeholders

Ultimately, the way forward requires a concerted effort from all stakeholders—geologists, policymakers, educators, industry leaders, and the public—to embrace innovation, prioritize sustainability, and foster collaboration across disciplines. By doing so, we can ensure that the mining and geological sectors contribute positively to society and the environment, paving the way for a sustainable future. The socio-environmental dimensions of mining must be at the forefront of this transition, ensuring that technological advancements not only enhance operational efficiency but also promote social equity and environmental justice in the communities affected by mining activities. Together, we can harness the power of geology and technology to create a resilient and sustainable future for all.

Emphasizing the Importance of Education and Outreach

To further support this transition, it is crucial to invest in education and outreach initiatives that promote geoscience literacy among students and the general public. By fostering a deeper understanding of geological processes and the importance of sustainable resource management, we can cultivate a culture that values environmental stewardship. Educational programs that highlight the role of geology in addressing contemporary challenges—such as climate change, natural disasters, and resource scarcity—can empower individuals to make informed decisions and advocate for sustainable practices within their communities.

Engaging Local Communities

Engaging local communities in the mining process is essential for ensuring that their needs and concerns are addressed. Mining companies should actively seek input from community members, particularly Indigenous populations, who often have a profound connection to the land and its resources. By incorporating traditional knowledge and local perspectives into mining operations, companies can develop more culturally sensitive and environmentally responsible practices. This engagement not only fosters trust but also enhances the social license to operate, which is increasingly important in today's socially conscious market.

The Role of Policy and Regulation

Effective policy and regulation are vital for guiding the mining industry towards sustainable practices. Policymakers must work closely with geologists, environmental scientists, and industry leaders to develop regulations that promote responsible resource extraction while protecting the environment and local communities. This includes establishing clear guidelines for environmental assessments, land use planning, and community engagement. By creating a regulatory framework that prioritizes sustainability, governments can incentivize mining companies to adopt best practices and invest in innovative technologies.

Innovations in Resource Management

As the mining industry evolves, innovations in resource management will play a critical role in ensuring sustainability. The adoption of a circular economy approach, which emphasizes recycling and reusing materials, can significantly reduce the environmental impact of mining. Technologies that facilitate the recovery of valuable minerals from waste products can help minimize the need for new extraction, thereby conserving natural resources. Additionally, advancements in data analytics and AI can optimize resource allocation and improve operational efficiency, further reducing the ecological footprint of mining activities.

Conclusion: A Vision for the Future

In summary, the future of mining and geology is intertwined with the principles of sustainability, social equity, and environmental responsibility. By embracing emerging technologies and fostering collaboration among stakeholders, the mining industry can navigate the complexities of modern resource extraction while contributing positively to society and the environment. The socio-environmental dimensions of mining must remain at the forefront of this transition, ensuring that technological advancements not only enhance operational efficiency but also promote social justice and environmental stewardship.

As we look ahead, let us commit to a shared vision of a sustainable future—one where the mining industry operates in harmony with the environment and local communities. By leveraging geological knowledge, innovative technologies, and collaborative approaches, we can create a resilient and sustainable future for all. Together, we can harness the power of geology and technology to address the pressing challenges of our time, ensuring that both people and the planet thrive for generations to come.

Building Resilience Through Collaboration

To achieve this vision, it is essential to build resilience through collaboration across various sectors. The mining industry, academia, government, and civil society must work together to share knowledge, resources, and best practices. Collaborative research initiatives can lead to innovative solutions that address the multifaceted challenges of mining, including environmental degradation, social inequities, and economic pressures. By pooling expertise and resources, stakeholders can develop comprehensive strategies that not only enhance mining operations but also contribute to broader sustainability goals.

The Importance of Monitoring and Accountability

Monitoring and accountability mechanisms are crucial for ensuring that mining companies adhere to sustainable practices and fulfill their commitments to local communities and the environment. Implementing robust monitoring systems that track environmental impacts, resource usage, and community engagement can provide valuable insights into the effectiveness of mining operations. These systems should be transparent and accessible to all stakeholders, allowing for independent assessments and public scrutiny. By fostering a culture of accountability, the mining industry can build trust with communities and demonstrate its commitment to responsible resource management.

Leveraging Technological Advancements for Sustainability

As technological advancements continue to reshape the mining landscape, it is vital to leverage these innovations for sustainability. For example, the use of blockchain technology can enhance supply chain transparency, ensuring that minerals are sourced responsibly and ethically. Additionally, advancements in artificial intelligence and machine learning can optimize operational processes, reducing waste and energy consumption. By embracing these technologies, the mining sector can not only improve its efficiency but also minimize its environmental impact and enhance its social license to operate.

Fostering a Culture of Sustainability

Creating a culture of sustainability within the mining industry requires a shift in mindset at all levels of operation. Companies must prioritize sustainability in their corporate strategies, integrating it into their core values and decision-making processes. This includes investing in employee training and development programs that emphasize the importance of sustainable practices and environmental stewardship. By fostering a culture that values sustainability, mining companies can empower their workforce to contribute to innovative solutions and drive positive change within the industry.

Conclusion: A Collective Responsibility

In conclusion, the path toward a sustainable future in mining and geology is a collective responsibility that requires the commitment and collaboration of all stakeholders. By embracing emerging technologies, prioritizing socio-environmental considerations, and fostering a culture of sustainability, the mining industry can navigate the complexities of modern resource extraction while contributing positively to society and the environment.

As we move forward, let us remain steadfast in our commitment to leveraging geological knowledge and innovative practices to create a resilient and sustainable future. Together, we can harness the power of collaboration, transparency, and accountability to address the pressing challenges of our time, ensuring that both people and the planet thrive for generations to come. The journey toward sustainability in mining is not just an industry imperative; it is a shared vision for a better world, where responsible resource management and environmental stewardship go hand in hand.

Emphasizing the Role of Education and Capacity Building

To sustain the momentum toward a more responsible mining industry, education and capacity building must be prioritized. Educational institutions should develop curricula that focus on sustainable mining practices, environmental science, and community engagement. By equipping the next generation of geologists, engineers, and industry leaders with the knowledge and skills necessary to address sustainability challenges, we can foster a workforce that is committed to responsible resource management.

Moreover, ongoing professional development for current industry professionals is essential. Workshops, seminars, and training programs can help mining personnel stay informed about the latest advancements in technology and sustainability practices. By investing in human capital, the mining industry can ensure that its workforce is prepared to implement innovative solutions and adapt to changing environmental and social expectations.

Encouraging Public Participation and Advocacy

Public participation is a critical component of sustainable mining practices. Mining companies should actively engage with local communities, stakeholders, and advocacy groups to solicit feedback and address concerns. This engagement can take many forms, including public forums, community meetings, and collaborative decision-making processes. By fostering an inclusive dialogue, mining companies can better understand the needs and priorities of the communities they impact, leading to more equitable and sustainable outcomes.

Additionally, advocacy for responsible mining practices should extend beyond local communities to include national and international platforms. By participating in global discussions on sustainability, mining companies can contribute to the development of best practices and standards that promote responsible resource extraction worldwide. This collective effort can help raise awareness of the importance of sustainable mining and encourage other industries to adopt similar practices.

The Role of Innovation in Addressing Climate Change

As the world grapples with the impacts of climate change, the mining industry has a unique opportunity to lead by example. Innovations in mining technology can significantly reduce greenhouse gas emissions and enhance energy efficiency. For instance, the adoption of renewable energy sources, such as solar and wind power, can help mining operations transition away from fossil fuels. Additionally, advancements in carbon capture and storage technologies can mitigate the environmental impact of mining activities.

By prioritizing climate resilience and sustainability, the mining industry can not only reduce its carbon footprint but also contribute to global efforts to combat climate change. This proactive approach can enhance the industry's reputation and strengthen its social license to operate, as stakeholders increasingly demand accountability and action on climate issues.

Conclusion: A Vision for Sustainable Mining

In conclusion, the future of mining and geology depends on our commitment to sustainability as a fundamental principle. The mining industry has faced significant criticism for its environmental impacts and social consequences, but by integrating emerging technologies, fostering collaboration, and prioritizing socio-environmental considerations, it can effectively navigate the complexities of modern resource extraction while making positive contributions to society and the environment. Innovations such as remote sensing, artificial intelligence, and blockchain technology can enhance operational efficiency, reduce waste, and improve transparency in supply chains. For example, remote sensing technologies enable real-time monitoring of land use changes and environmental impacts, allowing companies to respond swiftly to potential issues. Artificial intelligence can optimize resource extraction processes, minimizing energy consumption and waste generation, while blockchain can ensure traceability in the supply chain, confirming that minerals are sourced responsibly and ethically. By adopting these technologies, mining companies can significantly reduce their ecological footprint and ensure that their operations are both economically viable and environmentally responsible.

As we envision the future, we must commit to a shared vision of sustainable mining—one that respects the rights of local communities, protects the environment, and promotes social equity. Achieving this vision requires a fundamental shift in how mining companies engage with stakeholders, including indigenous populations, local governments, and environmental organizations. By fostering open dialogue and collaboration, the industry can better understand the needs and concerns of those affected by mining activities, which is crucial for identifying potential conflicts and finding mutually beneficial solutions. Implementing fair compensation practices and ensuring that local communities benefit from mining operations can help build trust and support for sustainable practices. This commitment to social responsibility is essential for creating a mining sector that not only extracts resources but also uplifts communities and contributes to their long-term well-being. Additionally, investing in community development initiatives—such as education, healthcare, and infrastructure—can further enhance the positive impact of mining operations on local populations.

Together, we can harness the power of geology and technology to address the pressing challenges of our time, ensuring that both people and the planet thrive for generations to come. The path forward is clear: by prioritizing sustainability, transparency, and community engagement, we can transform the mining industry into a model of responsible resource management that benefits everyone. This transformation will require ongoing education and training for industry professionals, as well as the development of policies that incentivize sustainable practices. Educational programs focused on sustainable mining practices, environmental stewardship, and community engagement can equip the next generation of mining professionals with the skills and knowledge necessary to drive change. Furthermore, governments and regulatory bodies must establish frameworks that encourage sustainable practices, such as tax incentives for companies that invest in green technologies or penalties for those that fail to meet environmental standards.

By working collaboratively across sectors and disciplines, we can create a mining landscape that not only meets the demands of a growing global population but also safeguards the environment and promotes social equity. This collaborative approach can involve partnerships between mining companies, governments, non-governmental organizations, and academic institutions to share knowledge, resources, and best practices. In doing so, we will pave the way for a future where mining is synonymous with stewardship, innovation, and resilience. Ultimately, the vision for sustainable mining is not merely about minimizing harm; it is about creating a positive legacy that enhances the quality of life for all stakeholders involved. By embracing this vision, we can ensure that the mining industry plays a vital role in building a sustainable future—one that balances economic growth with environmental protection and social responsibility.

CHAPTER 11

ROCK BLASTING CONSTRAINTS

Introduction

Blasting is a critical operation in mining and construction that involves the use of explosives to fragment rock masses for easier handling and processing. Effective blast design is essential for achieving optimal fragmentation, minimizing environmental impacts, and ensuring safety. This chapter explores the various constraints and considerations that must be taken into account when designing a blast, including the characteristics of the rock mass, properties of the explosives, initiation techniques, material size, haulage logistics, safety measures, and environmental requirements.

11.1. Background Knowledge for Blast Design

Before delving into the specifics of blast design, it is essential to understand the fundamental principles that govern blasting operations. A successful blast requires a comprehensive understanding of the geological and physical properties of the rock mass, the behavior of explosives, and the desired outcomes of the blasting operation. Knowledge of these principles allows for the optimization of blast parameters to achieve the best possible results while minimizing risks and environmental impacts. This foundational knowledge serves as the bedrock for effective blast design, ensuring that all subsequent decisions are informed by a thorough understanding of the underlying science.

11.2. Characteristics of the Rock Mass

The geological characteristics of the rock mass play a pivotal role in determining the blast design. Key factors include rock type, rock structure, and in-situ stress conditions. Different rock types, such as igneous, sedimentary, and metamorphic, exhibit varying degrees of hardness, density, and fracture patterns, which influence how they respond to blasting. For instance, igneous rocks may require different explosive energy compared to softer sedimentary rocks.

Additionally, the presence of joints, faults, and bedding planes can significantly affect the propagation of shock waves and the fragmentation of the rock. Understanding the orientation and spacing of these features is crucial for effective blast design, as they can either enhance or hinder the effectiveness of the blast. Furthermore, the existing stress state within the rock mass can impact the effectiveness of the blast. High-stress conditions may require adjustments in explosive loading and timing to achieve the desired fragmentation. By thoroughly assessing these geological characteristics, blast designers can tailor their approaches to maximize efficiency and minimize unintended consequences.

11.3. Properties of Explosives

The selection of explosives is a critical aspect of blast design. Key properties to consider include the velocity of detonation (VOD), density, water resistance, and sensitivity. The velocity of detonation refers to the speed at which the detonation wave travels through the explosive material, affecting the energy release and fragmentation pattern. Higher VOD explosives typically produce finer fragmentation, which can be advantageous in certain applications where smaller rock sizes are desired.

The density of the explosive influences its energy output, while water resistance is essential for maintaining performance in wet conditions. Explosives with higher density generally provide more energy, which can be beneficial in harder rock formations. Sensitivity is another crucial factor, as it determines how easily an explosive can be initiated. This is particularly important in areas where safety is a concern, as more sensitive explosives may pose a higher risk of accidental detonation. Selecting the appropriate sensitivity level is crucial for ensuring safe blasting operations, as it directly impacts the overall safety and effectiveness of the blast.

11.4. Initiation Techniques

The method of initiating explosives is crucial for controlling the timing and sequence of the blast. Common initiation techniques include shock tube systems, electronic detonators, and non-electric systems. Shock tube systems utilize a thin plastic tube filled with a reactive powder to transmit a signal from a detonator to the explosive charge. They are widely used for their reliability and safety, allowing for precise control over the initiation sequence.

Electronic detonators offer advanced capabilities, allowing for precise timing and sequencing of blasts. This enables more complex blast designs that can optimize fragmentation and minimize environmental impacts. The ability to program electronic detonators to initiate at specific times enhances the effectiveness of the blast and allows for better control over the resulting fragmentation. Non-electric systems are employed in environments where electrical initiation poses a risk, such as in the presence of flammable gases. These systems provide a safer alternative while still allowing for effective blast initiation, ensuring that safety remains a top priority in blasting operations.

11.5. Size of Material to be Fragmented

The desired size of the fragmented material is a key consideration in blast design. Factors influencing this decision include processing requirements and haulage considerations. The size of the material must align with the requirements of the crushing and processing plant. Oversized fragments can lead to increased wear on equipment and higher operational costs, while undersized fragments may not meet processing specifications.

Moreover, the size of the fragmented material affects haulage logistics. Larger fragments may require more robust haulage equipment and can increase transportation costs. Properly sizing the material can enhance efficiency in both processing and transportation, ultimately contributing to the overall success of the operation. By carefully considering the desired size of the fragmented material, blast designers can ensure that the resulting product meets operational needs while minimizing costs and maximizing efficiency.

11.6. Haulage of Material to the Crushing Plant

Efficient haulage of fragmented material is essential for maintaining productivity and minimizing costs. Key considerations include transport routes and material handling. The design of transport routes should consider the size and weight of haulage vehicles, as well as the potential for congestion and delays. Optimizing transport routes can reduce travel time and improve overall efficiency, ensuring that fragmented material reaches the crushing plant in a timely manner.

Additionally, the method of loading and unloading fragmented material must be efficient to minimize downtime and ensure a smooth flow of material to the crushing plant. Implementing effective material handling systems can enhance operational efficiency, reducing the time and resources required for transporting fragmented material. By focusing on efficient haulage practices, operations can maintain productivity and reduce costs associated with transportation and processing.

11.7. Safety Measures Involved

Safety is paramount in blasting operations. Key safety measures include blast area security, monitoring and communication, and training and protocols. Establishing a secure blast zone and ensuring that all personnel are at a safe distance during detonation is critical to preventing accidents. Clear signage and barriers should be used to delineate the blast area, ensuring that unauthorized personnel do not enter during blasting operations.

Implementing monitoring systems to detect any anomalies during blasting and maintaining clear communication among team members can enhance safety. Regular safety briefings and drills can prepare personnel for emergency situations, ensuring that everyone is aware of their roles and responsibilities. Furthermore, regular training for personnel involved in blasting operations and adherence to established safety protocols are essential for minimizing risks. Continuous education on safety practices can help reinforce a culture of safety within the organization, ultimately contributing to safer blasting operations.

11.8. Environmental Requirements

Blasting operations can have significant environmental impacts, and it is essential to consider these factors in the blast design. Key environmental requirements include flyrock control, noise management, vibration control, and wall damage prevention. Measures must be taken to minimize the risk of flyrock, which can pose hazards to personnel and infrastructure. This includes proper blast design, such as the use of appropriate stemming materials, controlled explosive loading, and the establishment of safe blast zones. Additionally, monitoring the blast area for potential hazards and ensuring that all personnel are at a safe distance are critical components of flyrock management.

Noise generated by blasting operations can impact nearby communities and wildlife. Strategies to mitigate noise include controlled blasting techniques, timing of blasts, and the installation of noise barriers. Using techniques such as pre-splitting or smooth blasting can help reduce noise levels by minimizing the energy released during detonation. Scheduling blasts during times when fewer people are present, such as during the day, can also help reduce the impact of noise on local communities. Furthermore, the installation of barriers or buffers around the blast site can help absorb and deflect sound waves, further reducing noise pollution.

Ground vibrations generated by blasting can affect nearby structures and ecosystems. To manage vibrations, monitoring systems can be installed to assess the impact of blasting on surrounding areas and ensure compliance with regulatory limits. Modifying the charge weight, timing, and pattern of the blast can help control vibration levels and minimize damage to nearby structures. Protecting the integrity of surrounding rock walls and structures is essential to prevent damage, which can be achieved through controlled blasting techniques and regular inspections of nearby structures and geological features before and after blasting.

11.9. Importance of Proper Fragmentation

Proper fragmentation is a critical outcome of effective blast design, directly influencing the efficiency and cost-effectiveness of subsequent operations in mining and construction. The size and distribution of fragmented material can significantly impact loading, hauling, crushing, and grinding processes, ultimately determining the overall success of the operation. Achieving the desired fragmentation not only enhances operational efficiency but also reduces environmental impacts by minimizing waste and optimizing resource use (Figure 18).

In summary, effective blast design requires a comprehensive understanding of various constraints and considerations, including the characteristics of the rock mass, properties of explosives, initiation techniques, material size, haulage logistics, safety measures, and environmental requirements. By addressing these factors, blasting operations can be conducted safely, efficiently, and with minimal environmental impact, contributing to the overall success of mining and construction projects. The integration of these considerations into blast design is essential for achieving optimal results while ensuring the safety of personnel and the protection of the environment.

.

Figure 18: Soapstone Material – Boulder size

11.9.1Optimized Loading

Efficient loading is a crucial aspect of the blasting process, as it directly impacts the overall productivity of mining and construction operations. Properly fragmented rock allows for more efficient loading into haul trucks or conveyor systems. When the material is of the right size, it can be loaded quickly and effectively, reducing the time and effort required for this operation. Conversely, oversized fragments can lead to increased wear on loading equipment and may necessitate additional handling to break them down further. This not only prolongs the loading process but also increases operational costs due to the need for extra equipment or labor. Moreover, when fragmentation is optimal, the loading equipment operates more smoothly, leading to less wear and tear on machinery. This reduction in wear translates to lower maintenance costs and an increased lifespan for the equipment. By ensuring that the rock is fragmented to the appropriate size, operators can enhance the efficiency of the loading process, ultimately contributing to a more streamlined operation and improved profitability.

11.9.2 Efficient Hauling

The efficiency of hauling fragmented material is significantly influenced by the size of the fragments. Transport logistics play a vital role in determining the choice of haulage equipment and the overall logistics of transporting the material to processing facilities. Properly sized fragments can be efficiently transported using standard haul trucks, while oversized or irregularly shaped fragments may require specialized equipment, which can increase operational costs. This consideration is essential for optimizing the entire supply chain from the blast site to the processing plant. Cost-effectiveness is another critical factor in efficient hauling. When well-fragmented material is transported, it reduces fuel consumption and transportation costs. Consistent fragment sizes allow for optimal loading of haul trucks, maximizing payloads and minimizing the number of trips required. This efficiency not only lowers operational costs but also contributes to a reduced environmental footprint by decreasing fuel usage and emissions associated with transportation.

11.9.3 Effective Crushing

The efficiency of the crushing process is heavily influenced by the size and shape of the fragmented material. Properly sized fragments can be fed directly into crushers without the need for further reduction, leading to lower energy consumption and increased throughput. When the material is already at an optimal size, the crushing equipment can operate more effectively, resulting in a smoother and more efficient processing operation. Additionally, the quality of end products is directly affected by the size distribution of the fragmented material. For example, in aggregate production, achieving the desired gradation of material is essential for meeting specifications for concrete and asphalt. Proper fragmentation ensures that the right proportions of different sizes are available for blending, which is crucial for producing high-quality construction materials. This attention to detail in the fragmentation process ultimately enhances the overall quality of the final product.

11.9.4 Grinding Processes

In mineral processing, the size of the material entering the grinding circuit significantly impacts energy consumption. Properly fragmented material requires less energy to grind to the desired size, leading to cost savings and improved efficiency in the milling process. When the material is already at an appropriate size, the grinding equipment can operate more effectively, reducing the overall energy requirements for processing. Moreover, the effectiveness of grinding operations is also influenced by the size of the feed material. Proper fragmentation can enhance the liberation of valuable minerals, leading to higher recovery rates and improved overall processing efficiency. By ensuring that the material is adequately fragmented before entering the grinding circuit, operators can maximize the recovery of valuable resources, ultimately improving the economic viability of the operation.

11.10. Material Size Considerations

The desired size of the fragmented material varies depending on the intended use, and different applications require specific size distributions. Understanding these requirements is essential for optimizing the blasting process and ensuring that the resulting material meets the necessary specifications. For aggregate production, the size of the fragmented material is critical. Gravel-sized fragments are typically required for use in concrete, asphalt, and other construction applications. The ideal size range for gravel aggregates is usually between 4.75 mm and 19 mm, depending on the specific requirements of the project. Proper fragmentation ensures that the gravel produced meets the gradation specifications necessary for optimal performance in construction applications. Well-graded aggregates provide better workability, strength, and durability in concrete mixes, making it essential to achieve the desired size distribution during the blasting process.

For dimension stone applications, such as granite or marble, larger boulder-sized fragments are often required. These fragments are typically cut and shaped into specific dimensions for use in building facades, countertops, and other architectural features. In ornamental stone applications, the size and aesthetic quality of the boulders are crucial. Proper fragmentation allows for the extraction of high-quality stone that can be polished and finished for decorative purposes. Ensuring that the boulders are of the appropriate size and quality is essential for meeting the demands of the market and achieving customer satisfaction.

11.10.1 Degree of Fragmentation

The degree of fragmentation refers to the extent to which the rock mass is broken down into smaller, manageable pieces during a blast. Achieving the desired degree of fragmentation is crucial for the efficiency of subsequent operations, including loading, hauling, and processing. Several factors influence the degree of fragmentation achieved during blasting operations. The choice of explosive type and quantity directly affects the energy released during detonation. Higher energy explosives can produce finer fragmentation, while lower energy explosives may result in larger fragments. The quantity of explosives used must be carefully calculated to achieve the desired fragmentation without causing excessive flyrock or vibrations.

Blast design parameters also play a significant role in determining the degree of fragmentation. Key parameters include burden, spacing, and stemming. Burden is the distance from the blast hole to the nearest free face, and a well-calculated burden ensures that the explosive energy is effectively utilized to break the rock. Spacing refers to the distance between adjacent blast holes, and proper spacing allows for optimal energy transfer and reduces the risk of unfragmented rock. Stemming is the material used to fill the upper portion of the blast hole, and adequate stemming helps contain the explosive gases, enhancing fragmentation. The inherent properties of the rock mass, including its hardness, density, and structural features, also influence how well it fragments. For instance, highly fractured or weathered rock may fragment more easily than dense, intact rock. Understanding these characteristics is essential for optimizing blast design and achieving the desired degree of fragmentation.

The degree of fragmentation can be assessed using various methods, including visual inspection, sieve analysis, and digital imaging. Visual inspection involves observing the size distribution of the fragmented rock immediately after blasting, providing a quick assessment of fragmentation quality. However, this method is subjective and may not provide precise data. Sieve analysis allows for a quantitative determination of the size distribution by collecting samples of the blasted material and conducting sieve tests. This method provides a more accurate assessment of fragmentation and can help identify whether the desired size distribution has been achieved. Advanced imaging techniques can also be employed to analyze the size and shape of fragments, providing detailed data on fragmentation quality. This method can offer insights into the uniformity of fragmentation and help identify areas for improvement in blast design.

11.11. Typical Layout

The layout of a blast site is critical for ensuring safety, efficiency, and effectiveness. A well-planned layout considers the geological characteristics, operational requirements, and safety protocols. The arrangement of blast holes is typically designed in a grid or staggered pattern, depending on the rock mass characteristics and desired fragmentation. Common patterns include square or rectangular patterns, which are often used in uniform rock masses to achieve consistent fragmentation, and triangular patterns, which are useful in irregular rock formations to optimize energy distribution. The depth and diameter of blast holes must also be determined based on the rock type and the desired fragmentation. Properly designed blast hole layouts contribute to the overall effectiveness of the blasting operation.

Safety is a paramount concern in blasting operations. Establishing buffer zones around the blast area is essential for protecting personnel and equipment from flyrock and vibrations. These zones should be clearly marked and monitored during blasting operations to ensure that unauthorized personnel do not enter the area. Additionally, the layout should include safe access routes for personnel and equipment, ensuring that all workers can evacuate the area quickly in case of an emergency. Proper planning and adherence to safety protocols are critical for minimizing risks associated with blasting operations.

11.12. Blasting Sequence

The blasting sequence refers to the timing and order in which individual blast holes are detonated. A well-planned blasting sequence is crucial for controlling fragmentation, minimizing vibrations, and reducing environmental impacts. Sequential blasting involves detonating holes in a specific order to control the direction of rock movement and optimize fragmentation. This technique can help reduce ground vibrations and flyrock by allowing for a more gradual release of energy. By controlling the timing of detonations, operators can achieve a more uniform fragmentation pattern, leading to better processing efficiency and reduced operational costs. The use of delays between detonations can enhance fragmentation and control the overall impact of the blast. Delays can be achieved using electronic detonators or shock tube systems, allowing for precise control over the timing of each blast hole. This level of control is essential for optimizing the effectiveness of the blast and minimizing potential negative impacts on the surrounding environment.

A carefully designed blasting sequence can help minimize environmental impact by reducing noise, vibrations, and flyrock, thereby lessening the impact on surrounding communities and ecosystems. Improved fragmentation is another significant benefit of a well-planned sequence. By controlling the timing and order of detonations, operators can achieve a more uniform fragmentation pattern, which leads to better processing efficiency and reduced operational costs.

11.13. Geological Influence on Design

The geological characteristics of the rock mass have a profound impact on blast design. Understanding these characteristics is essential for developing effective blasting strategies that optimize fragmentation and minimize risks. In the context of blasting operations, understanding the geological characteristics of the rock mass is crucial for effective blast design. Weak zones within the rock can significantly influence fragmentation, stability, and overall success of a blast. These weak zones may arise from various geological features, including tectonic joints, cooling joints, natural fissures and cracks, and previous blasting practices. Tectonic activity can create joints in the rock mass, which are fractures that occur due to stress and strain in the Earth's crust. These joints can serve as planes of weakness, allowing for easier fracturing during blasting. The orientation and spacing of these joints can affect how the explosive energy is distributed and how the rock will fragment. In igneous rocks, cooling joints form as the molten rock cools and contracts, creating natural planes of weakness that may influence the direction of rock movement during a blast. Understanding the orientation of these cooling joints is essential for optimizing blast design to ensure effective fragmentation and minimize unwanted rock displacement.

Natural fissures and cracks can occur due to weathering, erosion, or other geological processes. These features can create zones of weakness that may lead to preferential fracture patterns during blasting. Identifying and mapping these fissures is important for predicting how the rock will respond to explosive forces. Additionally, historical blasting activities can contribute to the formation of weak zones. The residual effects of previous blasts may create additional fractures or alter the existing geological structure, leading to planes of weakness that can be exploited or avoided in future blasting operations.

The presence of weak zones in the rock mass has several implications for blast design. Weak zones can lead to preferential fracture orientations, meaning that the rock may fracture more easily along certain planes. This can be advantageous for achieving desired fragmentation but may also pose risks if not properly accounted for. Understanding the orientation of these weak zones relative to the blasthole's axis is essential for optimizing blast parameters. Furthermore, the presence of weak zones can affect how explosive energy is distributed throughout the rock mass. If the energy is concentrated along weak planes, it may lead to excessive fragmentation in some areas while leaving other areas unfragmented. This uneven distribution can impact the efficiency of loading and hauling operations.

Weak zones can also influence the stability of the rock mass after blasting. If a blast creates significant fractures along weak planes, it may lead to instability in the surrounding rock, increasing the risk of rockfalls or other hazards. Proper assessment of weak zones is crucial for ensuring the safety of personnel and equipment. The identification of weak zones can inform the placement of blast holes. By strategically locating blast holes in relation to these zones, operators can enhance fragmentation and control the direction of rock movement, leading to more efficient and effective blasting outcomes.

In summary, geological influences, particularly weak zones, play a critical role in the design and execution of blasting operations. Understanding the nature and orientation of these weak zones allows for more effective blast design, optimizing fragmentation, enhancing safety, and improving overall operational efficiency. A thorough geological assessment is essential for identifying these features and integrating them into the blast design process. By considering geological characteristics and their implications, operators can develop blasting strategies that maximize productivity while minimizing risks and environmental impacts. This comprehensive approach to blast design not only improves operational outcomes but also contributes to the sustainability of mining and construction practices.

11.13.1 Rock Mass Properties

The properties of the rock mass play a crucial role in determining the effectiveness of blasting operations. Among these properties, strength and hardness are particularly significant. The compressive strength and hardness of the rock influence the choice of explosives and the design of the blast. Harder rocks may require higher energy explosives and more careful planning to achieve the desired fragmentation. Additionally, the presence of natural fractures, joints, and faults can significantly affect how the rock responds to blasting. Understanding the orientation and spacing of these features is crucial for optimizing blast design and minimizing the risk of unfragmented rock.

Dynamic compressive strength is a critical parameter in the context of blasting operations. It refers to the ability of rock to withstand compressive forces generated by explosive detonations. When an explosive charge is detonated, it produces a strain wave that travels through the rock. If the intensity of this strain wave exceeds the dynamic compressive breaking strength of the surrounding rock, several significant effects occur. One of the primary consequences is the formation of a crushed rock zone around the charge. This crushed rock annulus is characterized by a significant reduction in rock integrity and strength, leading to a highly fragmented area surrounding the blast hole. The presence of this crushed rock zone is detrimental to the transmission of strain waves in the surrounding rock, as the crushed material has different mechanical properties compared to intact rock. This inefficiency can result in suboptimal fragmentation and may affect the overall effectiveness of the blast.

Determining the dynamic compressive strength of rocks in situ before conducting blasting operations is essential. This assessment provides valuable information about how the rock will respond to explosive forces and helps in selecting appropriate blasting parameters. If the rocks exhibit low compressive strength values, it is advisable to use explosives with specific characteristics. In such cases, explosives with low density and low sonic velocity should be considered, as these types generate lower strain waves, reducing the risk of exceeding the rock's dynamic compressive strength and minimizing the formation of a detrimental crushed zone.

Understanding the dynamic compressive strength of the rock allows for better optimization of fragmentation. By selecting the right explosive type and charge weight, operators can achieve the desired fragmentation while minimizing the formation of crushed zones. Additionally, using explosives that are too powerful for the rock's compressive strength can lead to excessive fragmentation and instability in the surrounding rock mass, posing safety risks such as rockfalls and other hazards. Proper assessment of rock strength is crucial for ensuring safe blasting practices. Furthermore, selecting the appropriate explosive based on the dynamic compressive strength of the rock can lead to more efficient blasting operations. By avoiding over-blasting, operators can reduce costs associated with excessive fragmentation and improve the overall efficiency of loading and hauling processes.

Elastic Modulus

The elastic modulus, also known as the modulus of elasticity, is a fundamental property that describes the behavior of rocks under stress. It quantifies the relationship between stress (force per unit area) and strain (deformation) in a material. In the context of blasting operations, understanding the elastic modulus of the rock is crucial for optimizing blast design and achieving desired fragmentation. The elastic modulus can be determined using sonic techniques, which involve measuring the velocity of sound waves traveling through the rock. These measurements allow for the calculation of dynamic moduli, including the dynamic Young's modulus, which reflects the rock's response to rapid loading conditions, such as those produced by blasting.

Research has shown that the explosion pressure generated by a blast should not exceed 5% of the dynamic Young's modulus of the rock to achieve optimal results. This threshold is critical for ensuring that the rock responds appropriately to the explosive forces without causing excessive fragmentation or instability. After detonation, the gas pressure generated within the blasthole exerts force on the surrounding rock. This pressure acts on the cracks and fractures between the blasthole and the free face, pushing the burden (the rock mass above the blast) forward and producing heave. Understanding the elastic modulus helps predict how the rock will behave under this pressure, allowing for better control of the heave effect. The bulk modulus of a rock, which measures its resistance to uniform compression, is also essential for calculating heave. By knowing the bulk modulus, operators can estimate how much the burden will move as a result of the post-detonation gas pressure.

By understanding the elastic modulus and its relationship to explosion pressure, operators can optimize the charge weight and placement of explosives. This optimization helps ensure that the explosive energy is effectively utilized to fragment the rock without exceeding the elastic limits of the material. Proper assessment of the elastic modulus contributes to the safety and stability of the rock mass after blasting. If the explosion pressure is too high relative to the elastic modulus, it can lead to excessive heave, instability, and potential hazards such as rockfalls. Knowledge of the elastic modulus allows for better control over the fragmentation process. By tailoring the explosive parameters to the rock's elastic properties, operators can achieve more uniform fragmentation, which is essential for efficient loading and hauling operations.

Density

The density of a rock is a fundamental property that is closely correlated with its strength and overall behavior during blasting operations. Density is defined as the mass of the rock per unit volume and plays a crucial role in determining how the rock will respond to explosive forces. Understanding the density of the rock is essential for optimizing blast design and achieving effective fragmentation. Generally, an increase in rock density is associated with an increase in the rock's compressive strength. Denser rocks tend to have more tightly packed mineral grains, which contributes to their overall strength and resistance to fragmentation.

When blasting occurs, denser rocks often exhibit less displacement of the fragmented rock mass. This reduced displacement can be attributed to the rock's higher strength, which makes it more resistant to movement under explosive forces. As a result, achieving adequate displacement in denser rocks can be more challenging. To effectively fragment and displace denser rock masses during blasting, several strategies can be employed. One approach is to increase the blasthole diameter. By increasing the diameter of the blasthole, more explosive material can be placed in the hole, which can enhance the energy output of the blast. A larger blasthole diameter allows for a greater volume of explosive to be used, leading to increased energy release and improved displacement of the rock mass.

Another strategy involves reducing the blasthole pattern. A tighter blasthole pattern can increase the concentration of explosive energy in a given area. By reducing the spacing between blastholes, operators can create a more effective blast that maximizes the energy transfer to the surrounding rock. This approach can help overcome the resistance of denser rocks and achieve better fragmentation and displacement. Additionally, changing to an explosive with stronger heave energy can significantly impact the displacement of denser rock. Explosives designed to produce greater gas volume and pressure can create more effective heave, pushing the burden forward and facilitating the movement of the fragmented rock mass. This adjustment can be particularly beneficial when dealing with high-density rocks that require additional energy to achieve adequate displacement.

Understanding the density of the rock allows for better optimization of blast design parameters. By tailoring the blasthole diameter, pattern, and explosive selection to the specific density of the rock, operators can enhance fragmentation and improve overall blasting efficiency. Proper assessment of rock density is essential for ensuring safety during blasting operations. If the explosive energy is insufficient to displace dense rock, it may lead to unintended consequences, such as fly rock or inadequate fragmentation, which can pose risks to personnel and equipment. Effective blast design that accounts for rock density can lead to more efficient loading and hauling operations. By achieving optimal fragmentation and displacement, operators can reduce costs associated with re-blasting or excessive handling of fragmented material.

Porosity

Porosity is a critical property of rocks that refers to the volume of voids or spaces within the rock material. It is expressed as a percentage of the total volume and plays a significant role in determining how rocks respond to blasting operations. High porosity can adversely affect the efficiency of blasting, leading to challenges in achieving optimal fragmentation. Research has shown that the lengths of strain-wave-induced cracks in highly porous rocks are significantly shorter—approximately 25% of those in non-porous rocks with identical mineralogy. This reduction in crack length indicates that the energy from the explosive is less effectively transmitted through porous rocks, leading to less efficient fragmentation.

In highly porous rocks, fragmentation is primarily achieved through heave energy rather than through the propagation of strain waves. The presence of voids within the rock structure means that the explosive energy is not fully utilized to create fractures, resulting in a reliance on the expansion of post-detonation gases to achieve rock displacement. To enhance the efficiency of blasting operations in porous rocks, it is essential to implement strategies that maximize the effectiveness of heave energy and ensure that post-detonation gases are effectively utilized. One such strategy is bottom priming, which involves placing a small charge of explosive at the bottom of the blasthole. This technique can help initiate the main charge more effectively, enhancing the energy transfer to the surrounding rock and improving the overall fragmentation process.

Adequate stemming is another critical factor in optimizing blasting in porous rocks. Stemming refers to the inert material placed above the explosive charge in the blasthole. Adequate stemming is crucial for trapping the post-detonation gases at high pressure until they have performed their task of fragmenting the rock. By preventing premature venting of gases, stemming allows for a more effective expansion of gases, which can enhance the heave effect and improve rock displacement. Understanding the porosity of the rock allows for better optimization of charge configuration and placement. By adjusting the amount of explosive and the stemming length, operators can tailor the blast design to the specific characteristics of the porous rock, maximizing fragmentation efficiency.

Proper assessment of porosity is essential for ensuring safety during blasting operations. If the explosive energy is not effectively utilized due to high porosity, it may lead to inadequate fragmentation and potential hazards, such as fly rock or unfragmented material. Effective blast design that accounts for porosity can lead to more efficient loading and hauling operations. By achieving optimal fragmentation in porous rocks, operators can reduce costs associated with re-blasting or excessive handling of fragmented material. In conclusion, porosity is a significant factor that influences the efficiency of blasting operations. The reduced effectiveness of strain-wave-induced cracking in highly porous rocks necessitates the use of strategies such as bottom priming and adequate stemming to optimize the use of heave energy. By understanding the implications of porosity on blast design, operators can enhance fragmentation, improve safety, and increase the overall efficiency of blasting operations in mining and construction projects.

Conclusion

In summary, the properties of rock mass—specifically dynamic compressive strength, elastic modulus, density, and porosity—are critical factors that influence the design and execution of blasting operations. Each of these properties plays a unique role in determining how rock will respond to explosive forces, and understanding these relationships is essential for optimizing blast design.

Dynamic compressive strength is vital for selecting appropriate explosives and charge weights, ensuring that the explosive energy is effectively utilized without causing excessive fragmentation or instability. The elastic modulus provides insights into how the rock will behave under stress, particularly in relation to explosion pressure and post-detonation gas effects. By understanding the elastic properties of the rock, operators can better control fragmentation and minimize risks associated with heave and instability.

Density is closely correlated with rock strength and behavior during blasting. Denser rocks may require specific strategies, such as increasing blasthole diameter or using explosives with higher heave energy, to achieve adequate displacement and effective fragmentation. Porosity, on the other hand, presents unique challenges, as highly porous rocks may not transmit explosive energy efficiently. Implementing techniques like bottom priming and ensuring adequate stemming can help optimize the use of heave energy in these situations.

By thoroughly assessing these rock mass properties and integrating this knowledge into blast design, operators can enhance fragmentation, improve safety, and increase the overall efficiency of blasting operations. This comprehensive understanding not only contributes to successful and responsible mining practices but also supports the sustainability of construction projects. Ultimately, the careful consideration of rock mass properties in blast design is essential for achieving optimal results while minimizing environmental impacts and ensuring the safety of personnel and equipment.

Internal Friction

Overview

Conclusion

In summary, the properties and characteristics of rock mass, including internal friction, water content, in situ stress, and bedding, are critical factors that significantly influence the effectiveness of blasting operations. Each of these elements plays a unique role in determining how rocks respond to explosive forces, affecting fragmentation, stability, and overall blast performance.

Internal Friction is essential for understanding how strain waves propagate through rock. Factors such as porosity, permeability, jointing, and rock type can significantly impact internal friction, which in turn affects the selection of explosives and blast design. High internal friction may necessitate the use of more powerful explosives to achieve effective fragmentation, while water-filled pores can reduce internal friction and enhance energy transfer.

Water Content is another critical factor that influences rock behavior during blasting. While water saturation can increase the velocity of strain wave propagation, it can also weaken the rock structure, leading to challenges in achieving effective fragmentation. Understanding the effects of water on rock behavior is essential for optimizing blast design and ensuring safety in mining operations. Implementing effective dewatering strategies can help mitigate risks associated with water saturation, leading to improved blasting outcomes and enhanced stability of rock slopes.

In Situ Stress refers to the existing static stresses within a rock mass, which can significantly affect crack propagation and fragmentation during blasting. High in situ stresses can lead to complications such as the curvature of blastholes, closure of microcracks, and resistance to crack extension. By understanding the orientation and magnitude of in situ stresses, operators can optimize drill patterns and select appropriate explosives, enhancing safety and effectiveness in blasting operations.

Bedding and joint characteristics are also vital considerations in blast design. The orientation of bedding planes and joints can dictate how fractures propagate and how effectively explosive energy is utilized. Maximum fragmentation is typically achieved when joint planes are aligned with the free face of the rock. By tailoring blast patterns to the geological features present, operators can maximize fragmentation efficiency and minimize the risk of overbreak or instability.

Recommendations for Practice

1. **Comprehensive Geological Assessment**: Conduct thorough geological surveys to assess rock mass properties, including internal friction, water content, in situ stress, and bedding characteristics. This information is crucial for informed decision-making in blast design.

2. **Tailored Blast Design**: Utilize the insights gained from geological assessments to tailor blast designs. Adjust charge weights, explosive

types, and blasthole patterns based on the specific characteristics of the rock mass to optimize fragmentation and minimize risks.

3. **Monitoring and Adjustments**: Implement monitoring systems to assess the effectiveness of blasting operations. Use data collected from blasts to make real-time adjustments and improve future blasting strategies.

4. **Safety Protocols**: Establish safety protocols that account for the unique challenges posed by rock mass properties. Ensure that personnel are trained to recognize and respond to potential hazards associated with blasting operations.

5. **Research and Development**: Encourage ongoing research into the effects of rock mass properties on blasting efficiency. Innovations in blasting technology and techniques can lead to improved outcomes and enhanced safety in mining and construction projects.

Final Thoughts

A thorough understanding of rock mass properties and their implications for blasting operations is crucial for achieving optimal results. By incorporating insights on internal friction, water content, in situ stress, and bedding characteristics into blast planning, operators can significantly enhance fragmentation, improve safety, and increase the overall efficiency of blasting activities. This comprehensive approach not only fosters successful and responsible mining practices but also promotes the sustainability of construction projects, ensuring that operations are conducted safely and effectively.

Conclusion

In conclusion, bedding and joint characteristics are critical factors that influence the behavior of rocks during blasting operations. The orientation of bedding planes and joints plays a pivotal role in determining the effectiveness of explosive energy in achieving desired fragmentation patterns. By gaining a deeper understanding of these geological features, operators can optimize blast design, enhance fragmentation efficiency, and improve safety in both mining and construction projects. A thorough assessment of bedding and joint characteristics is essential for achieving successful and efficient blasting outcomes.

Figures

Below are Figures 19 to 22, illustrating the relationship between blasthole positioning and joint planes.

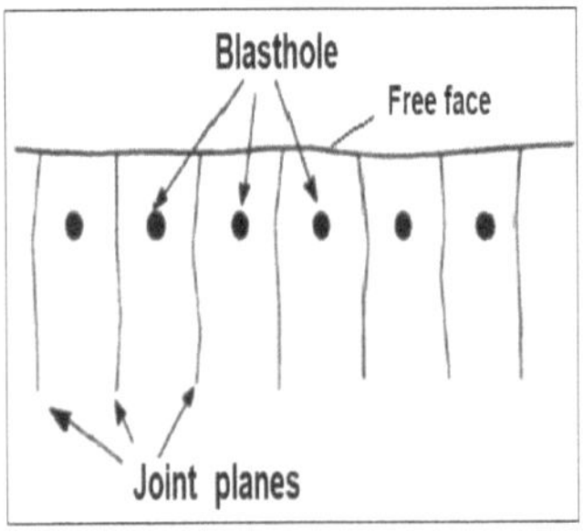

Figure 19

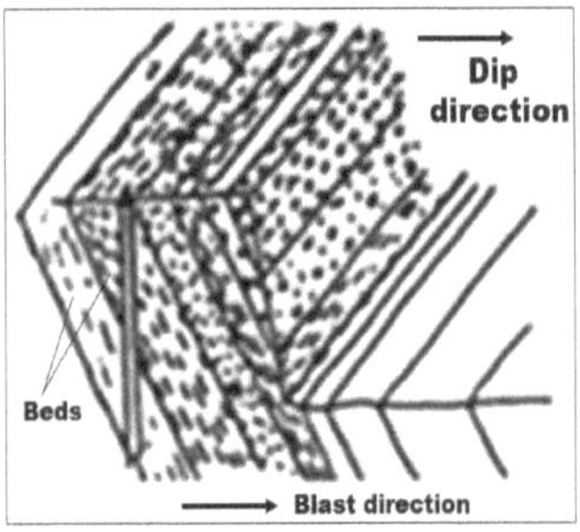

Figure 20

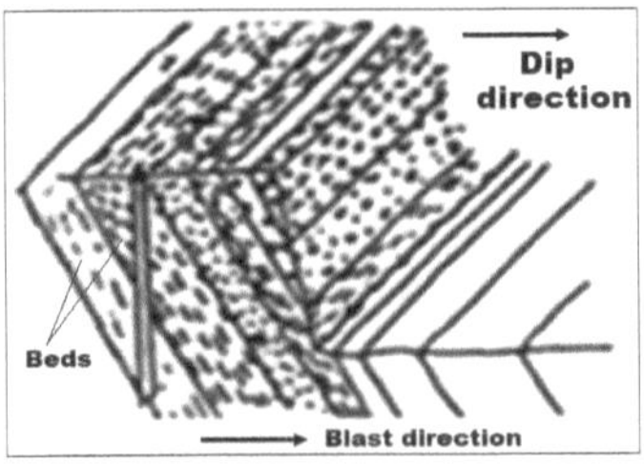

Figure 21

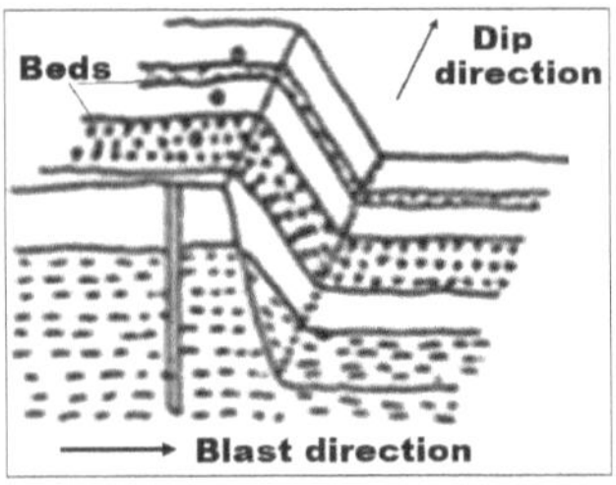

Figure 22

Bedding Orientation

Overview

The orientation of major bedding planes within a rock mass is a critical factor that can significantly influence the outcomes of blasting operations. Understanding the relationship between bedding orientation and blasting results is essential for optimizing blast design and ensuring effective fragmentation. There are three primary scenarios to consider regarding bedding orientation during blasting: shooting in the dip direction of the strata, shooting against the dip direction of the strata, and shooting along the strike of the strata.

1. Shooting in the Dip Direction of the Strata

Characteristics: When blasting is conducted in the dip direction of the strata, the explosive energy is aligned with the natural inclination of the bedding planes. This orientation often results in more back break and fewer toe problems, as the energy effectively utilizes the existing geological structure.

Implications: While this method may reduce toe issues, it can lead to the creation of an unstable highwall. The back break can compromise the integrity of the rock face, increasing the risk of rockfalls and instability. Additionally, the muckpile may be poorly displaced, resulting in more work during excavation. The irregularities in the muckpile can complicate loading and hauling operations, leading to inefficiencies in the overall process. Operators must carefully consider these factors to mitigate potential hazards and optimize the blasting outcome.

Visual Reference: Refer to Figure 7 for a visual representation of shooting in the dip direction.

2. Shooting Against the Dip Direction of the Strata

Characteristics: Blasting against the dip direction of the strata tends to result in less back break since the strata is dipping into the wall. This orientation allows for more controlled fragmentation and reduces the risk of instability, as the explosive energy is directed away from the highwall.

Implications: However, this method may lead to rougher floor conditions, as the explosive energy may not effectively displace the rock material away from the face. The muckpile may become higher, with less movement of blasted material from the face, potentially complicating subsequent excavation efforts. Operators need to balance the benefits of reduced back break with the challenges of managing muckpile characteristics to ensure efficient operations.

Visual Reference: Refer to Figure 8 for a visual representation of shooting against the dip direction.

3. Shooting Along the Strike of the Strata

Characteristics: When blasting is conducted along the strike of the strata, it often results in a highly saw-toothed profile. This irregularity can complicate the working face and affect the efficiency of subsequent operations, as the explosive energy may not be effectively utilized.

Implications: To achieve a smoother pit floor and facilitate greater movement of blasted material away from the face, it may be beneficial to reorient the working face to a more favorable direction. This adjustment can lead to a lower muckpile profile and improved excavation conditions, ultimately enhancing operational efficiency. Operators should consider the orientation of the bedding planes when planning blasts to minimize complications and optimize fragmentation.

Visual Reference: Refer to Figure 9 for a visual representation of shooting along the strike of the strata.

Conclusion

In conclusion, the orientation of bedding planes plays a significant role in determining the outcomes of blasting operations. Each orientation—shooting in the dip direction, against the dip direction, or along the strike—has distinct implications for back break, muckpile characteristics, and overall stability of the rock face. By understanding these effects, operators can optimize blast design, enhance fragmentation efficiency, and improve safety in mining and construction projects. Careful consideration of bedding orientation is essential for achieving successful and efficient blasting outcomes.

Geological Mapping

Overview

Geological mapping is a critical component of the blasting process in mining and construction operations. It involves the systematic investigation and documentation of the geological features of a site, including the distribution of rock types, structural characteristics, and in-situ stress conditions. Thorough geological mapping provides essential information that informs blast design, enhances safety, and optimizes operational efficiency.

Importance of Geological Surveys

Conducting comprehensive geological surveys is fundamental to understanding the subsurface conditions of a site. These surveys typically involve:

- **Mapping Rock Types:** Identifying and classifying the various rock types present in the area is crucial for understanding their physical and mechanical properties. Different rock types exhibit varying behaviors under stress, which can significantly influence blasting outcomes. For example, igneous rocks may have different fragmentation characteristics compared to sedimentary rocks.
- **Assessing Structural Features:** Geological mapping involves documenting structural features such as faults, joints, bedding planes, and folds. These features can act as planes of weakness or strength within the rock mass, affecting how explosive energy propagates during blasting. Understanding the orientation and density of these features is essential for optimizing blast design.
- **Evaluating In-Situ Stress Conditions:** In-situ stress refers to the existing static stresses within the rock mass prior to any excavation or blasting. Mapping these stress conditions helps identify potential challenges related to rock stability and fragmentation. High in-situ stresses can influence crack propagation and may require adjustments to blast design to ensure effective fragmentation.

Hydrogeological Survey Report for Giftmart Supermarket In Kericho County, Kenya

Location: Kapkatet Area, Kapkatet Location, Bureti Division, Bureti Sub-County, Kericho County, Kenya

This report presents the findings of a hydrogeological and geophysical survey conducted for Giftmart Supermarket on a plot of land located in the Kapkatet Area of Kericho County. The primary objective of the survey was to identify a suitable site for drilling a productive borehole to serve as the sole water source for domestic purposes. The estimated water demand for the supermarket is approximately 25 cubic meters per day (m^3/day).

The project area is situated in a zone characterized by medium to low groundwater potential. The climate in Kericho is classified as warm and temperate, with colder seasons experiencing higher rainfall. The average temperature is around 14°C, and the average annual rainfall is approximately 1951 mm. The topography of the area features undulating terrain, located on the slopes of the Mau Escarpment, which is a significant geological feature formed by the East African Rift System. The Mau Forest lies to the southeast of the project area, contributing to the region's unique ecological characteristics.

Geologically, the project area is underlain by superficial deposits of volcanic soils, which are found above layers of vitric pumice tuffs, ignimbrites, and welded tuffs. These layers are further underlain by lacustrine sediments, graded tuffs, diatomites, and phonolitic trachytes. Groundwater is typically found at the contact zones between different volcanic lava flows, at old land surfaces, and within weathered and fractured zones of the volcanic rocks. This geological setting is crucial for understanding the potential for groundwater extraction in the area.

Based on the survey findings, it is recommended that the borehole be drilled at Point 4, reaching a maximum depth of 230 meters to ensure full penetration of the deeper aquifer. The yield of surrounding boreholes in the area ranges from 2.2 to 4.7 cubic meters per hour (m^3/hr), and it is anticipated that the yield of the borehole at the recommended location will fall within this range. Careful construction and development practices will be essential to maximize productivity and ensure the longevity of the borehole.

The geophysical measurements conducted within the plot indicated that a suitable site was located where the rock is deeply weathered and fractured to depths exceeding 180 meters below ground level. This geological condition is favorable for groundwater accumulation and extraction, making it an ideal location for the proposed borehole.

To achieve the project objectives, a construction summary has been proposed. This includes drilling the borehole to a maximum depth of 230 meters at Point 4, installing appropriate casing to prevent collapse and contamination, and implementing proper development techniques to enhance yield. Additionally, water quality testing should be conducted post-drilling to ensure the safety of the water for domestic use, along with pumping tests to assess the yield and sustainability of the borehole. Establishing a monitoring plan for water levels and quality over time will also be crucial for long-term management.

In conclusion, the hydrogeological survey conducted for Giftmart Supermarket has successfully identified a suitable site for drilling a productive borehole within the Kapkatet Area. The geological and hydrological conditions indicate that a borehole drilled at the recommended location will meet the water demand for domestic purposes. By following the outlined construction summary and recommendations, the project can achieve its objectives effectively and sustainably.

Techniques for Geological Mapping

Geological mapping is a fundamental process in understanding the Earth's subsurface and is essential for various applications, including mining, construction, and environmental management. Several techniques are employed to gather data on geological features, each contributing unique insights into the composition and structure of the Earth.

Field Surveys

Field surveys are the cornerstone of geological mapping, allowing geologists to collect firsthand data on rock types, structural features, and other geological characteristics. During these surveys, geologists conduct visual inspections of outcrops, collect rock samples, and take measurements of various rock properties, such as hardness, color, and texture. This hands-on approach enables geologists to develop a comprehensive understanding of the geological landscape, identify mineral deposits, and assess the potential for natural hazards. Field surveys also facilitate the mapping of geological formations, which is crucial for planning construction projects and resource extraction.

Geophysical Methods

Non-invasive geophysical techniques play a vital role in geological mapping by providing valuable insights into subsurface conditions without the need for extensive drilling. Methods such as seismic surveys, ground-penetrating radar (GPR), and electrical resistivity imaging allow geologists to visualize the subsurface structure and identify geological features, such as faults, fractures, and stratigraphic layers. Seismic surveys utilize sound waves to detect variations in rock density and composition, while GPR employs radar pulses to image the subsurface. Electrical resistivity imaging measures the electrical conductivity of the ground, helping to delineate different geological materials. These techniques are particularly useful in areas where direct access to the subsurface is limited or where drilling may be cost-prohibitive.

Drilling and Sampling

Drilling and sampling provide a direct means of examining the rock mass, allowing geologists to obtain core samples for laboratory analysis. Core drilling involves extracting cylindrical samples of rock from various depths, which can then be analyzed to determine physical and mechanical properties, such as compressive strength, porosity, and mineral composition. This detailed examination is crucial for understanding the geological conditions at a site and for making informed decisions regarding construction and resource extraction. The data obtained from drilling and sampling can also inform geotechnical analyses, which assess the stability and behavior of the rock mass during activities such as blasting.

Geotechnical Analysis

Importance of Geotechnical Analysis

Geotechnical analysis is a critical component of geological mapping, focusing on the study of the physical and mechanical properties of rock and soil. This analysis is essential for assessing the stability of the rock mass and predicting its behavior during and after blasting operations. Key aspects of geotechnical analysis include stability assessment and understanding the behavior of the rock mass during blasting.

Stability Assessment

Evaluating the stability of the rock mass is paramount for ensuring safety during blasting operations. Geotechnical analysis can identify potential failure mechanisms, such as rockfalls or slope instability, that may arise due to blasting activities. By understanding these risks, engineers can design blasting operations that minimize the likelihood of failure and ensure the safety of personnel and equipment.

Behavior During Blasting

Understanding how the rock mass will respond to explosive forces is vital for optimizing blast design. Geotechnical analysis provides insights into factors such as fragmentation patterns, vibration propagation, and the potential for overbreak. This knowledge allows engineers to tailor blast parameters to achieve desired outcomes, such as effective fragmentation and minimal environmental impact.

Key Components of Geotechnical Analysis

Laboratory Testing

Laboratory testing of rock samples is a fundamental aspect of geotechnical analysis. Tests can determine key properties such as compressive strength, tensile strength, elasticity, and porosity. These properties are essential for predicting how the rock will behave under stress and during blasting. For example, understanding the compressive strength of a rock can help engineers determine the appropriate charge weight and blast pattern to achieve optimal fragmentation.

Numerical Modeling

Advanced numerical modeling techniques are increasingly used to simulate the behavior of the rock mass under various blasting scenarios. These models can predict fragmentation outcomes, ground vibrations, and the potential for damage to surrounding structures. By utilizing numerical modeling, engineers can refine blast designs and enhance safety measures, ultimately leading to more efficient and effective blasting operations.

Safety Measures

The insights gained from geotechnical analysis are instrumental in guiding the selection of appropriate blast parameters, such as charge weight, blast pattern, and timing. Implementing safety measures based on geotechnical findings can help mitigate risks associated with blasting operations. For instance, adjusting the timing of blasts can reduce the impact of vibrations on nearby structures, while optimizing charge weights can enhance fragmentation efficiency.

Environmental Impacts of Blasting Operations

Blasting operations can have significant environmental impacts that must be carefully considered and managed. One of the primary concerns is noise pollution. The loud noises generated during blasting can disturb local communities and wildlife, leading to prolonged exposure to high noise levels that may result in health issues for nearby residents. Additionally, the stress induced in animals can disrupt their natural behaviors and habitats, further impacting local ecosystems.

Another critical impact of blasting is vibration. The shockwaves produced by detonations can cause vibrations that may damage nearby structures, disrupt ecosystems, and affect the stability of the surrounding rock mass. It is essential to monitor and control vibration levels to minimize these risks, as uncontrolled vibrations can lead to structural damage and pose safety hazards.

Air quality degradation is also a significant concern associated with blasting operations. The detonation of explosives can release dust and gases into the atmosphere, potentially degrading air quality. This degradation can have adverse effects on human health, particularly in populated areas, as well as on the environment. The release of particulate matter and harmful gases can contribute to respiratory issues and other health problems for residents living near blasting sites.

Blasting can also have detrimental effects on local ecosystems. The process can disrupt habitats and lead to the displacement of flora and fauna. The removal of vegetation and alteration of landforms can have long-term ecological consequences, affecting biodiversity and the essential ecosystem services that these habitats provide. The loss of plant and animal species can disrupt food chains and reduce the resilience of ecosystems to environmental changes.

Furthermore, water quality can be compromised by blasting operations. Runoff from blasting sites can introduce sediments and contaminants into nearby water bodies, impacting aquatic life and overall water quality. Proper management practices are necessary to prevent pollution and protect water resources, as contaminated water can have far-reaching effects on both ecosystems and human health.

Mitigation Strategies

To address these environmental impacts, operators must implement effective mitigation strategies. One approach is noise and vibration control. Utilizing sound barriers, scheduling blasts during off-peak hours, and employing controlled blasting techniques can help minimize the impacts of noise and vibrations on surrounding communities and ecosystems.

Dust suppression is another critical strategy. The use of water sprays and dust suppressants during blasting operations can significantly reduce airborne particles, thereby improving air quality for nearby communities and ecosystems. This proactive measure helps to mitigate the health risks associated with dust exposure.

Environmental monitoring is essential for identifying potential issues early and allowing for timely interventions. Regular monitoring of air and water quality, as well as noise and vibration levels, can help operators assess the effectiveness of their mitigation strategies and make necessary adjustments to minimize environmental impacts.

Ecosystem management is also crucial in mitigating the effects of blasting. Conducting environmental impact assessments (EIAs) before blasting can help identify sensitive areas and inform strategies to protect local ecosystems. Additionally, rehabilitating disturbed areas post-blasting is vital for restoring habitats and ensuring that ecosystems can recover from the impacts of blasting activities.

By incorporating these environmental considerations into blast design and execution, operators can significantly reduce the ecological footprint of their activities. This commitment to environmental stewardship not only enhances the sustainability of blasting operations but also fosters positive relationships with local communities and stakeholders. Ultimately, responsible management practices can lead to successful outcomes in resource extraction while safeguarding the environment for future generations.

Conclusion

In conclusion, geological mapping and a comprehensive understanding of various constraints are essential for successful blasting operations in mining and construction. Thorough site investigations and geotechnical analyses provide critical insights into the geological characteristics of the rock mass, informing blast design to optimize parameters, improve fragmentation efficiency, and ensure stability during and after blasting. Proper fragmentation is vital for subsequent processes like loading, hauling, crushing, and grinding, aligning the size of fragmented material with its intended use. By integrating geological insights into blast design and selecting appropriate explosives and initiation techniques, operators can enhance efficiency, reduce costs, and improve product quality.

Incorporating safety measures and environmental considerations into blast design is crucial for promoting sustainable practices, as blasting can significantly impact the environment through noise, vibration, and air quality degradation. By assessing and mitigating these impacts, operators can minimize their ecological footprint and promote responsible resource management. Ongoing research and advancements in blasting technology, including innovations in explosives and real-time monitoring systems, will further enhance the effectiveness and safety of operations. This chapter emphasizes the importance of a multidisciplinary approach that considers geological, operational, safety, and environmental factors.

By adhering to these principles, practitioners can optimize blasting operations and contribute to responsible natural resource management, leading to successful outcomes in mining and construction while safeguarding the environment for future generations. Ultimately, the successful execution of blasting operations relies on a thorough understanding of geological conditions, careful planning, and the implementation of best practices that prioritize safety and environmental stewardship. As the industry evolves, embracing advancements in technology and methodologies will foster a culture of continuous improvement and sustainability in resource extraction and construction.

CHAPTER 12

STRATEGIC METALS, MINERALS, AND RARE EARTH ELEMENTS (REES) IN THE MINING SECTOR IN KENYA

Introduction

Kenya is known for its rich mineral resources, which include a variety of strategic metals and minerals, as well as rare earth elements (REEs). These resources play a crucial role in various industries, including technology, renewable energy, and defense. The mining sector in Kenya has the potential to significantly contribute to the country's economic growth and development.

12.1. Importance of Strategic Metals and Minerals

Strategic metals and minerals play a crucial role in the modern economy, serving as the backbone for the production of high-tech devices, renewable energy technologies, and advanced materials. Their significance is underscored by several key areas. First and foremost, technological advancements rely heavily on these materials. Metals such as lithium, cobalt, and nickel are vital for the production of batteries, particularly for electric vehicles and renewable energy storage systems. As the world shifts towards cleaner energy solutions, the demand for these metals is expected to surge, making them indispensable for future innovations.

In addition to their role in technology, strategic metals and minerals are essential for defense and aerospace applications. Certain metals, including titanium and rare earth elements, are crucial for manufacturing advanced military and aerospace technologies. These materials are used in everything from aircraft components to missile systems, highlighting their importance in national security and defense capabilities. The unique properties of these metals, such as strength, lightweight, and resistance to corrosion, make them ideal for high-performance applications.

Furthermore, the transition to a green economy is heavily reliant on strategic minerals. Renewable energy technologies, such as solar panels and wind turbines, require a variety of minerals for their production. The shift towards sustainable development necessitates a robust supply of these materials to support the infrastructure needed for clean energy solutions. As countries strive to meet their climate goals, the importance of strategic metals and minerals will only continue to grow, positioning them as critical components in the global effort to achieve sustainability.

12.2. Current State of Mining in Kenya

The mining sector in Kenya has experienced significant developments in recent years, driven by government initiatives and foreign investment. The Kenyan government has established a legal and regulatory framework aimed at promoting responsible mining practices and attracting investment. The Mining Act of 2016 is a pivotal piece of legislation that seeks to streamline licensing processes and enhance transparency within the sector. This regulatory framework is designed to create a conducive environment for both local and international investors, fostering growth and development in the mining industry.

Investment opportunities in Kenya's mining sector have attracted both local and international stakeholders, with a particular focus on the exploration and production of strategic minerals and rare earth elements (REEs). The country's rich mineral resources, including deposits of titanium, gold, and gemstones, present significant potential for economic growth. As the global demand for strategic metals and minerals continues to rise, Kenya is well-positioned to capitalize on its mineral wealth. The government's commitment to improving the mining sector's regulatory environment and promoting sustainable practices is expected to further enhance investment prospects, ultimately contributing to the country's economic development and job creation.

12.3. Key Strategic Resources in Kenya

12.3.1 Rare Earth Elements (REEs)

Occurrence:

Rare Earth Elements (REEs) are a group of 17 chemically similar elements that are critical in various high-tech applications, including electronics, renewable energy technologies, and defense systems. In Kenya, REEs are primarily found in geological formations such as the carbonatites of Mrima Hill in Kwale County. This area is particularly notable for its deposits of niobium and other rare earth minerals, which have garnered interest from both local and international mining companies. The geological setting of Mrima Hill, characterized by its unique mineralogy, provides a promising environment for the extraction of these valuable resources.

Potential: The global demand for REEs has been steadily increasing, driven by their essential role in modern technology and green energy solutions. As countries seek to reduce their reliance on traditional energy sources and transition to more sustainable alternatives, the need for REEs is expected to rise significantly. By developing its REE resources, Kenya has the potential to position itself as a key player in the global supply chain. This development could not only enhance the country's economic growth through increased foreign investment and job creation but also contribute to the diversification of its mineral resource base. Furthermore, establishing a robust REE mining sector could help Kenya leverage its strategic geographical location to serve as a hub for REE processing and distribution in East Africa.

Mrima Hill Carbonatites

Mrima Hill is not only significant for its mineral deposits but also for its unique geological features, including the presence of hot springs in the surrounding area. One of the most notable hot springs is the Dzombo hot spring, commonly referred to as 'Maji-moto hot springs,' located to the north of Mrima Hill.

Geothermal Activity: Research conducted by Tole (1990b) indicated that the Monchiquite dykes associated with Mrima Hill are linked to two of the three hot springs in the vicinity, exhibiting maximum surface discharge temperatures ranging from 56°C to 76°C (Tole, 2003). This geothermal activity is indicative of the underlying magmatic processes and the shallow depth of the magmatic bodies that form the Jombo alkali igneous complex, of which Mrima Hill is a part.

Chemical Analysis: Several studies, including those by Mulusa (2014), have undertaken chemical analyses of the hot springs to characterize the hot water and compare it with geothermal waters found in the Kenya Rift, particularly at Lake Bogoria. These analyses reveal important insights into the geochemical processes at play and the potential for geothermal energy development in the region.

Hydrothermal Fluids and Mineral Distribution: The presence of hot springs is also a manifestation of hydrothermal activity, which plays a crucial role in the distribution of minerals within the carbonatites. It is believed that hydrothermal fluids contributed to the upward movement of minerals in the Buru carbonatites, and similar processes may apply to the Jombo igneous complex and its associated formations. However, the seepage of ocean water into the Maji-moto hot springs suggests a reduction in the associated hydrothermal activities. Analyses by Mulusa (2014) indicated that approximately 3% of seawater was infiltrating the Maji-moto hot springs, which has implications for the geochemical characteristics of the hot spring waters.

Geothermometry Calculations: Geothermometry temperature calculations have shown a 5°C temperature difference between the original and recalculated Maji-moto hot spring water, indicating the influence of seawater on the thermal regime of the springs. The Na-K-Mg ternary model used in these analyses further supports the correlation between the original chemistry of the Maji-moto hot springs and seawater, highlighting the complex interactions between geothermal systems and external water sources.

Uses of Rare Earth Elements

1. Electronics

Rare Earth Elements (REEs) are integral to the production of a wide range of electronic devices that are ubiquitous in modern life. Their unique chemical and physical properties make them essential for various components. For instance, REEs are used in **capacitors**, which store and release electrical energy in electronic circuits. They help improve the efficiency and miniaturization of electronic devices, allowing for more compact designs without sacrificing performance. Additionally, **neodymium**

(Nd) and **samarium (Sm)** are key components in the production of high-performance permanent magnets. These magnets are crucial in various applications, including hard disk drives, speakers, and headphones, where strong magnetic fields are required in a small form factor.

Moreover, REEs such as **europium (Eu)** and **terbium (Tb)** are used in phosphors for displays and lighting. They are essential for producing vibrant colors in LED screens, televisions, and fluorescent lamps, enhancing the visual experience for consumers. REEs are also found in other electronic components, such as lasers, semiconductors, and optical devices, contributing to advancements in technology and communication.

2. Green Technologies

The transition to renewable energy sources has significantly increased the demand for REEs, which are vital in the development of green technologies. For example, **neodymium** and **dysprosium** are used to create high-strength permanent magnets for wind turbine generators. These magnets enable more efficient energy conversion and contribute to the overall performance of wind energy systems, making them a cornerstone of sustainable energy production.

In the realm of electric vehicles (EVs), REEs are critical in the production of electric motors. The use of neodymium magnets in these motors enhances their efficiency and power output, allowing for longer driving ranges and improved performance. Additionally, REEs are used in batteries and other components that support the electrification of transportation. Furthermore, certain REEs are utilized in the manufacturing of solar cells, improving their efficiency and energy conversion rates. This application is crucial for advancing solar technology and increasing the viability of solar energy as a mainstream power source.

3. Defense Applications

The unique properties of REEs make them indispensable in various military technologies, enhancing performance and operational capabilities. For instance, REEs are used in the guidance systems of **precision-guided munitions**, improving targeting accuracy and effectiveness. Their ability to function reliably under extreme conditions is critical for military operations.

Additionally, REEs are employed in **radar systems**, where they enhance signal processing and detection capabilities. This application is vital for surveillance, reconnaissance, and defense systems, providing military forces with a strategic advantage. REEs are also used in communication technologies, including satellite systems and secure communication devices. Their properties help improve signal clarity and transmission efficiency, which are essential for modern military operations.

4. Catalysts

Certain Rare Earth Elements play a significant role in catalysis, particularly in reducing harmful emissions from vehicles. **Cerium (Ce)** is widely used as a catalyst in automotive catalytic converters, which are essential for reducing harmful emissions from internal combustion engines. Cerium helps facilitate the conversion of toxic gases such as carbon monoxide (CO), hydrocarbons (HC), and nitrogen oxides (NOx) into less harmful substances, thereby improving air quality and meeting environmental regulations.

REEs are also used in various industrial catalytic processes, including petroleum refining and chemical manufacturing. Their ability to enhance reaction rates and selectivity makes them valuable in producing cleaner fuels and chemicals. Beyond automotive uses, REEs are being explored for their potential in environmental remediation processes, such as removing pollutants from water and air, further underscoring their importance in sustainable practices.

Conclusion

The diverse applications of Rare Earth Elements across electronics, green technologies, defense, and catalysis highlight their critical role in modern society. As technology continues to advance and the demand for sustainable solutions grows, the importance of REEs will only increase, making their responsible sourcing and development a priority for the future. The strategic management of these resources will be essential for ensuring that they can meet the needs of various industries while minimizing environmental impacts.

12.3.2 Titanium

Significance

Kenya is recognized as one of the leading producers of titanium, primarily extracted from mineral sands located along its coastal regions. The country's rich mineral deposits, particularly in Kwale County, have positioned it as a key player in the global titanium market. Titanium is a versatile metal known for its exceptional properties, including high strength, low density, and excellent corrosion resistance. These characteristics make it invaluable across various industries, including aerospace, medical, and manufacturing. The growing demand for titanium in these sectors underscores its significance to both Kenya's economy and the global supply chain.

Major Projects

One of the most notable projects contributing to Kenya's titanium production is the Kwale Mineral Sands Project, operated by Base Resources. This project involves the extraction of titanium and zircon from mineral sands, which are rich in valuable minerals. The Kwale project has not only boosted local employment and economic development but has also positioned Kenya as a reliable supplier of titanium to international markets. The success of this project has led to increased investment in mineral exploration and extraction in the region, further enhancing Kenya's role in the global titanium industry. By establishing a strong foundation for titanium production, the Kwale project exemplifies the potential for mineral resources to drive economic growth and development in Kenya.

Uses of Titanium

Titanium is extensively used in the aerospace industry due to its remarkable strength-to-weight ratio, which is crucial for improving fuel efficiency and performance in aircraft. Key applications include the manufacturing of critical aircraft components such as airframes, engine parts, and landing gear. The lightweight nature of titanium helps reduce overall aircraft weight, leading to lower fuel consumption and increased payload capacity. Additionally, titanium can withstand high temperatures, making it suitable for components exposed to extreme conditions, such as turbine blades and exhaust systems in jet engines. Its resistance to corrosion ensures the longevity and reliability of aerospace components, which is vital for safety and performance in aviation.

In the medical field, titanium's biocompatibility and resistance to corrosion make it an ideal material for various applications. It is widely used in the production of medical implants, including hip and knee replacements, dental implants, and spinal fusion devices. The compatibility of titanium with human tissue minimizes the risk of rejection and promotes healing, making it a preferred choice for medical applications. Furthermore, the durability and corrosion resistance of titanium make it suitable for surgical instruments, ensuring they maintain their integrity and performance over time. Titanium is also utilized in prosthetic devices, providing strength and lightweight properties that enhance patient comfort and mobility.

Titanium's unique properties make it valuable in various industrial sectors as well. In chemical processing, titanium is used in equipment and piping due to its resistance to corrosion from aggressive chemicals. This application is particularly important in industries such as petrochemicals and pharmaceuticals, where durability and reliability are essential. In the oil and gas sector, titanium is utilized in offshore drilling equipment and pipelines, where its strength and corrosion resistance are crucial for durability in harsh marine environments. Additionally, titanium is employed in marine applications, including shipbuilding and underwater equipment, where exposure to saltwater and other corrosive elements is a concern.

Another significant use of titanium is in pigments and coatings, particularly titanium dioxide (TiO_2), which is one of the most widely used white pigments in the world. TiO_2 serves as a key ingredient in paints and coatings, providing opacity, brightness, and durability. Its ability to reflect UV light also enhances the longevity of exterior coatings. Moreover, titanium dioxide is used in plastics and paper products to improve whiteness and brightness, making it a crucial component in consumer goods. In cosmetics and personal care products, TiO_2 serves as a pigment and provides sun protection due to its UV-blocking properties.

Conclusion

In conclusion, titanium is a critical resource for Kenya, with significant implications for various industries, including aerospace, medical, and manufacturing. The country's strategic position as a leading producer of titanium, particularly through projects like the Kwale Mineral Sands Project, highlights its potential to contribute to the global supply chain. As demand for titanium continues to grow, especially in high-tech and sustainable applications, Kenya's role in the titanium market is likely to expand. This expansion offers opportunities for economic development and technological advancement, positioning Kenya as a key player in the global titanium industry.

12.3.3 Cobalt and Lithium

Emerging Resources

Cobalt and lithium are emerging as critical resources in the context of the global transition to renewable energy and electric mobility. Both minerals are essential for the production of batteries, particularly lithium-ion batteries, which power electric vehicles (EVs), portable electronics, and renewable energy storage systems. In Kenya, potential deposits of cobalt and lithium have been identified, particularly in regions with pegmatite formations, which are known to host these valuable minerals. The exploration and development of these resources present significant opportunities for economic growth and technological advancement in the country.

Investment Interest

The global shift towards electric vehicles and renewable energy technologies has sparked considerable interest in the exploration and development of cobalt and lithium resources in Kenya. As countries and companies strive to meet climate goals and reduce carbon emissions, the demand for batteries is surging. This has led to increased investment in mining and extraction projects focused on cobalt and lithium, positioning Kenya as a potential player in the global supply chain for these critical minerals. The government and private sector are exploring partnerships and initiatives to attract investment, enhance infrastructure, and develop sustainable mining practices.

Uses of Cobalt

Cobalt is a key component in lithium-ion batteries, which are widely used in various applications. In the realm of **electric vehicles (EVs)**, cobalt enhances the performance, energy density, and stability of these batteries, making them suitable for powering electric vehicles. The inclusion of cobalt in battery cathodes improves energy storage capacity and extends battery life, which is crucial for the widespread adoption of EVs. Additionally, cobalt is utilized in batteries for **portable electronics** such as smartphones, laptops, and tablets, where high energy density and longevity are essential for consumer satisfaction. Furthermore, cobalt-containing batteries are employed in **renewable energy storage systems**, enabling the efficient storage of energy generated from solar and wind sources for later use, thereby supporting the transition to sustainable energy solutions.

In addition to its role in batteries, cobalt is utilized in the production of high-performance alloys, particularly in aerospace and industrial applications. Cobalt-based alloys are known for their exceptional strength, heat resistance, and durability, making them ideal for **aerospace applications**. These alloys are commonly used in components exposed to extreme conditions, such as turbine blades and engine parts in aircraft. Moreover, cobalt alloys find applications in various **industrial settings**, including cutting tools, wear-resistant coatings, and medical devices, where high performance and reliability are required to meet demanding operational standards.

Cobalt also serves as a catalyst in several chemical processes, contributing to advancements in energy diversification and sustainability. In the production of **synthetic fuels**, cobalt is used in the Fischer-Tropsch synthesis process, which converts natural gas into liquid hydrocarbons. This process is essential for producing synthetic fuels and chemicals, thereby supporting efforts to diversify energy sources. Additionally, cobalt catalysts are employed in various **chemical reactions**, enhancing reaction rates and selectivity, which is vital for efficient industrial processes.

Uses of Lithium

Lithium is a primary component in rechargeable lithium-ion batteries, which are critical for various applications. In the context of **electric vehicles (EVs)**, lithium-ion batteries serve as the backbone of electric mobility, providing the necessary energy storage for these vehicles. The demand for lithium is surging as automakers transition to electric vehicles to meet environmental regulations and consumer preferences. Furthermore, lithium batteries power a wide range of **portable electronic devices**, including smartphones, laptops, and tablets, where lightweight and high-capacity energy storage is essential. Lithium-ion batteries are also increasingly used in **renewable energy storage systems**, allowing for the efficient storage of energy generated from solar and wind sources, thereby enhancing grid stability and energy reliability.

Beyond batteries, lithium compounds are utilized in the production of **glass and ceramics**. Lithium is added to glass formulations to improve strength, reduce thermal expansion, and enhance durability. This is particularly important in applications such as glassware, ceramics, and tiles, where thermal shock resistance is critical. Additionally, lithium is used in the production of **specialty glass**, including glass-ceramics and high-performance glass used in electronics and optics, further showcasing its versatility.

Lithium salts have important applications in the **pharmaceutical industry** as well. Lithium carbonate is a common medication prescribed for the treatment of bipolar disorder and other mental health conditions. It acts as a mood stabilizer, helping to manage the symptoms of mood disorders and improve the quality of life for patients. Ongoing research into the therapeutic effects of lithium continues to explore its potential benefits for various neurological and psychiatric conditions, further solidifying its importance in medicine and highlighting its multifaceted role in both health and technology.

Conclusion

Cobalt and lithium are emerging as vital resources in the context of the global transition to electric mobility and renewable energy. With Kenya's potential deposits of these minerals, there is significant opportunity for economic development and investment in the mining sector. The increasing global demand for batteries, driven by the rise of electric vehicles and renewable energy storage solutions, positions Kenya favorably in the international market for cobalt and lithium.

As the country seeks to attract investment and develop its mineral resources, it is essential to implement sustainable mining practices that minimize environmental impacts and promote social responsibility. This includes engaging local communities, ensuring fair labor practices, and adhering to environmental regulations. By fostering a responsible mining sector, Kenya can not only capitalize on its mineral wealth but also contribute to global sustainability efforts.

Furthermore, the development of cobalt and lithium resources can stimulate technological innovation and infrastructure development within Kenya. Investments in mining can lead to improvements in transportation, energy, and communication infrastructure, benefiting not only the mining sector but also the broader economy.

In summary, cobalt and lithium represent significant opportunities for Kenya as it navigates the challenges and benefits of a rapidly changing global landscape. By leveraging its mineral resources, Kenya can enhance its economic prospects, support the transition to a low-carbon future, and play a crucial role in the global supply chain for these essential minerals.

12.3.4 Other Strategic Minerals

Gold

Kenya has a long-standing history of gold mining, with significant deposits located in regions such as Western Kenya, particularly in areas like Kakamega and Migori. Gold mining has been a vital part of the local economy, providing employment and contributing to national revenue through exports. Despite fluctuations in global gold prices, gold remains a valuable export commodity for Kenya, attracting both local and foreign investment. The enduring demand for gold underscores its importance as a financial asset and a driver of economic activity in the region.

Gold is highly prized in the jewelry industry due to its aesthetic appeal, malleability, and resistance to tarnish. It can be easily shaped into intricate designs and is often alloyed with other metals, such as copper or silver, to enhance its strength and durability. The demand for gold jewelry remains strong, particularly in emerging markets and among affluent consumers. Additionally, gold is considered a safe-haven asset, especially during times of economic uncertainty. Investors purchase gold in various forms, including coins, bars, and exchange-traded funds (ETFs), as a means of preserving wealth and hedging against inflation. The global demand for gold as an investment continues to drive its market value. Furthermore, gold's excellent conductivity and resistance to corrosion make it an essential material in the electronics industry, where it is commonly used in components such as connectors, switches, and circuit boards.

Cotant

Cotant, a lesser-known mineral, has garnered attention due to its potential applications and the geological significance of its deposits in Embu, Kenya. This region, characterized by its diverse geological formations, presents a unique opportunity for mineral exploration and extraction. Understanding the properties, uses, and implications of cotant is essential for assessing its role in the local economy and the broader mining sector in Kenya.

Embu County is located in the central part of Kenya and is characterized by a variety of geological formations, including volcanic rocks, sedimentary layers, and metamorphic terrains. The presence of cotant in this region suggests a complex geological history that may have contributed to its formation. Cotant is believed to form under specific conditions, often in association with other minerals. Its occurrence in Embu may be linked to hydrothermal processes, where mineral-rich fluids circulate through rock formations, depositing minerals as they cool and crystallize. Understanding the geological context of cotant deposits can aid in exploration efforts and inform mining strategies.

The discovery and potential extraction of cotant in Embu could have several economic implications for the region and the country as a whole. The mining of cotant can create job opportunities for local communities, contributing to economic development and improving livelihoods. If cotant is found to have significant commercial value, it could become an important export commodity for Kenya, with potential applications in industries such as electronics, construction, or agriculture. The presence of cotant in Embu may also attract investment from mining companies and exploration firms, leading to the development of infrastructure that benefits not only the mining sector but also other industries in the region.

While specific applications of cotant may vary based on its chemical and physical properties, potential uses could include industrial applications, agricultural uses, and technological applications. Depending on its composition, cotant may be utilized in various industrial processes, serve as a raw material in the production of ceramics or glass, and even find applications in the electronics industry if it exhibits unique electrical or thermal properties. However, the extraction of cotant must be approached with caution to minimize environmental impacts, emphasizing the need for sustainable mining practices and community involvement.

Soda Ash

The Lake Magadi area in Kenya is renowned for its production of soda ash, a key industrial mineral. Soda ash, or sodium carbonate, is extracted from the lake's alkaline waters and is used in various applications across multiple industries. One of the primary uses of soda ash is in **glass manufacturing**, where it serves as a fundamental ingredient in the production of container glass, flat glass, and fiberglass. By lowering the melting point of silica, soda ash makes the glass-making process more energy-efficient and cost-effective. The glass industry is one of the largest consumers of soda ash, and its demand is closely tied to the construction and packaging sectors.

In addition to glass manufacturing, soda ash is utilized in the production of various chemicals, including sodium bicarbonate, sodium silicate, and detergents. It serves as a pH regulator and a cleaning agent in many industrial processes, contributing to the production of household and industrial cleaning products. Furthermore, in water treatment processes, soda ash is used to soften water and adjust pH levels, helping to remove impurities and improve water quality for both industrial and municipal applications. This makes soda ash an essential component in maintaining safe and clean water supplies.

12.4. Challenges Facing the Mining Sector

Despite the potential for growth in Kenya's mining sector, several challenges hinder its development. Regulatory hurdles remain a significant issue, as bureaucratic processes can delay project approvals and licensing. Investors often face challenges navigating the regulatory landscape, which can lead to increased costs and uncertainty. Additionally, environmental concerns associated with mining activities can have significant impacts, including habitat destruction, water pollution, and soil degradation. These concerns can lead to community opposition and increased regulatory scrutiny, making it essential for mining companies to adopt sustainable practices and engage with local stakeholders. The need for responsible mining practices is paramount to ensure that the benefits of mining do not come at the expense of the environment or local communities.

Another challenge facing the mining sector in Kenya is infrastructure deficiencies. Inadequate infrastructure, including poor road networks and unreliable energy supply, can limit the efficiency of mining operations and increase operational costs. The lack of proper transportation routes can hinder the movement of minerals to markets, while unreliable energy sources can disrupt production processes. Investments in infrastructure development are crucial to support the growth of the mining sector and facilitate the transportation of minerals, ultimately enhancing the overall competitiveness of the industry.

12.5. Opportunities for Growth

Despite these challenges, the mining sector in Kenya presents several opportunities for growth. Increased investment in exploration activities can lead to the discovery of new mineral deposits, particularly for strategic metals and rare earth elements (REEs). This can enhance Kenya's position in the global mining landscape and attract foreign investment, which is essential for the sector's development. As global demand for minerals continues to rise, Kenya has the potential to become a significant player in the mining industry.

Embracing technological advancements is another opportunity for growth in the mining sector. The adoption of modern mining technologies, such as automation, data analytics, and environmentally friendly extraction methods, can enhance operational efficiency, reduce environmental impacts, and improve safety standards in mining operations. By leveraging technology, mining companies can optimize their processes, minimize waste, and ensure more sustainable practices.

Furthermore, building strong relationships with local communities and stakeholders is essential for fostering support for mining projects. Engaging communities in decision-making processes and ensuring that they benefit from mining activities can lead to more sustainable practices and a social license to operate. By prioritizing community involvement and addressing local concerns, mining companies can create a positive impact on the communities in which they operate, ultimately contributing to social and economic development.

Conclusion

In conclusion, Kenya's mining sector holds significant potential for economic growth and development, driven by strategic minerals such as gold, cotant, and soda ash. While challenges such as regulatory hurdles, environmental concerns, and infrastructure deficiencies exist, there are also numerous opportunities for growth through investment in exploration, technological advancements, and community engagement. By adopting sustainable mining practices and prioritizing the well-being of local communities, Kenya can harness its mineral wealth to foster economic development, create jobs, and enhance its position in the global mining industry. Continued research and exploration efforts will be essential to fully understand the potential of these minerals and their role in shaping the future of Kenya's mining sector.

Conclusion

Kenya's mining sector holds significant potential for the extraction of strategic metals, minerals, and rare earth elements. By addressing the challenges and leveraging opportunities, the country can enhance its position in the global mining landscape, contribute to economic growth, and support sustainable development. Continued investment, technological innovation, and responsible mining practices will be key to unlocking the full potential of Kenya's mineral resources. As the global demand for minerals continues to rise, Kenya has the opportunity to become a significant player in the mining industry, benefiting both its economy and its communities. By fostering a conducive environment for investment and ensuring that mining activities are conducted responsibly, Kenya can harness its rich mineral resources to drive economic development and improve the quality of life for its citizens.

To achieve this, the government, private sector, and local communities needs to collaborate effectively. This collaboration can lead to the establishment of policies that promote sustainable mining practices, protect the environment, and ensure that the benefits of mining are equitably shared.

Moreover, as global trends shift towards sustainability and the circular economy, Kenya can position itself as a leader in responsible mining practices. This includes exploring opportunities for recycling and reusing minerals, investing in cleaner technologies, and developing a skilled workforce that can support the mining sector's growth.

In summary, the future of Kenya's mining sector is promising, with significant opportunities for growth and development. By addressing existing challenges and capitalizing on emerging trends, Kenya can enhance its role in the global mining industry, contribute to economic resilience, and support sustainable development goals. The way forward will require commitment, innovation, and a focus on responsible practices that benefit both the economy and the environment.

Summary: Environmental Geology of Strategic Metals, Minerals, and Rare Earth Elements (REEs) in the Mining Sector in Kenya

Introduction

The mining sector in Kenya is increasingly recognized for its potential to contribute to economic growth through the extraction of strategic metals, minerals, and rare earth elements (REEs). This chapter explores the significance of these resources, their applications, and the environmental implications associated with their extraction and processing. Understanding the environmental geology of these materials is crucial for promoting sustainable mining practices and mitigating adverse impacts on ecosystems and communities. As global demand for these resources rises, particularly in the context of renewable energy and technological advancements, the need for responsible mining practices becomes paramount.

Strategic Metals and Minerals

Kenya is endowed with various strategic metals and minerals, including titanium, cobalt, lithium, gold, and soda ash. Each of these resources plays a vital role in global industries:

1. **Titanium:** Extracted primarily from mineral sands along the coast, titanium is essential in aerospace, medical implants, and as a pigment in paints. Its high strength-to-weight ratio makes it ideal for reducing aircraft weight and improving fuel efficiency.

2. **Cobalt and Lithium:** These minerals are critical for battery production, particularly in electric vehicles (EVs) and renewable energy storage systems. Cobalt enhances battery performance, while lithium is a primary component in rechargeable lithium-ion batteries. The global shift towards electric mobility has sparked interest in exploring and developing these resources in Kenya.

3. **Gold:** With a long history of gold mining, particularly in Western Kenya, gold remains a valuable export commodity. Its uses span jewelry, investment, and electronics, where it is prized for its conductivity and resistance to corrosion.

4. **Soda Ash:** The Lake Magadi area is known for its production of soda ash, which is used in glass manufacturing, chemical production, and water treatment. Soda ash plays a crucial role in various industrial applications, making it an important mineral for Kenya's economy.

Environmental Implications

The extraction and processing of strategic metals and minerals pose several environmental challenges that must be addressed to ensure sustainable development:

1. **Habitat Destruction:** Mining activities can lead to significant habitat loss and fragmentation, impacting local biodiversity. The removal of vegetation and soil can disrupt ecosystems, threaten wildlife, and lead to the extinction of vulnerable species. The alteration of land use can also affect local communities that rely on these ecosystems for their livelihoods.
2. **Water Pollution:** Mining operations often generate waste materials that can contaminate local water sources. Heavy metals, chemicals, and tailings from processing can leach into groundwater and surface water, posing risks to human health and aquatic life. The pollution of water bodies can have cascading effects on agriculture, fisheries, and drinking water supplies.
3. **Soil Degradation:** The disturbance of soil during mining can lead to erosion, loss of soil fertility, and increased sedimentation in water bodies. This degradation can affect agricultural productivity and local food security, particularly in rural areas where communities depend on farming for their livelihoods.
4. **Air Quality:** Dust and emissions from mining activities can degrade air quality, leading to respiratory issues for nearby communities and contributing to broader environmental concerns. The release of particulate matter and volatile organic compounds can have detrimental effects on public health and the environment.
5. **Community Impact:** Mining operations can lead to social conflicts, particularly when local communities feel excluded from decision-making processes or when they do not benefit from the economic gains of mining activities. Issues such as land displacement, inadequate compensation, and lack of access to resources can exacerbate tensions between mining companies and local populations.

Sustainable Mining Practices

To address these environmental challenges, the chapter emphasizes the importance of adopting sustainable mining practices. Key strategies include:

1. **Environmental Impact Assessments (EIAs):** Conducting thorough EIAs before initiating mining projects can help identify potential environmental risks and inform mitigation strategies. EIAs should

involve stakeholder consultations to ensure that community concerns are addressed.

2. **Community Engagement:** Involving local communities in the planning and decision-making processes can foster trust and ensure that their needs and concerns are addressed. Building strong relationships with stakeholders can lead to more sustainable practices and social license to operate.

3. **Rehabilitation and Restoration:** Implementing plans for land rehabilitation and ecosystem restoration post-mining can help mitigate long-term environmental impacts and promote biodiversity recovery. This includes reforestation, soil stabilization, and the restoration of natural habitats.

4. **Technological Innovation:** Embracing modern mining technologies, such as automation, data analytics, and environmentally friendly extraction methods, can enhance operational efficiency, reduce waste, and minimize environmental footprints. Innovations in processing techniques can lead to more sustainable practices and lower energy consumption.

5. **Regulatory Frameworks:** Strengthening regulatory frameworks and enforcement mechanisms is essential for ensuring compliance with environmental standards. Governments should work to streamline permitting processes while maintaining rigorous environmental protections.

6. **Sustainable Supply Chains:** Promoting sustainable supply chains for strategic metals and minerals can enhance transparency and accountability in the mining sector. This includes traceability of minerals from extraction to end-use, ensuring that they are sourced responsibly and ethically. Companies can adopt certification schemes that verify sustainable practices, thereby appealing to environmentally conscious consumers and investors.

Conclusion

The mining sector in Kenya presents significant opportunities for economic development through the extraction of strategic metals, minerals, and rare earth elements (REEs). However, the environmental implications of mining activities necessitate a careful and responsible approach. By prioritizing sustainable practices, engaging with local communities, and implementing effective environmental management strategies, Kenya can harness its mineral wealth while protecting its natural resources and promoting social equity.

The integration of environmental geology principles into mining operations will be essential for ensuring that the benefits of mineral extraction are realized without compromising the health of ecosystems and communities. As global demand for strategic metals and minerals continues to rise, Kenya has the opportunity to position itself as a leader in responsible mining practices, contributing to both local and global sustainability efforts.

In summary, the future of Kenya's mining sector is promising, with significant opportunities for growth and development. By addressing existing challenges and capitalizing on emerging trends, Kenya can enhance its role in the global mining industry, contribute to economic resilience, and support sustainable development goals. The way forward will require commitment, innovation, and a focus on responsible practices that benefit both the economy and the environment.

Through collaborative efforts among government, industry, and local communities, Kenya can unlock the full potential of its mineral resources while ensuring that environmental and social considerations remain at the forefront of mining activities. This holistic approach will not only safeguard the environment but also foster a more equitable and sustainable future for all Kenyans.

Natural Earth Materials Hosting Rare-Earth Elements

Rare-Earth Elements (REEs) are critical components in various high-tech applications, including electronics, renewable energy technologies, and advanced materials. These elements are primarily extracted from large carbonatite bodies, which are igneous rocks rich in carbonate minerals. Notable global mines for REE extraction include:

- **Mountain Pass, California, USA:** The Mountain Pass mine, operated by Molycorp, Inc., was the only active producer of REEs in the United States as of 2014. This mine is situated in a carbonatite intrusion and was the leading REE resource in the country at that time (see the quarry in Figure 23).

Figure 23: Bastnäsite mineral specimen (A) and Mount Pass Mine Oct.2020, California (B) (Adopted from USGS, 2014)

- **Bayan Obo, China:** This is one of the largest REE mines in the world, located in Inner Mongolia. It is known for its significant reserves of both rare-earth elements and iron.
- **Maoniuping, China:** Another major REE mining site, Maoniuping is recognized for its large-scale production of various rare-earth elements.
- **Daluxiang and Weishan, China:** These mines also contribute to China's dominance in the global REE market, providing substantial quantities of these critical materials.

Uses and Genesis of Rare-Earth Elements in Relation to Carbonatites of Kenya

In Kenya, significant occurrences of REEs are associated with carbonatite formations, particularly at Buru Hill, Homa Mountain, and Mrima Hill. These locations are notable for their potential to host economically viable deposits of rare-earth elements.

- **Buru Hill:** This site is characterized by its unique geological features and has been identified as a potential source of REEs.
- **Homa Mountain:** Similar to Buru Hill, Homa Mountain exhibits geological formations conducive to the presence of rare-earth elements.
- **Mrima Hill:** This location has garnered attention for its rich deposits of REEs, which are associated with the carbonatite complex.

The genesis of REEs in these carbonatite bodies is linked to the processes of magmatic differentiation and hydrothermal alteration, which concentrate these elements in economically recoverable forms. The exploration and potential mining of REEs in these areas could significantly contribute to Kenya's mineral resource sector and provide opportunities for economic development.

In Kenya, rare-earth elements (REEs) have been found to be enriched in carbonatite magmas. However, they did not receive significant attention until the past two decades, when detailed exploration efforts began on carbonatite bodies in the coastal and western regions of the country. To date, various researchers continue to undertake petrological and geochemical characterization of the Nyanza Rift carbonatites.

For instance, Onuonga (1996), as a scholar, studied carbonatites under academic scholarship and through JICA initiatives. In 2000, he performed X-ray fluorescence (XRF) analyses on drill cores obtained from Buru Hill to determine lanthanide concentrations. The results obtained from this analysis are presented in the following Table 1.2:

Table 1.2: Major Elements for Buru Hill Carbonatite (Onuonga, 1996)

Constituent	**Major Elements by Wt. (%)**								**Average (wt. (%)**
	BR-1	**BR-5**	**BR-8**	**BR-13**	**BR-16**	**BR-28**	**BR-59**	**BR-150**	
SiO_2	1.75	3.38	1.43	12.86	2.23	2.39	1.28	2.46	3.47
Al_2O_3	0.39	0.42	0.24	2.55	0.36	0.14	0.41	0.35	0.61
Fe_2O_3	30.48	62.8	59.37	14.98	21.88	19.38	38.54	48.39	36.98
MgO	0.22	0.28	0.16	2.04	1.31	2.44	2.32	1.96	1.34
CaO	26.91	2.40	7.89	28.71	36.41	26.36	11.6	8.56	18.61
Na_2O	0.12	0.06	0.06	0.85	0.27	0.38	0.16	0.26	0.27
K_2O	0.12	0.20	0.09	1.55	0.29	0.35	0.29	0.23	0.39
TiO_2	0.40	0.51	0.51	0.37	0.24	0.16	0.15	0.14	0.31
P_2O_5	1.23	0.64	0.50	0.94	0.29	2.80	0.21	0.44	0.88

MnO	7.22	9.48	8.9	2.37	2.67	4.17	5.65	5.68	5.77
Cr_2O_3	n.d	n.d	n.d	n.d	n.d	n.d	n.d	n.d	-
Total	**68.72**	**80.18**	**79.15**	**67.22**	**65.95**	**58.57**	**60.61**	**68.47**	
	n.d – not detected								

Similarly, the Buru hill carbonatite was enriched with Cr_2O_3 oxide but lacked P_2O_5 oxide while the silicate rocks were enriched with P_2O_5 oxide but lacked Cr_2O_3 oxide.

Nyanza Rift Carbonatites

The Nyanza Rift Carbonatites are notable geological formations characterized by a diverse array of primary mineral constituents, which include calcite, dolomite, ankerite, siderite, and, in rare instances, magnesite. These minerals play a crucial role in the overall composition and economic potential of the carbonatites, as they can influence the extraction and processing methods employed in mining operations. Calcite and dolomite, for example, are commonly used in various industrial applications, including construction and agriculture, while ankerite and siderite are significant sources of iron. The presence of these minerals indicates the potential for economic exploitation, but it also necessitates careful consideration of the environmental implications associated with their extraction.

In addition to the primary minerals, the Nyanza Rift Carbonatites contain variable amounts of secondary minerals, which further enrich their geological complexity. These secondary minerals include magnetite, biotite, apatite, phlogopite, K-feldspar, quartz, pyrochlore, nepheline, clinopyroxene, and amphibole. Each of these minerals contributes unique properties and potential applications, such as magnetite's use in iron production and apatite's role as a phosphate source in fertilizers. The diversity of mineral constituents not only enhances the economic viability of the carbonatites but also presents challenges in terms of processing and environmental management, as different minerals may require distinct extraction techniques and have varying environmental impacts.

Chemically, the Nyanza Rift Carbonatites are enriched with incompatible elements, including barium (Ba), strontium (Sr), niobium (Nb), thorium (Th), titanium (Ti), phosphorus (P), and rare earth elements (REEs). The presence of these elements is significant, as they are often sought after for their applications in high-tech industries, including electronics, renewable energy, and advanced materials. The enrichment of incompatible elements indicates that these carbonatites may serve as valuable resources for the extraction of critical minerals, which are essential for modern technological advancements. However, the extraction of these elements must be balanced with environmental considerations, as their mining can lead to significant ecological disturbances.

Environmental Impacts of Carbonatites

The environmental impact of mining and processing carbonate minerals, such as siderite and ankerite, can be profound and multifaceted. Habitat destruction is one of the most immediate consequences of mining activities, as the removal of vegetation and soil can lead to the loss of critical ecosystems that support diverse flora and fauna. This disruption can result in the fragmentation of habitats, making it difficult for species to thrive and reproduce, ultimately diminishing biodiversity in the affected areas.

Soil degradation is another significant concern associated with the extraction of carbonate minerals. The removal of topsoil and alteration of landforms can lead to erosion, reduced soil fertility, and impaired water retention capabilities. These changes can adversely affect agricultural productivity and the health of local ecosystems, as soil quality is essential for sustaining plant life and maintaining ecological balance.

Water contamination is a critical environmental issue linked to mining operations, particularly due to the release of heavy metals and other pollutants during the extraction and processing of carbonate minerals. Contaminated runoff can seep into local water bodies, posing risks to aquatic life and compromising the quality of drinking water for nearby communities. The introduction of pollutants can disrupt aquatic ecosystems, leading to declines in fish populations and other wildlife that depend on clean water for survival.

CHAPTER 13

FUNDAMENTALS OF GROUND INVESTIGATION

Introduction

Ground investigation is a critical process in engineering, construction, and environmental management. It involves the systematic collection and analysis of data regarding subsurface conditions to inform decision-making and design practices. The primary objectives of ground investigation are to gather essential data about the subsurface environment, assess the suitability of the ground for proposed projects, and inform design and construction practices. This chapter outlines the key objectives of ground investigation, focusing on the characteristics of the ground, variations in those characteristics, qualitative descriptions, the response of the ground to imposed conditions, and environmental issues.

13.1. Characteristics of the Ground

13.1.1 Definition of Ground Characteristics

Ground characteristics refer to the physical, chemical, and mechanical properties of the soil and rock that make up the subsurface environment. These characteristics are fundamental to understanding how the ground will behave under various conditions, including loading, moisture changes, and environmental impacts. A comprehensive understanding of ground characteristics is essential for evaluating the suitability of the ground for construction, infrastructure development, and environmental management.

The assessment of ground characteristics involves a multi-faceted approach that includes laboratory testing, field investigations, and geophysical surveys. By analyzing these properties, engineers and geologists can make informed decisions regarding site selection, design parameters, and construction methodologies.

13.1.2 Key Characteristics to Investigate

Ground investigations focus on several key characteristics that are critical for understanding subsurface conditions. These include:

Soil Type: The identification and classification of various soil types (e.g., clay, sand, silt, gravel) are fundamental to ground investigation. Each soil type exhibits distinct physical and mechanical properties, such as grain size distribution, plasticity, and compaction behavior. For instance, clay soils may have high plasticity and low permeability, while sandy soils typically exhibit good drainage characteristics. Understanding soil type is essential for predicting behavior under load and for determining appropriate construction techniques.

Rock Type: The classification of rock types (e.g., igneous, sedimentary, metamorphic) is equally important in ground investigation. Each rock type has unique mechanical properties, including compressive strength, tensile strength, and durability. For example, granite, an igneous rock, is known for its high strength and durability, making it suitable for heavy construction. In contrast, sedimentary rocks may exhibit varying degrees of weathering and porosity, influencing their stability and suitability for construction.

Moisture Content: The assessment of moisture content in soils and rocks is critical, as it significantly affects their behavior under load. Moisture content influences soil cohesion, friction angle, and overall stability. For example, saturated soils may experience reduced shear strength, leading to potential failure under load. Accurate measurement of moisture content is essential for effective design and construction practices.

Density and Porosity: The measurement of soil density and porosity provides insights into the material's ability to drain and its overall stability. Density is a key factor in determining the load-bearing capacity of soils, while porosity affects water retention and movement. High porosity may indicate good drainage but can also lead to instability if not properly managed. Conversely, low porosity may result in waterlogging and reduced strength.

13.2. Variation of the Characteristics

13.2.1 Importance of Variability

Ground characteristics can vary significantly over short distances due to geological processes, human activities, and environmental factors. This variability is a critical consideration in ground investigation, as it can have profound implications for project design and construction.

Geological Processes: Natural geological processes, such as sedimentation, erosion, and tectonic activity, can lead to significant variations in soil and rock properties. For example, a site may transition from sandy soils in one area to clay-rich soils just a few meters away, affecting drainage and load-bearing capacity. Understanding these geological variations is essential for accurate site characterization and risk assessment.

Human Activities: Human activities, including construction, excavation, and land use changes, can also introduce variability in ground characteristics. For instance, the compaction of soil during construction can alter its density and porosity, impacting its behavior under load. Additionally, contamination from industrial activities can affect soil chemistry and stability, necessitating thorough investigation and remediation.

Environmental Factors: Environmental factors, such as climate, vegetation, and hydrology, can influence ground characteristics over time. Changes in precipitation patterns can affect soil moisture content and drainage, while vegetation can impact soil structure and organic matter content. Understanding these environmental influences is crucial for predicting how ground characteristics may change over time and for developing effective management strategies.

Conclusion

In conclusion, ground investigation is a fundamental aspect of engineering and environmental management that provides critical insights into subsurface conditions. By understanding the characteristics of the ground, including soil and rock types, moisture content, and density, professionals can make informed decisions that enhance the safety and sustainability of construction projects. Recognizing the variability of these characteristics is equally important, as it allows for accurate assessments and tailored design solutions that account for the complexities of the subsurface environment. As such, thorough ground investigation is essential for successful project outcomes and the responsible management of natural resources.

13.2.2 Factors Influencing Variation

Understanding the factors that influence variation in soil and rock types, as well as their structural features, is crucial for fields such as geology, agriculture, environmental science, and civil engineering. This section delves into three primary factors: geological history, hydrological conditions, and human activities. Each of these factors plays a significant role in shaping the characteristics of the ground and can lead to substantial variations in soil and rock properties.

Geological History: The geological history of an area encompasses the processes and events that have shaped the Earth's crust over millions of years. This history is fundamental in determining the types of soil and rock present in a given location, as well as their layering and structural features. Formation processes such as sedimentation, volcanic activity, metamorphism, and erosion contribute to the formation of different rock types. For instance, sedimentary rocks, formed from the accumulation of sediments, often contain layers that reflect changes in environmental conditions over time. In contrast, igneous rocks, formed from cooled magma, may exhibit different characteristics based on their cooling history and mineral composition.

Hydrological Conditions: Hydrological conditions refer to the distribution and movement of water within the soil and rock layers, which can significantly influence their physical and chemical properties.

1. **Groundwater Levels**: Fluctuations in groundwater levels can lead to variations in soil saturation. High groundwater levels can result in waterlogged soils, which may affect their strength and stability. Conversely, low groundwater levels can lead to drier soils, which may become more susceptible to erosion and compaction.

2. **Soil Saturation and Strength**: The degree of soil saturation directly impacts its mechanical properties. Saturated soils tend to have lower shear strength, making them more prone to failure under load. This is particularly important in construction and engineering, where understanding soil strength is critical for foundation design.

3. **Flow Patterns**: The flow of water through soil and rock layers can influence the distribution of nutrients and contaminants. For example, areas with high water flow may experience leaching, where essential nutrients are washed away, leading to nutrient-poor soils. Conversely, areas with slow water movement may accumulate nutrients, resulting in more fertile soils.

4. **Surface Water Interaction**: The interaction between surface water bodies (such as rivers, lakes, and wetlands) and groundwater can also affect soil characteristics. For instance, the presence of a nearby river may influence the moisture content of adjacent soils, leading to variations in vegetation and soil composition.

Human Activities: Human activities have a profound impact on the natural characteristics of the ground, often leading to significant variations in soil and rock properties.

1. **Construction and Excavation**: Activities such as construction, mining, and excavation can alter the landscape and disturb existing soil and rock layers. The removal of topsoil during construction can lead to a loss of nutrients and organic matter, affecting soil fertility. Additionally, excavation can expose underlying rock layers, which may have different properties than the surface soils.

2. **Land Use Changes**: Changes in land use, such as converting forests to agricultural land or urban development, can significantly impact soil characteristics. Agricultural practices, including tillage and the use of fertilizers, can alter soil structure, compaction, and nutrient availability. Urbanization can lead to soil sealing, where impervious surfaces prevent water infiltration and disrupt natural hydrological processes.

3. **Pollution and Contamination**: Human activities can introduce pollutants into the soil and groundwater, leading to variations in soil quality and health. Contaminants from industrial activities, agricultural runoff, and waste disposal can affect soil chemistry and biology, resulting in degraded soil conditions.

4. **Soil Management Practices**: Agricultural and land management practices, such as crop rotation, cover cropping, and conservation tillage, can influence soil health and variability. Sustainable practices can enhance soil structure, increase organic matter, and improve water

retention, while poor management can lead to soil degradation and erosion.

5. **Climate Change**: Human-induced climate change is altering precipitation patterns, temperature regimes, and extreme weather events, which can further influence soil and rock characteristics. Changes in climate can affect soil moisture levels, erosion rates, and the distribution of vegetation, leading to variations in soil properties over time.

13.3. Hydrogeology Case Study

The hydrogeological characteristics of an area can significantly influence the availability and quality of groundwater resources. In this case study, we examine the various aquifers associated with the rock formations in the project area, highlighting their properties, the implications for groundwater management, and the environmental impacts associated with these resources.

13.3.1 Aquifer Types

Several aquifers in the project area are associated with distinct rock formations, each exhibiting unique characteristics:

- **Gravel and Sands**: These are principal aquifers found within the Tuffaceous series. They are extensive and highly variable, providing significant groundwater resources. The porosity and permeability of these materials allow for substantial water storage and movement, making them crucial for local water supply.
- **Porous Tuffs and Lapilli Tuffs**: These aquifers are well developed in the Kedowa tuffs and are sporadically distributed. They behave similarly to the gravel and sands, offering moderate groundwater yields. Their variability can affect local water availability, particularly during dry seasons.
- **Solution Cavities in Tuffs**: Although not widespread in the project area, these cavities serve as excellent aquifers. Their formation through dissolution processes creates spaces that can store and transmit water effectively, providing a reliable source of groundwater in areas where they are present.

- **Lateritic Zones**: These zones consist of murram that mantles high ground and lower depressions, forming perched water aquifers. They feed most of the shallow wells dug in the higher grounds, providing accessible water sources for local communities. However, their limited extent can lead to competition for water resources during periods of drought.
- **Fractured and Fissured Zones**: Typically encountered within phonolites and welded tuffs, these zones contain water within fractures and joints. The presence of water in these zones is crucial for groundwater supply, especially in areas with limited aquifer development. However, the variability in fracture density can lead to inconsistent water yields.
- **Old Land Surfaces**: These surfaces form excellent aquifers, particularly where developed between lava flows. They are common in the highlands north and east of the project area, especially within various phonolite flows and at the contact between the Kedowa tuffs and phonolites. Their ability to store water makes them vital for local agriculture and domestic use.

13.3.2 Implications for Groundwater Management

Understanding the characteristics of these aquifers is essential for effective groundwater management. The variability in aquifer types and their properties necessitates tailored management strategies to ensure sustainable use of groundwater resources. Key considerations include:

1. **Water Quality Monitoring**: Regular monitoring of groundwater quality is crucial, especially in areas where human activities may introduce contaminants. Protecting groundwater from pollution is essential for maintaining safe drinking water supplies and supporting agricultural practices.
2. **Sustainable Extraction Practices**: Implementing sustainable extraction practices is vital to prevent over-extraction and depletion of aquifers. This includes setting limits on groundwater withdrawal based on aquifer recharge rates and promoting water conservation measures among local communities.

3. **Recharge Enhancement**: Strategies to enhance aquifer recharge, such as rainwater harvesting and the restoration of natural drainage patterns, can help maintain groundwater levels. Encouraging vegetation cover and minimizing soil erosion can also contribute to improved recharge rates.

4. **Community Engagement**: Involving local communities in groundwater management decisions fosters a sense of ownership and responsibility. Educating communities about the importance of sustainable water use and the impacts of over-extraction can lead to more effective management practices.

13.3.3 Environmental Impacts

The extraction and management of groundwater resources can have significant environmental impacts. Key considerations include:

1. **Ecosystem Health**: Over-extraction of groundwater can lead to reduced water levels in nearby rivers, lakes, and wetlands, negatively impacting aquatic ecosystems. Maintaining ecological balance is essential for preserving biodiversity and ecosystem services.

2. **Land Subsidence**: Excessive groundwater withdrawal can result in land subsidence, where the ground sinks due to the loss of support from underground water. This phenomenon can damage infrastructure and alter surface water flow patterns.

3. **Salinization**: In coastal areas, over-extraction of groundwater can lead to saltwater intrusion, where seawater contaminates freshwater aquifers. This salinization can render groundwater unsuitable for drinking and irrigation, posing a significant threat to agricultural productivity.

4. **Soil Degradation**: Changes in groundwater levels can affect soil moisture content, leading to soil degradation and reduced agricultural yields. Sustainable groundwater management practices are essential to mitigate these impacts and ensure long-term soil health.

Conclusion

In summary, the hydrogeological characteristics of the project area, including the various aquifer types and their properties, play a crucial role in groundwater availability and quality. Understanding these characteristics is essential for effective groundwater management, which must consider the implications for local ecosystems and communities. By implementing sustainable practices and engaging local stakeholders, it is possible to harness groundwater resources responsibly while minimizing environmental impacts. Continued research and monitoring will be vital to adapt management strategies to changing conditions and ensure the long-term sustainability of groundwater resources in the region.

13.3.4 Groundwater Resources

A notable feature of the project area is the impermeability of certain rock types. For instance, the trachytes exhibit a massive, un-jointed nature, which limits water movement. Similarly, the Kedowa tuffs are characterized by ignimbrites and decomposed clayey pumice tuffs, contributing to their impermeability. This characteristic may necessitate drilling to considerable depths below the water table to access permeable layers that can yield sufficient water.

Recharge of the aquifers primarily occurs from local sources, with rainfall percolating down through fissures until it reaches a porous stratum. In some cases, surfac3.4 e water bodies also contribute to aquifer recharge. The hydrological dynamics in the area are influenced by the alignment of rivers and streams along fault lines. These fault zones act as conduits for groundwater movement, creating trough-like depressions in the water table, particularly in lowland areas.

Water moves through these fault zones into flat-lying aquifers, generally flowing southward from the Mau Ranges to the plains in the north. The faults, while obscured by later pyroclastics, remain significant as open fracture zones devoid of impervious clay fillings. This allows them to function as conduits rather than barriers, facilitating groundwater movement.

The aquifers within older faulted formations, including sedimentary intercalations and old land surfaces, are interconnected through these fault zones. This creates an irregular series of enclosed flat-lying aquifers, with faults serving as the connecting links that integrate them into a single hydrostatic system. The outlet of one aquifer often provides the inlet to the next, although this system does not favor artesian conditions due to the disruption of aquicludes at intervals along the fault zones.

13.3.5 Existing Boreholes

The project area has a limited number of boreholes, with only one borehole drilled within the vicinity. Unfortunately, data regarding this borehole could not be traced from the database. The discharge data available does not show a direct relationship with the drill depth, indicating that the borehole output may be influenced by the complex geomorphology of the area, hydrogeological conditions, and the methods used during borehole development and construction.

Permits for drilling are typically issued based on the nature of the aquifer and any special conditions that may apply, which can override radial distance requirements. This highlights the importance of understanding the hydrogeological context when planning for groundwater extraction and management.

13.3.6 Environmental Impacts

The extraction and management of groundwater resources in the project area can have significant environmental impacts, which must be carefully considered to ensure sustainable practices.

1. **Depletion of Aquifers**: Over-extraction of groundwater can lead to the depletion of aquifers, resulting in reduced water availability for both human use and ecological systems. This can lead to a decline in water levels, affecting the health of nearby rivers, lakes, and wetlands that rely on groundwater contributions.

2. **Land Subsidence**: Excessive groundwater withdrawal can cause land subsidence, where the ground sinks due to the collapse of underground voids. This phenomenon can damage infrastructure, disrupt drainage patterns, and lead to increased flooding in low-lying areas.

3. **Water Quality Deterioration**: The movement of water through fractured and fissured zones can lead to the contamination of groundwater supplies. Pollutants from surface activities, such as agricultural runoff or industrial discharges, can infiltrate aquifers, compromising water quality and posing health risks to communities relying on these sources.

4. **Alteration of Natural Hydrology**: Groundwater extraction can alter the natural hydrological cycle, affecting surface water bodies and

ecosystems. Changes in groundwater levels can impact the flow of rivers and streams, leading to reduced habitat availability for aquatic species and altering the ecological balance of the area.

5. **Impact on Vegetation**: Changes in groundwater availability can affect local vegetation, particularly in areas dependent on shallow groundwater for survival. Reduced water levels can lead to stress on plant communities, potentially resulting in shifts in species composition and loss of biodiversity.

6. **Perched Water Aquifers**: The presence of perched water aquifers in lateritic zones can be particularly sensitive to changes in land use and groundwater extraction. Over-extraction can lead to the drying up of these aquifers, impacting local communities that rely on shallow wells for drinking water and irrigation.

Conclusion

The hydrogeological characteristics of the project area, including the various aquifers and their interactions, play a crucial role in determining groundwater availability and quality. Understanding these factors is essential for effective water resource management, particularly in regions where groundwater is a primary source of water for communities and agriculture.

The environmental impacts associated with groundwater extraction and management must be carefully considered to promote sustainable practices. By recognizing the complexities of the hydrogeological system and the potential consequences of human activities, stakeholders can make informed decisions that protect groundwater resources and ensure their availability for future generations. Sustainable management practices, including monitoring groundwater levels, implementing conservation measures, and engaging local communities, are vital for maintaining the health of both the aquifers and the ecosystems they support.

13.3.7 Methods for Assessing Variation

Sampling and Testing

Sampling and testing are fundamental methods for assessing variations in subsurface conditions. This process involves the systematic collection of soil and rock samples from various depths and locations across a site. By analyzing these samples, engineers and geologists can gain critical insights into the physical and chemical properties of subsurface materials. The data obtained from these analyses can reveal variations in soil type, moisture content, and potential contamination levels, all of which are essential for risk assessment and engineering design. For instance, understanding the variations in soil composition can significantly influence the design of foundations and other structural elements, ensuring that they are adequately supported by the underlying materials.

Geophysical Surveys

Geophysical surveys are essential tools in the field of environmental geology, employing non-invasive techniques to map subsurface conditions and identify variations without the need for extensive drilling or excavation. These methods, which include seismic, resistivity, magnetic, and electromagnetic techniques, provide a broader spatial understanding of geological features, enabling engineers and geoscientists to detect anomalies such as voids, fractures, or changes in material properties.

By utilizing geophysical surveys, professionals can minimize the environmental impact associated with traditional drilling methods, preserving the integrity of the surrounding ecosystem while still obtaining valuable data about subsurface conditions. This non-invasive approach not only reduces the physical disturbance to the land but also lowers the risk of contaminating groundwater resources, making it a preferred choice in sensitive environmental contexts.

Case Study: Hydrogeological/Geophysical Survey for Giftmart Supermarket

This case study presents the hydrogeological and geophysical survey conducted on the plot of land designated for Giftmart Supermarket, located in the Kapkatet Area of Bureti Division, Bureti Sub-County, Kericho County. The primary objective of the study was to identify a suitable site for drilling a productive borehole that would serve as the sole water source for domestic purposes within the supermarket premises. The estimated water demand for the facility is approximately 25 m^3 per day.

Field Exploration Program

The field exploration program involved a comprehensive assessment of the hydrogeological conditions in the area, utilizing both hydrogeological and geophysical methods to ensure the identification of an optimal borehole location. The following steps were undertaken during the survey:

1. **Site Assessment**:
    - Initial site visits were conducted to evaluate surface conditions and gather preliminary data on existing water sources, land use, and potential environmental impacts.
2. **Geophysical Surveys**:
    - Non-invasive geophysical techniques, such as resistivity measurements, were employed to map subsurface geological features. This approach allowed for the identification of water-bearing formations and the assessment of their potential productivity without the need for extensive drilling.
3. **Hydrogeological Analysis**:
    - The hydrogeological characteristics of the area were analyzed, including the lithology, porosity, and permeability of the subsurface materials. This analysis was crucial for understanding groundwater flow patterns and the sustainability of the proposed borehole.
4. **Data Interpretation**:
    - The data collected from the geophysical surveys and hydrogeological assessments were integrated to create a comprehensive model of the subsurface conditions. This model facilitated the identification of the most promising locations for borehole drilling.
5. **Recommendations**:
    - Based on the findings of the survey, recommendations were made regarding the optimal site for drilling the borehole, taking into consideration factors such as water quality, yield potential, and environmental sustainability.

This case study underscores the importance of thorough hydrogeological and geophysical investigations in ensuring the successful establishment of a reliable water source for domestic use. The insights gained from this survey will not only support the operational needs of Giftmart Supermarket but also contribute to sustainable water resource management in the Kapkatet area.

1.3 Project Site Location, Geographical Location, and Topography

The proposed site for the Giftmart Supermarket is situated within the Kapkatet Shopping Centre, located on the right-hand side of the Kaplong-Litein Road, approximately 5 kilometers south of Litein Town (Figure 24). Administratively, the plot falls within the Kapkatet Area, Kapkatet Location, Bureti Division, Bureti Sub-County, Kericho County.

Geographical Coordinates

The geographical coordinates for the proposed site are as follows:

- **Grid Reference**: 36M 744503 E
- **UTM Reference**: 9930602 S
- **Elevation**: 1,748 meters above sea level

These coordinates correspond to the topographical map sheet 131/1 of the Chemegel Area, as provided by the Survey of Kenya.

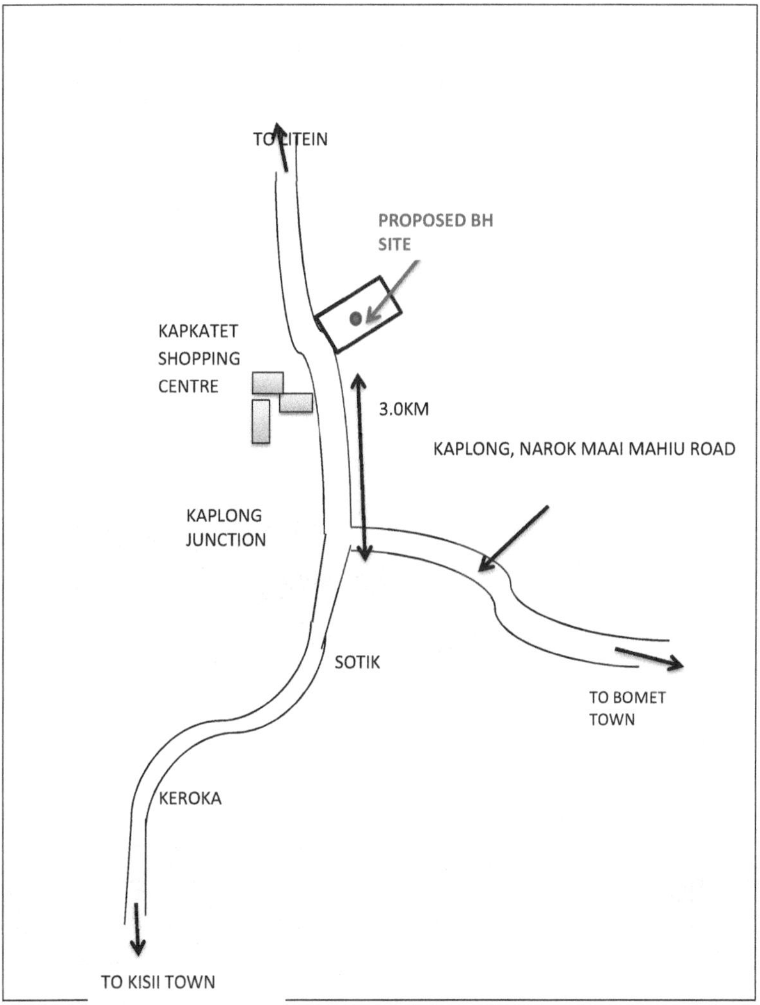

Figure 24: Sketch of the proposed site (NOT TO SCALE)

Topography

The topography of the area is characterized by gently rolling hills and valleys, typical of the Kericho County region. The elevation of 1,748 meters above sea level indicates that the site is situated in a relatively elevated area, which may influence local hydrology and groundwater availability. The landscape is predominantly covered with agricultural land (Figure 25), interspersed with patches of natural vegetation, which contributes to the region's biodiversity and ecological balance.

The topographical features of the site and its surroundings play a crucial role in determining water drainage patterns, soil characteristics, and potential groundwater recharge areas. Understanding these factors is essential for the successful implementation of the borehole project and for ensuring sustainable water resource management in the area.

Figure 25: Google Earth Image showing location of the proposed project site

Agricultural Context and Water Requirements

The project site is situated in an agricultural zone where mixed farming is the predominant activity. The region primarily cultivates tea as the main cash crop, which significantly influences local economic activities and water usage patterns. Given the agricultural nature of the area, understanding the water requirements for both domestic use and agricultural activities is essential for sustainable resource management.

Objective of the Investigation

The primary objective of the investigation was to establish the optimal location for a borehole that would serve as the main source of water for domestic use within the project parcel. This borehole is intended to meet the water needs of the Giftmart Supermarket and its associated facilities, ensuring a reliable supply for daily operations.

13.4 Rainfall and Climate

The project area experiences relatively high rainfall in the highlands, with a significant decrease in precipitation as one moves toward the lowlands. On average, the region receives more than 1,000 mm of rainfall annually, which plays a crucial role in supporting both agricultural activities and groundwater recharge with temperatures ranging between 10- 25 °C (Figures 26 and 27).

The remarkable topographical diversity of this area is primarily attributed to the tectonic and volcanic disturbances associated with the Kavirondo and Rift Valley regions. These geological processes have disrupted the peneplained surfaces, resulting in the formation of distinct ridges and troughs that predominantly trend north-south. The complex history of dislocation and volcanic activity is reflected in the physiography of the landscape, which showcases multiple periods of geological activity.

Overall, the topography of the project area can be characterized as undulating, featuring a combination of valleys and ridges in the highlands, while the lowlands are relatively flat.

Key topographical features in the vicinity include the forested Mau Hills to the east and the Transmara Forest to the southwest. These features significantly influence groundwater movement in the region, as numerous fault fissures act as drainage zones for groundwater, facilitating its discharge into the plains to the west. This area serves as a critical catchment for groundwater and river systems, underscoring the importance of understanding the hydrological dynamics at play in the project area.

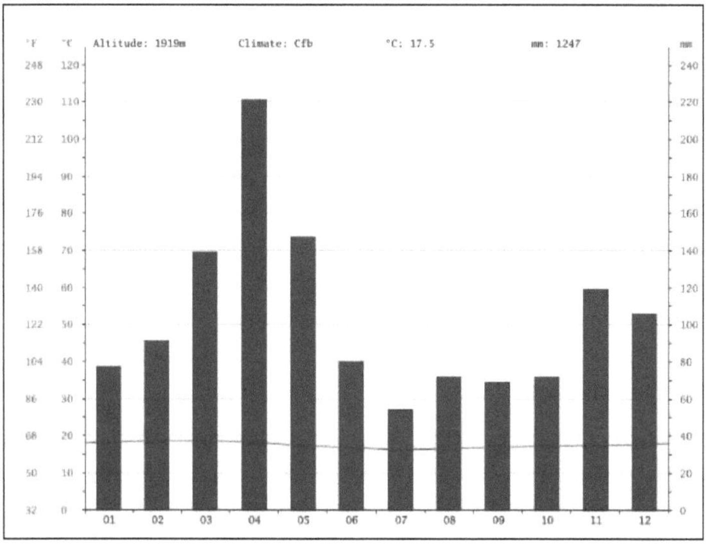

Figure 26: Graph showing average rainfall in Litein

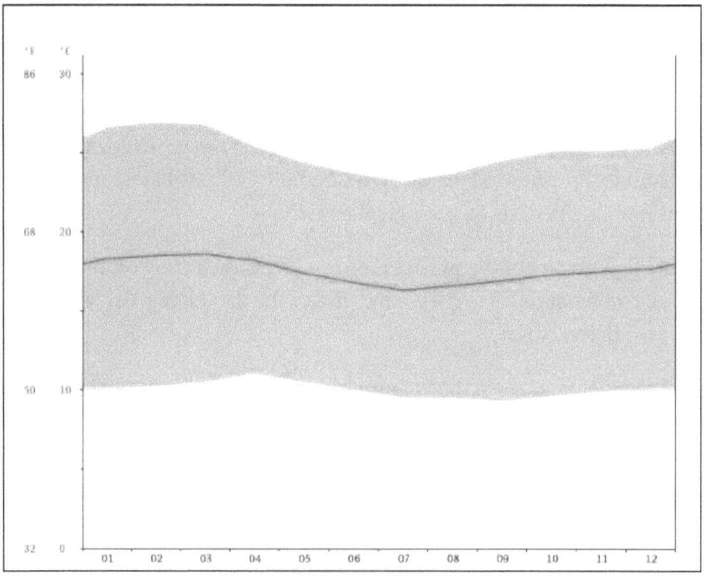

Figure 27: Graph showing average temperatures in Litein

13.5 Drainage

The drainage patterns in the project area are primarily influenced by the elevation of Lake Victoria to the west. The Kipsonoi River, which is the main river in the region, is fed by several youthful streams originating from the Mau Ranges to the east. Most of these streams are perennial, ensuring a consistent flow of water throughout the year.

At the base of the scarp, these streams have carved deep channels through poorly sorted lenticular deposits of coarse gravels, sands, and sandy clays. This geological composition plays a significant role in the hydrology of the area, affecting both surface water flow and groundwater recharge.

The Kipsonoi River flows westward, eventually draining into the Sondu River before reaching Lake Victoria, which is situated to the northwest of the project area. The majority of these rivers originate from the Mau Forest, located to the east of the project site, highlighting the forest's critical role in sustaining the local hydrological system.

In summary, the drainage system in the project area is characterized by the prominent Kipsonoi River and its tributaries, which are vital for maintaining the ecological balance and supporting agricultural activities in the region. Understanding these drainage dynamics is essential for effective water resource management and planning for the proposed borehole project.

Water Requirement Needs

The following table outlines the estimated water requirements for the project:

Table 0.1: Water requirement needs for the client

Population to be supplied with water	**100**
Expected Rate of water use (m^3/day)	**25**

13.6 Field Exploration Program

13.6.1 Prospecting Methods

13.6.1.1 Resistivity (Basic Principles)

Resistivity methods are grounded in the electrical properties of rocks found in the upper part of the Earth's crust. These properties are influenced by several factors, including lithology, porosity, the degree of pore space saturation, and the salinity of the pore water. A thorough understanding of these characteristics is essential for accurately interpreting subsurface conditions.

The resistivity of rocks varies significantly based on their composition and moisture content. Saturated rocks typically exhibit lower resistivity compared to unsaturated or dry rocks. This phenomenon occurs because the presence of water, particularly saline water, enhances the conductivity of the material. Additionally, higher porosity in saturated rocks correlates with lower resistivity, as the increased volume of pore space filled with conductive fluids facilitates the flow of electric current. Conversely, the salinity of the saturating fluids plays a critical role; higher salinity results in lower resistivity. This aspect is particularly important in hydrogeological studies, where the salinity of groundwater can indicate the presence of contaminants or the overall quality of water resources.

The presence of clays and other conductive minerals further reduces the resistivity of rocks. Clays, in particular, can retain water and enhance conductivity due to their ion-exchange properties. This characteristic is vital for identifying clay-rich zones, which may significantly influence groundwater flow and the transport of contaminants.

To study the resistivity of Earth materials, electrical potential distribution is measured at the Earth's surface when an electric current is passed through the ground. This is typically accomplished using an array of electrodes placed in the ground, allowing for the measurement of voltage differences that occur as current flows through the subsurface materials. Various configurations, such as Wenner, Schlumberger, or dipole-dipole arrays, can be employed depending on the specific objectives of the survey and the geological conditions present. Each configuration has its advantages and limitations, influencing the depth of investigation and the resolution of the data obtained.

Resistivity surveys have a wide range of applications, including groundwater exploration, mineral exploration, environmental assessments, and geotechnical investigations. By providing insights into subsurface conditions, resistivity methods inform decisions related to resource management, site remediation, and infrastructure development. The resistance R of a certain material is directly proportional to its length L and cross-sectional area A, expressed as:

$$R = \rho * L.A \qquad (1)$$

Where ρ is known as the specific resistivity, characteristic of the material and independent of its shape or size, With Ohm's Law;

$$R = \Delta V/I \qquad (2)$$

Where V is the potential difference across the resistor and I is the electric current through the resistor, the specific resistivity may be determined by:

$$\rho = (A/L)\ \Delta\ V/I) \qquad (3)$$

Survey Design

Two primary categories of field techniques exist for conventional resistivity analysis of the subsurface: Vertical Electrical Sounding (VES) and Horizontal Electrical Profiling (HEP).

Vertical Electrical Sounding (VES) aims to deduce the variation of resistivity with depth below a specific point on the ground surface. This method correlates resistivity data with available geological information to infer the depths and resistivities of the layers present. In VES using the Wenner configuration, the array spacing (a) is increased in steps while keeping the midpoint fixed (e.g., (a = 2, 6, 18, 54 ...etc)). In the Schlumberger method, the potential electrodes are moved occasionally, while the current electrodes are systematically moved outward in steps, with (**AB > 5 MN**).

Horizontal Electrical Profiling (HEP) focuses on detecting lateral variations in the resistivity of the ground, such as lithological changes and near-surface faults. In the Wenner procedure of HEP, the four electrodes with a defined array spacing (a) are moved as a whole in suitable steps, typically ranging from 10 to 20 meters, with all four electrodes moving after each measurement. In the Schlumberger method of HEP, the current electrodes remain fixed at a relatively large distance, often a few hundred meters apart, while the potential electrodes, separated by a small constant distance (MN), are moved between points A and B.

Field Work

The fieldwork for this project was conducted on March 17, 2025. Prior to the actual survey, a reconnaissance site visit was undertaken, which formed an essential part of the desk study for the project site. This preliminary visit allowed for an initial assessment of the area, helping to identify key features and potential challenges that could influence the survey process.

During the reconnaissance, various factors such as topography, land use, and existing water sources were evaluated. This information was crucial in planning the survey methodology and ensuring that the fieldwork would be conducted efficiently and effectively. The insights gained from the reconnaissance visit informed the selection of specific locations for resistivity measurements and helped establish a clear understanding of the geological and hydrological context of the site.

The fieldwork itself involved the deployment of geophysical techniques, specifically resistivity measurements, to assess subsurface conditions. The data collected during this phase will be instrumental in determining the optimal location for the borehole, ensuring that it meets the water demand requirements for the Giftmart Supermarket while also considering environmental sustainability.

13.6.2 Surface Ground Potential Measurement

For the surface ground potential measurement, the equipment employed consists of two electrodes (designated as MN) that are inserted into the ground at a predetermined distance of ten (10) meters apart (Figure 28). The midpoint between these two electrodes is considered the reference point for the measurement, allowing for accurate assessment of the electrical potential at that specific location.

This method is essential for determining the ground potential distribution, which provides valuable insights into the subsurface electrical properties. By measuring the voltage difference between the two electrodes, we can infer information about the resistivity of the surrounding materials, which is crucial for understanding groundwater conditions and identifying suitable locations for borehole drilling.

The setup is designed to ensure that the electrodes are properly grounded and that the measurements are not influenced by external factors, such as nearby electrical sources or environmental noise. This careful calibration and placement of the electrodes are vital for obtaining reliable data that can be used in conjunction with other geophysical methods to create a comprehensive picture of the subsurface conditions.

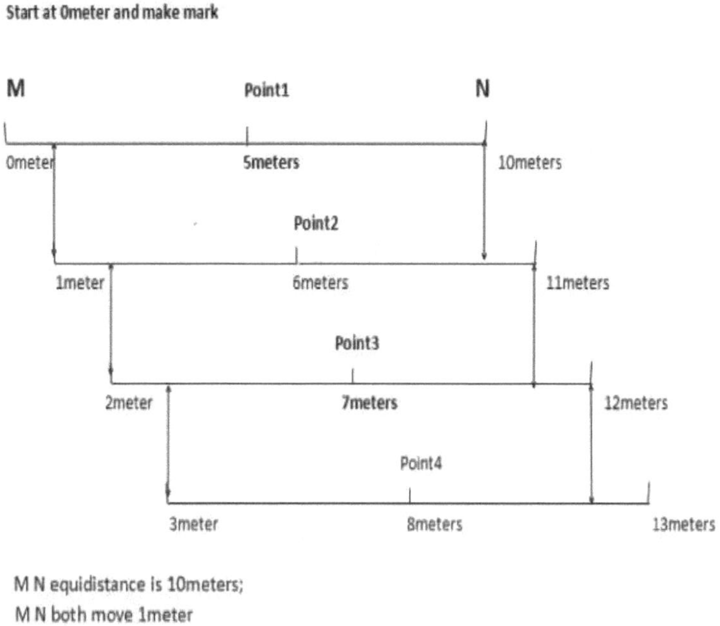

Figure 28: Demonstration of Surface Ground Potential Measurement

In this figure, a visual representation illustrates the setup for surface ground potential measurement using two electrodes (MN) placed ten (10) meters apart. The midpoint between the electrodes is highlighted as the reference point for measuring the electrical potential. This diagram serves to clarify the methodology employed during the fieldwork, showcasing the arrangement of the equipment and the process of obtaining measurements on site.

Data Collection and Analysis

Once the electrodes are securely inserted into the ground, readings are taken using the resistivity measurement equipment. The resistivity machine records the electrical potential differences, allowing for the identification of areas with abnormal resistivity values, which may indicate variations in subsurface materials or water content. High resistivity values could suggest the presence of dry or less conductive materials, while low resistivity values may indicate saturated zones or the presence of conductive minerals.

After completing the data collection on site, the next step involves analyzing the gathered data to interpret the subsurface conditions accurately. This analysis will include the following steps:

1. **Data Processing**: The raw data collected from the resistivity measurements will be processed to remove any noise or anomalies that could affect the results. This may involve applying filtering techniques and correcting for any systematic errors.

2. **Resistivity Mapping**: The processed data will be used to create resistivity maps, which visually represent the variations in subsurface resistivity across the survey area. These maps will help identify patterns and anomalies that warrant further investigation.

3. **Interpretation of Results**: The resistivity maps will be analyzed in conjunction with geological and hydrological information from the reconnaissance phase. This interpretation will help determine the likely composition and saturation levels of the subsurface materials, as well as identify potential locations for borehole drilling.

4. **Identification of Target Zones**: Based on the analysis, specific target zones will be identified where the resistivity values indicate favorable conditions for groundwater extraction. These zones will be prioritized for further exploration and potential borehole development.

5. **Reporting**: Finally, the findings from the data analysis will be compiled into a comprehensive report, detailing the methodology, results, and recommendations for the next steps in the project.

This systematic approach to data analysis ensures that the insights gained from the resistivity measurements are effectively utilized to inform decision-making regarding groundwater resource management and borehole placement.

Sample Data Analysis

First observe the curve to find the "V" "L" "W" "A" (measuring point display regular a lot of falling curves, lower potential difference data, and mark the position number (Figure 29). (The horizontal line of curve is measurement point; the vertical line of the curve is underground material's potential difference data which could be regarded as resistivity value).

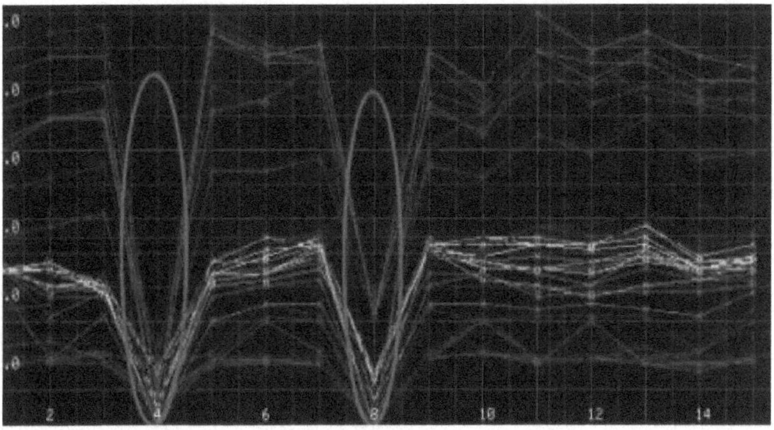

Figure 29: From the Curve, finding the "V", which is a lot of falling, the low value is low potential difference data and high probability for water. The below case is at point 4 and point 8.

Results Analysis and Profile Mapping

In addition to the resistivity measurements, we will analyze the profile map generated from the data, utilizing a color-coded system to interpret the results effectively. The colour scheme is as follows (Figure 30):

- **Blue to Light Blue to Light Yellow**: These colors indicate low resistivity values, which are typically associated with materials such as water, soft mud, and some metal minerals. While low resistivity often suggests the presence of water, it is important to note that not all low-resistivity readings indicate water saturation.
- **Yellow**: This color represents medium resistivity, which may correspond to materials such as solid rock or other geological formations that do not exhibit extreme conductivity or resistivity.

- **Red**: High resistivity is indicated by red, which may signify the presence of cavities, hard rock, or other dense materials that do not allow for easy electrical current flow.

Each color on the profile map corresponds to specific geological characteristics, aiding geologists in analyzing the subsurface structure. The map serves as a valuable tool for visualizing resistivity variations and understanding the geological context of the area. It is essential to combine these resistivity findings with local geological knowledge to enhance the accuracy of interpretations.

From a geological perspective, compact rock formations that are neither fractured nor weathered typically lack the ability to transmit or store groundwater. However, the presence of cracks or fractures within these compact rocks significantly increases the likelihood of encountering water at depth.

The results from the profile map indicate a higher probability of finding water at points 4 and 8, which have been selected as proposed drilling locations. These points exhibit resistivity values consistent with the presence of water-bearing formations, making them prime candidates for further exploration and borehole development. By focusing on these areas, we can optimize our efforts to access groundwater resources effectively.

From Profile Map, blue to light yellow stands for low-value, representing high probability for water; Yellow stands for middle value, high probability for rock; Red stand for high-value, high probability for a cave or hard rock.

13.6.3 Field Work

Field reconnaissance and a detailed field visit were conducted on March 19, 2025. During this phase, the Portable Groundwater Detection Tool (PQWT) was utilized to identify points of anomaly in the subsurface conditions. The data collected during this fieldwork is crucial for understanding the geological and hydrological characteristics of the area, and the results are discussed in the following sections.

13.6.4 The Line Curve of Data Collected on Site

The line curve generated from the data collected on site provides a graphical representation of the resistivity measurements taken during the survey. This curve illustrates the variations in resistivity across the surveyed area, highlighting areas of interest where anomalies were detected (Figures 30 to 31).

The line curve typically displays resistivity values on the vertical axis, while the horizontal axis represents the distance along the survey line. As the curve fluctuates, it indicates changes in subsurface materials and conditions. Key features of the line curve include:

- **Peaks and Valleys**: High peaks in the curve correspond to areas of high resistivity, which may indicate the presence of hard rock, cavities, or other dense materials. Conversely, valleys in the curve represent low resistivity zones, suggesting the presence of water, soft sediments, or conductive minerals.

- **Anomalies**: Specific points along the curve where significant deviations occur are identified as anomalies. These points warrant further investigation, as they may indicate potential groundwater sources or other geological features of interest.

- **Interpretation**: By analyzing the shape and characteristics of the line curve, geologists can infer the subsurface structure and identify potential drilling locations. The data collected will be compared with existing geological information to enhance the accuracy of interpretations and guide future exploration efforts.

Overall, the line curve serves as a vital tool in the analysis of subsurface conditions, providing insights that will inform decisions regarding groundwater extraction and resource management. The results from this analysis will be integrated with other findings to develop a comprehensive understanding of the geological framework in the project area.

The line curve of data collected on site

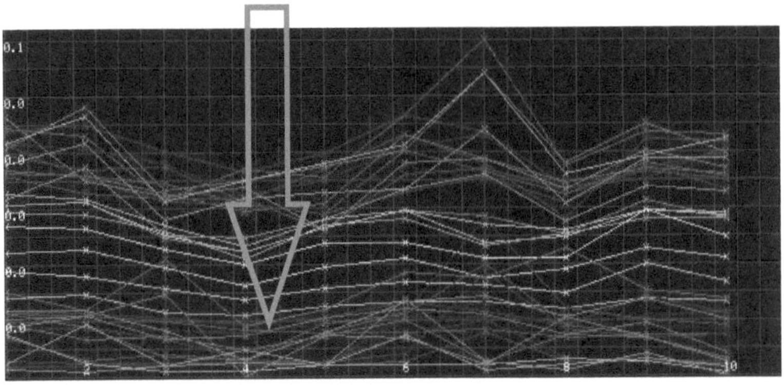

Figure 30: Line Curves from Data Collected on Site

The line curves generated from the data collected during the fieldwork provide a visual representation of the resistivity measurements across the surveyed area. These curves are instrumental in identifying subsurface conditions and potential groundwater sources.

In particular, **Line 1** shows a notable feature at **Point 4**, which has been pre-marked as having low resistivity. This low resistivity reading indicates a high probability of encountering water at this location. The significance of this finding is further underscored by the fact that Point 4 coincides with the selected site marked for the proposed drilling of the borehole.

The alignment of the low resistivity anomaly with the proposed drilling point indicates that this location is particularly favorable for groundwater extraction. Low resistivity values typically suggest the presence of water-saturated materials, making Point 4 an optimal choice for drilling. The data collected supports the decision to proceed with drilling at this site, as it is likely to yield positive results in terms of water availability, which is crucial for meeting the needs of local communities and supporting sustainable resource management.

This analysis underscores the significance of utilizing resistivity measurements to inform drilling decisions, ensuring that resources are allocated efficiently and effectively in the quest for groundwater. By relying on geophysical data, stakeholders can minimize the risks associated with drilling in less promising areas, thereby optimizing both time and financial investments. The insights gained from the resistivity line curves will be instrumental in guiding subsequent phases of the project, as they provide a robust foundation for further exploration and development efforts.

Moreover, the integration of resistivity data into the decision-making process enhances the overall reliability of groundwater assessments. As the project progresses, these insights will not only aid in identifying additional drilling locations but also contribute to a more comprehensive understanding of the aquifer system. This knowledge is essential for developing effective management strategies that ensure the sustainable use of groundwater resources, ultimately benefiting both the environment and the communities that depend on them. By leveraging advanced geophysical techniques, the project can achieve its objectives while promoting responsible water resource management practices.

Profile Curve from data collected on site

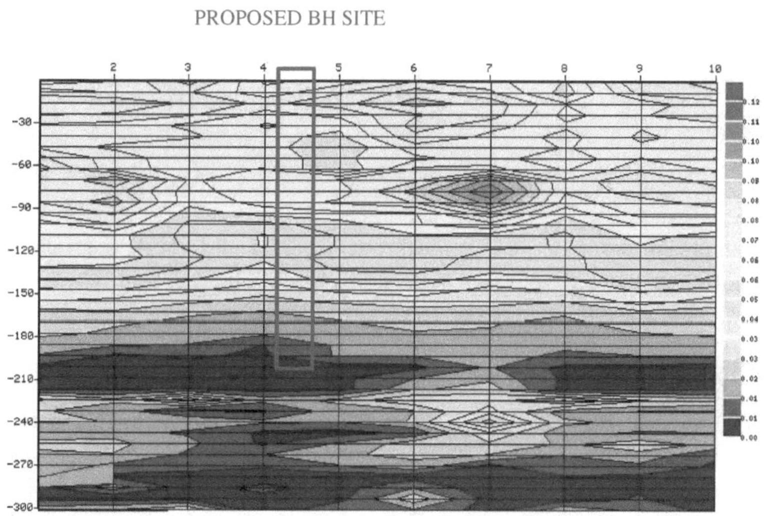

Figure 31: Line Curves from Data Collected on Site

The line curves generated from the data collected during the fieldwork provide a visual representation of the resistivity measurements across the surveyed area. These curves are instrumental in identifying subsurface conditions and potential groundwater sources.

In particular, **Line 1** shows a notable feature at **Point 4**, which has been pre-marked as having low resistivity. This low resistivity reading indicates a high probability of encountering water at this location. The significance of this finding is further underscored by the fact that Point 4 coincides with the selected site marked for the proposed drilling of the borehole.

At depths beyond 200 meters, **Point 4** exhibits a blue segment on the line curve, which indicates weathered rock formations. This observation further supports the likelihood of striking water at depth, as weathered rock can often serve as a reservoir for groundwater.

The results from the survey have effectively served their purpose, confirming the suitability of Point 4 for drilling. This conclusion was discussed on-site with the caretaker, ensuring that all stakeholders are aligned on the proposed drilling location. The data collected not only reinforces the decision to proceed with drilling at Point 4 but also provides confidence in the potential for successful groundwater extraction in this area.

Overall, the integration of resistivity data and geological insights will guide the next steps in the project, facilitating informed decision-making and resource management.

Interpretation of Field Curves

Interpreted results of the vertical electrical sounding with the inferred geological layering are shown in the tables below while the resistivity curves are as presented in the annexes.

Evaluation of the VES Data

The Vertical Electrical Sounding (VES) data collected during the survey provides valuable insights into the subsurface geological conditions. The interpretation of the VES results indicates a homogeneous strata, characterized primarily by volcanic rock formations.

Key Findings:

- **Homogeneous Volcanic Strata**: The VES data reveals a consistent layer of volcanic rocks throughout the surveyed area. This homogeneity suggests that the geological conditions are relatively uniform, which can simplify the assessment of groundwater potential.

- **Layer Composition**: The interpreted results indicate that the top layer consists of loam soils, which are typically conducive to water retention. Beneath this layer, there is a transition to weathered volcanic materials, followed by fractured volcanic layers extending to a depth of approximately 180 meters below ground level.
- **Groundwater Potential**: The presence of weathered and fractured volcanic rocks is significant, as these formations can enhance groundwater storage and movement. The highly weathered and fractured volcanic materials found below 180 meters are particularly promising for groundwater extraction, as they are likely to contain significant water reserves.

Based on these findings, the site has been assessed to have good groundwater potential. Consequently, it is recommended that drilling be conducted for the proposed borehole at this location. The combination of loam soils, weathered volcanic strata, and fractured rock layers creates an optimal environment for accessing groundwater, making this site a favorable choice for further exploration and development.

Overall, the VES data evaluation supports the decision to proceed with drilling, as it indicates a high likelihood of successfully encountering groundwater at the proposed depth (Table 1.3).

Table 1.3: Interpreted Survey Data

Depth (In Meters)	**Formation**	**Formation Status**
0.0 – 2.0	Sandy surficial deposits	Dry
2.0 – 12.0	Fractured sub-base	Dry
12.0– 75.0	Weathered volcanics	Dry
75.0 – 100.0	Fractured Volcanics	Dry
110.0 – 130.0	Highly fractured/Weathered volcanics	Wet
130.0-180.0	Highly fractured/Weathered volcanics	Wet to dry
>180.00	Fractured volcanics	Wet

Surface Ground Potential Measurement

The equipment utilizes to electrodes (MN) that are inserted into the ground at a predetermined distance of ten (10) meters where the center of the two electrodes is considered as the point to be measured as demonstrated below.

In summary, resistivity methods are a powerful tool in geophysical surveys, offering valuable insights into subsurface conditions while minimizing environmental impact. By understanding the electrical properties of rocks and the factors influencing resistivity, geoscientists can effectively map geological features and assess the potential for resource extraction or environmental remediation.

Impacts of Proposed Drilling Activity

Within the study area, there are currently no boreholes that have been drilled in the immediate neighborhood. This absence of existing boreholes minimizes the potential for significant drawdown effects, reduced radius of influence, and increased pumping costs associated with groundwater extraction. The proposed borehole will be drilled to the recommended depth, which has been determined based on the geological and hydrological assessments conducted during the study.

Impacts on Local Aquifer Quantity and Quality

The sustainability of the water quantity from the proposed borehole is contingent upon the balance between groundwater abstraction and the natural recharge rate of the aquifer. If groundwater is extracted at a rate that exceeds its natural replenishment, the project risks becoming unsustainable, potentially leading to dry pumping conditions.

Based on the yields observed in similar boreholes in the area, the proposed abstraction rate of **25 m^3/day** over a **10-hour pumping regime** is not anticipated to have any major adverse impacts on the local aquifers. The aquifer is expected to be quite productive, allowing for this level of abstraction without significant depletion.

Water quality is influenced by several factors, including the characteristics of the host rock, the design of the borehole construction, and the age of the water being extracted. It is essential to monitor these factors to ensure that the water quality remains within acceptable standards for use.

Overall, the expected impacts resulting from the borehole on the environment, including both quantity and quality of groundwater, will be thoroughly addressed in the Environmental Impact Assessment (EIA) study to be conducted. This assessment will outline potential impacts and propose mitigation measures to minimize any negative effects.

Impacts of Existing Boreholes in the Area

It is noteworthy that most existing boreholes are located more than a kilometer away from the proposed drilling site. Given this distance, no significant negative impacts are expected on the other existing boreholes within the vicinity. The spatial separation reduces the likelihood of interference with the hydraulic conditions of these boreholes, ensuring that their functionality and water supply remain unaffected.

In conclusion, the proposed drilling activity is expected to have manageable impacts on the local aquifer system, and the EIA will provide a comprehensive framework for monitoring and mitigating any potential environmental effects.

13.7 Qualitative Description

Importance of Qualitative Descriptions

Qualitative descriptions are vital for providing a narrative understanding of ground conditions, complementing the quantitative data obtained from tests and measurements. They offer context and insight that numerical data alone may not convey, allowing engineers and geologists to interpret findings more holistically. This narrative approach is essential for understanding the complexities of subsurface conditions and their implications for construction projects.

Elements of Qualitative Description

Key elements of qualitative description include visual observations, field notes, and contextual information. Visual observations involve detailed descriptions of the appearance, color, texture, and layering of soils and rocks, which can indicate the presence of specific minerals or organic materials. Field notes document observations made during site visits, capturing signs of instability, erosion, or other geological features that may not be evident through testing alone. Contextual information provides background about the site, including land use, vegetation, and historical data, enriching the qualitative description and highlighting potential environmental impacts.

Application of Qualitative Descriptions

Qualitative descriptions assist engineers and geologists in interpreting data, assessing risks, and effectively communicating findings to stakeholders. By integrating qualitative insights with quantitative data, professionals can develop a comprehensive understanding of site conditions, leading to more informed decision-making and enhanced project outcomes.

13.8 Response of the Ground to Certain Imposed Conditions

Understanding Ground Response

Understanding how the ground responds to imposed conditions—such as loading, excavation, and environmental changes—is critical for predicting behavior during construction and operation. This knowledge is essential for designing structures that can withstand various stresses and environmental factors, ensuring safety and stability throughout the lifecycle of a project.

Types of Imposed Conditions

Imposed conditions can be categorized into static loads, dynamic loads, and excavation and construction activities. Static loads refer to the effects of permanent structures, such as buildings and bridges, on the ground. Understanding these effects is crucial for ensuring stability and safety. Dynamic loads, which include transient forces from earthquakes, traffic, and machinery, can significantly impact ground response and must be considered in design. Additionally, excavation and construction activities can alter ground stability and behavior, necessitating careful assessment to maintain safety during construction.

Assessing Ground Response

Assessing ground response involves both laboratory testing and field monitoring. Laboratory tests, such as triaxial and unconfined compression tests, evaluate how soil and rock respond to different loading conditions, providing critical data for understanding material behavior under stress. Field monitoring, through the use of instrumentation like piezometers and inclinometers, allows for the observation of changes in ground conditions during and after construction. This real-time data collection is vital for assessing the effectiveness of mitigation measures and ensuring ongoing safety.

13.9 Environmental Issues

Importance of Environmental Considerations

Incorporating environmental considerations into ground investigations is essential to ensure that construction activities do not adversely affect the surrounding ecosystem and comply with regulatory requirements. A thorough understanding of environmental impacts can lead to more sustainable construction practices, ultimately benefiting both the project and the environment.

Key Environmental Issues to Address

Several key environmental issues must be addressed during ground investigations. Soil and water contamination is a primary concern, necessitating investigations into potential contamination from previous land use, industrial activities, or hazardous materials. Groundwater protection is also critical, as construction activities can impact groundwater quality and quantity. Additionally, understanding how ground disturbance may affect local flora and fauna is vital for mitigating habitat destruction and changes in surface water flow. Finally, incorporating sustainability practices into ground investigation and construction processes is essential for minimizing waste and promoting ecological balance.

Methods for Addressing Environmental Issues

To address environmental issues effectively, Environmental Impact Assessments (EIAs) should be conducted to evaluate the potential environmental effects of proposed projects and identify mitigation measures. Implementing monitoring programs to track environmental conditions before, during, and after construction ensures compliance with environmental regulations. Continuous monitoring allows for timely interventions if adverse effects are detected, thereby safeguarding the environment throughout the construction process.

Conclusion

The objectives of ground investigation are essential for ensuring the safety, stability, and success of construction projects. By thoroughly understanding the characteristics of the ground, assessing variations, providing qualitative descriptions, evaluating the response to imposed conditions, and addressing environmental issues, engineers and geologists can make informed decisions that mitigate risks and enhance project outcomes. A comprehensive ground investigation lays the foundation for effective design and construction practices, ultimately contributing to the responsible management of natural resources and the built environment.

Stages of Site Investigation

Introduction

Site investigation is a critical process in engineering and construction that involves a systematic approach to understanding the subsurface conditions of a site. This investigation is essential for informing design decisions, ensuring safety, and optimizing construction practices. The site investigation process can be divided into four main stages: Preliminary, Reconnaissance, Detailed, and Reporting. Each stage plays a vital role in gathering and analyzing data about the ground conditions.

1. Preliminary Stage

Interpretation of Geological Conditions

The preliminary stage aims to develop an initial understanding of the geological conditions at the site based on existing data. Activities during this stage include reviewing geological maps, reports, and studies relevant to the area, identifying potential geological hazards such as faults or landslides, and assessing the general soil and rock types expected at the site.

Acquisition of Additional Information

To enhance the understanding of site conditions, supplementary information must be gathered. This involves consulting local geological organizations, government agencies, and experts in the field, as well as reviewing historical data from previous site investigations and construction projects in the vicinity. Additionally, collecting information on groundwater levels, soil contamination, and environmental considerations is crucial.

2. Reconnaissance Stage

Confirmation of Interpreted Information

The reconnaissance stage focuses on validating the initial interpretations made during the preliminary stage. This involves conducting field visits to observe surface conditions and geological features, as well as engaging with local stakeholders to gather insights about the site and its history.

Acquisition of Camping Rights

Securing the necessary permissions for conducting site investigations is essential. This includes negotiating access rights with landowners or relevant authorities and ensuring compliance with legal and regulatory requirements for site access.

Selection of Sampling and Drilling Sites

Identifying specific locations for sampling and drilling based on preliminary findings is a critical activity in this stage. Evaluating potential sampling sites based on geological features and accessibility helps prioritize areas that are likely to provide critical information about subsurface conditions.

Selection of Best Investigation Methods and Equipment

Determining the most appropriate methods and equipment for the investigation is essential for effective data collection. This involves assessing various investigation techniques, such as drilling, trial pits, and geophysical surveys, and selecting equipment based on site conditions, budget, and project requirements.

3. Detailed Stage

Field Investigations

The detailed stage involves conducting comprehensive field investigations to gather data on actual geological conditions. Activities include excavating trial pits to visually inspect soil layers and collect samples, performing borehole drilling to obtain soil and rock samples at various depths, utilizing geophysical surveys to map subsurface conditions, and conducting in-situ tests to assess soil properties.

Laboratory Investigations

Analyzing collected samples in a controlled environment is crucial for determining their physical and mechanical properties. Laboratory investigations include petrological studies to analyze rock samples, drill core logging to document characteristics of drill cores, and performing laboratory tests to evaluate soil behavior under various conditions.

4. Reporting Stage

Preamble

The reporting stage begins with an introduction that outlines the purpose and scope of the report. This includes summarizing the objectives of the site investigation and the methodologies employed, as well as highlighting the significance of the investigation for the proposed project.

Description of the Site

A detailed description of the site and its surroundings is provided, including information on location, topography, land use, and any relevant historical context. Maps, photographs, and diagrams are included to illustrate site conditions.

Description of Sub-Soil

This section details the findings related to subsurface conditions, presenting the results of field and laboratory investigations, including soil and rock types, stratigraphy, and groundwater conditions. Any geological hazards or anomalies identified during the investigation are also discussed.

Borehole Journals

Borehole journals document the drilling process and findings from each borehole. This includes recording details such as borehole location, depth, soil/rock descriptions, and any encountered groundwater, as well as logs of in-situ tests and observations made during drilling.

Description of Laboratory Soil Tests

Summarizing the results of laboratory analyses is essential for understanding the implications of these results for project design and construction. This section presents test results, including physical and mechanical properties of soil samples.

Conclusion

The conclusion provides a summary of the investigation findings and their relevance to the project. Key insights gained from the investigation are highlighted, along with recommendations for design considerations, construction methods, and any necessary mitigation measures.

Final Conclusion

The stages of site investigation are essential for gathering comprehensive data about subsurface conditions, which informs design and construction practices. By following a systematic approach that includes preliminary assessments, reconnaissance, detailed investigations, and thorough reporting, engineers and geologists can ensure that projects are built on a solid understanding of the ground conditions. This process ultimately contributes to the safety, efficiency, and success of construction projects while addressing environmental concerns and promoting sustainable practices.

Factors Influencing the Choice of Site Investigation Methods

Introduction

The choice of site investigation methods is a critical decision that can significantly impact the success of a project. Various factors influence this choice, including the geological and topographical nature of the site, the type of information required, and considerations of cost and time. Understanding these factors is essential for selecting the most effective and efficient investigation methods, ensuring that the data collected meets the project's needs while optimizing resources.

1. Geological Nature of the Site

1.1 Soil and Rock Types

The geological composition of a site, including the types of soil and rock present, plays a crucial role in determining the appropriate investigation methods. Different soil types, such as clay, sand, and silt, exhibit varying physical and mechanical properties that may require specific sampling and testing techniques. For instance, cohesive soils, which tend to retain water and have a high plasticity, may necessitate different in-situ testing methods compared to granular soils, which are more permeable and less cohesive.

Understanding these differences is vital for selecting the right investigation techniques that will yield accurate and relevant data for engineering decisions.

1.2 Geological Structure

The geological structure of a site, characterized by features such as faults, fractures, and layering, plays a pivotal role in determining the appropriate methods for site investigation. Understanding these structural elements is essential for accurately assessing subsurface conditions, as they can significantly influence groundwater flow, resource availability, and the stability of the ground. Complex geological structures may necessitate the use of specialized investigation techniques, such as geophysical surveys or advanced drilling methods, to ensure that critical data is captured without extensive excavation.

Influence of Geological Features on Investigation Methods

Faults and Fractures: The presence of faults and fractures within the geological structure can create pathways for groundwater movement, affecting aquifer recharge and discharge. If a site is known to have fault lines, it may be essential to employ methods that can detect these features without extensive excavation. Geophysical techniques, such as seismic reflection or resistivity surveys, can be particularly useful in identifying subsurface discontinuities and assessing their implications for groundwater flow and resource management.

Layering: The stratification of geological materials can also influence the choice of investigation methods. Different layers may have varying physical and hydraulic properties, which can affect the behavior of groundwater. For instance, a layer of impermeable rock can act as a barrier to groundwater flow, while permeable layers can facilitate it. Understanding the stratigraphic succession is crucial for selecting optimal locations for boreholes or trial pits, ensuring that the investigation captures relevant data about the subsurface environment.

Case Study: Groundwater Assessment for Utafiti Primary School, Kiambu County, Kenya

This report summarizes the results of a groundwater assessment study conducted for Utafiti Primary School, located within the Nguirubi area of Thigio Sub-Location, Ndeiya Ward, Limuru Sub County, Kiambu County. The client intends to sink a borehole to provide water for domestic purposes within the school and the local community.

Site Location and Description

The site is represented on Topographic Map Sheet No. 148/1-LIMURU (1:50,000), with coordinates UTM 37M 0237349 E & 9865954 N (WGS 84) and an altitude of approximately 2027 meters above mean sea level. The land parcel is identified as Land Reference No. NGUIRUBI/NDIUNI/1380 and covers an area of 1.65 hectares. It is situated within Muguga Forest (Gacuthi and Bama Forest), approximately 1.8 kilometers northwest of Nderi Shopping Centre. Access to the project site is via Waiyaki Way, turning west at Zambezi Shopping Centre and continuing for about 1.8 kilometers towards Muguga Forest.

Hydrogeological Study Overview

The hydrogeological study was conducted to assess the groundwater potential of the area, with findings based on a combination of desk studies, field reconnaissance, and analysis of existing hydrogeological data. The study was carried out in July 2022, and the results indicate that the area has medium to high groundwater potential.

Subsurface Geology

The subsurface geology of the project area consists of a stratigraphic succession that includes recent soil deposits, Kerichwa Valley Tuffs, Kandizi Phonolytes, Limuru Trachytes, Ololua Trachytes, and Nairobi Trachytes. The main aquifer system is expected to be found within the highly weathered Limuru Trachytes, which are known for their capacity to store and transmit groundwater. Additionally, other aquifers may be encountered in the fractured zones of the Kandizi Phonolytes, which can also contribute to the overall groundwater yield.

Groundwater Yield and Recommendations

Based on the available data from existing boreholes in the vicinity, the study indicates variable and moderate yields. It is anticipated that sufficient water will be encountered within a depth of 250 meters to meet the client's water demand of 30,000 liters per day. Therefore, the report recommends drilling the borehole to a maximum depth of 250 meters, targeting the aquifers within the highly fractured and weathered Limuru Trachytes and Kandizi Phonolytes zones.

Environmental Considerations

The geological structure and the presence of aquifers in the area necessitate careful consideration of environmental impacts during the drilling process. Potential risks include groundwater contamination, habitat disruption, and changes to local hydrology. It is crucial to adhere to best practices in borehole drilling, construction, and completion to minimize these impacts. The report provides additional information on procedures and precautions to ensure maximum efficiency and longevity of the borehole.

Conclusion

Understanding the geological structure of a site is essential for effective groundwater assessment and resource management. The case study of Utafiti Primary School highlights the importance of employing appropriate investigation methods based on the geological features present. By accurately assessing subsurface conditions, stakeholders can make informed decisions regarding water resource development while minimizing environmental impacts. The findings of this study underscore the need for a comprehensive approach to groundwater management that considers both the geological context and the potential implications for local ecosystems and communities.

13.9 Groundwater Conditions

The presence and behavior of groundwater can significantly affect site investigation methods. Sites with high groundwater levels may require dewatering techniques during drilling or specific methods to assess groundwater quality and flow. For instance, if groundwater is encountered during drilling, it may necessitate the use of piezometers or other monitoring equipment to track water levels and quality over time. Additionally, understanding groundwater conditions is crucial for evaluating potential contamination risks and designing appropriate mitigation strategies.

Topographical Nature of the Site

Site Accessibility

The physical accessibility of the site can greatly influence the choice of investigation methods. Steep or rugged terrain may limit the use of heavy drilling equipment and necessitate the use of portable or lightweight equipment that can be easily maneuvered. Accessibility issues may also affect the selection of sampling locations and the overall logistics of the investigation. For example, if certain areas of the site are difficult to reach, alternative methods such as remote sensing or geophysical surveys may be employed to gather data without the need for extensive physical access.

Surface Features

The presence of surface features such as vegetation, buildings, and infrastructure can impact the choice of methods as well. Areas with dense vegetation may require preliminary clearing or the use of non-invasive geophysical methods to minimize disturbance to the environment. Furthermore, proximity to existing structures may necessitate careful planning to avoid damage and ensure safety during drilling or excavation. In urban settings, for instance, the presence of underground utilities may require additional precautions and specialized techniques to prevent disruptions.

13.10 Type of Information Required

Specific Data Needs

The type of information required for the project will dictate the choice of investigation methods. If detailed information about soil strength and behavior is needed, methods such as Standard Penetration Tests (SPT) or Cone Penetration Tests (CPT) may be appropriate. Conversely, for environmental assessments, methods that focus on soil and groundwater contamination may be prioritized, such as soil sampling and chemical analysis. Clearly defining the data needs at the outset of the investigation is essential for selecting the most effective methods.

Depth of Investigation

The required depth of investigation can also influence the choice of methods. Shallow investigations may be adequately addressed with trial pits or shallow drilling, while deeper investigations may require rotary drilling or specialized techniques to reach greater depths. The depth of groundwater and the geological profile will also affect the selection of drilling methods. For example, if the project requires information from deep geological layers, advanced drilling techniques that can penetrate hard rock formations may be necessary.

Cost and Time

Budget Constraints

The available budget for site investigation can significantly influence the choice of methods. More advanced and specialized investigation techniques may incur higher costs, which may not be feasible within the project budget. Therefore, cost-effective methods that provide sufficient data for decision-making may be prioritized, especially for preliminary investigations. It is essential to balance the need for comprehensive data with budgetary constraints to ensure that the investigation remains financially viable.

Project Timeline

The time available for site investigation can also impact the choice of methods. Rapidly changing project timelines may necessitate the use of quicker investigation methods, such as geophysical surveys or existing data reviews, rather than extensive drilling and laboratory testing. The urgency of the project may also affect the scheduling of fieldwork and laboratory analyses, requiring a flexible approach to ensure that critical deadlines are met without compromising data quality.

Groundwater Conditions

The presence and behavior of groundwater are critical factors that can significantly influence site investigation methods. Understanding groundwater conditions is essential for evaluating potential contamination risks, designing appropriate mitigation strategies, and ensuring the successful implementation of construction projects.

Influence of Groundwater Levels on Investigation Methods

Sites with high groundwater levels may require specific techniques to manage water during drilling operations. Dewatering techniques, such as well-point systems or sump pumps, may be necessary to lower the water table temporarily, allowing for safe and effective drilling. If groundwater is encountered during drilling, it may necessitate the use of piezometers or other monitoring equipment to track water levels and quality over time. Piezometers can provide valuable data on hydraulic head, groundwater flow direction, and potential contamination, which are crucial for understanding the hydrogeological dynamics of the site.

Additionally, understanding groundwater conditions is vital for assessing the potential for contamination from nearby sources, such as agricultural runoff or industrial activities. This knowledge can inform the design of protective measures, such as the installation of impermeable barriers or the implementation of monitoring programs to detect changes in water quality.

Environmental Impacts of Groundwater Conditions

1. **Contamination Risks**: High groundwater levels can increase the risk of contamination from surface activities. Pollutants from agricultural runoff, industrial discharges, or septic systems can easily infiltrate the groundwater, leading to widespread contamination. This can affect not only the immediate site but also surrounding areas, posing risks to public health and the environment.

2. **Altered Hydrology**: Dewatering activities can alter the natural hydrology of an area, potentially leading to reduced groundwater recharge and changes in surface water flow patterns. This can have downstream effects on ecosystems that rely on consistent water levels, such as wetlands and rivers.

3. **Soil Erosion and Instability**: The removal of groundwater through dewatering can lead to soil erosion and instability, particularly in areas with loose or unconsolidated materials. This can increase the risk of landslides or subsidence, particularly in steep or hilly terrains.

4. **Impact on Vegetation**: Changes in groundwater levels can affect local vegetation, particularly in riparian zones where plants depend on consistent moisture levels. A decline in groundwater can lead to stress

on these plants, potentially resulting in reduced biodiversity and habitat loss.

Topographical Nature of the Site

Site Accessibility

The physical accessibility of the site can greatly influence the choice of investigation methods. Steep or rugged terrain may limit the use of heavy drilling equipment, necessitating the use of portable or lightweight equipment that can be easily maneuvered. Accessibility issues may also affect the selection of sampling locations and the overall logistics of the investigation. For example, if certain areas of the site are difficult to reach, alternative methods such as remote sensing or geophysical surveys may be employed to gather data without the need for extensive physical access.

In the case of Utafiti Primary School, located in the Nguriunditu area of Kiambu County, the site is accessed through Waiyaki Way, turning west at Zambezi Shopping Centre and continuing towards Muguga Forest. The accessibility of the site is crucial for the planned borehole drilling, as it determines the type of equipment that can be used and the feasibility of transporting materials and personnel to the site.

Environmental Impacts of Site Accessibility

1. **Disturbance to Natural Habitats**: Limited accessibility may require the construction of access roads or paths, which can disturb natural habitats and lead to habitat fragmentation. This can negatively impact local wildlife and plant species.

2. **Soil Compaction and Erosion**: Increased traffic from construction vehicles can lead to soil compaction and erosion, particularly in sensitive areas. This can reduce soil fertility and disrupt local ecosystems.

3. **Pollution from Equipment**: The use of heavy machinery and vehicles can lead to soil and water pollution from fuel spills, oil leaks, and other contaminants. This can have long-term effects on local groundwater quality and surface water bodies.

Surface Features

The presence of surface features such as vegetation, buildings, and infrastructure can also impact the choice of investigation methods. Areas with dense vegetation may require preliminary clearing or the use of non-invasive geophysical methods to minimize disturbance to the environment. Furthermore, proximity to existing structures may necessitate careful planning to avoid damage and ensure safety during drilling or excavation. In urban settings, for instance, the presence of underground utilities may require additional precautions and specialized techniques to prevent disruptions.

Type of Information Required

Specific Data Needs

The type of information required for the project will dictate the choice of investigation methods. If detailed information about soil strength and behavior is needed, methods such as Standard Penetration Tests (SPT) or Cone Penetration Tests (CPT) may be appropriate. Conversely, for environmental assessments, methods that focus on soil and groundwater contamination may be prioritized, such as soil sampling and chemical analysis. Clearly defining the data needs at the outset of the investigation is essential for selecting the most effective methods.

In the context of the Utafiti Primary School borehole project, the specific data needs include understanding the groundwater yield, quality, and the characteristics of the aquifers present in the area. This information is critical for ensuring that the borehole can meet the school's water demand of 30,000 liters per day.

Environmental Impacts of Data Collection Methods

1. **Invasive Sampling Techniques**: Methods such as drilling and soil sampling can disturb the natural environment, leading to habitat destruction and soil erosion. Careful planning is required to minimize these impacts.

2. **Chemical Contamination**: The use of chemical agents for soil and water testing can pose risks of contamination if not handled properly. Proper disposal and management of these chemicals are essential to prevent environmental harm.

Depth of Investigation

The required depth of investigation can also influence the choice of methods. Shallow investigations may be adequately addressed with trial pits or shallow drilling, while deeper investigations may require rotary drilling or specialized techniques to reach greater depths. The depth of groundwater and the geological profile will also affect the selection of drilling methods. For example, if the project requires information from deep geological layers, advanced drilling techniques that can penetrate hard rock formations may be necessary.

In the case of Utafiti Primary School, the hydrogeological study indicates that sufficient water is expected to be encountered within a depth of 250 meters, targeting the highly weathered Limuru Trachytes and Kandizi Phonolytes. This depth requirement necessitates the use of appropriate drilling techniques to ensure successful borehole construction.

Environmental Impacts of Depth of Investigation

1. **Groundwater Depletion**: Drilling to significant depths can lead to the depletion of groundwater resources, particularly if the aquifer is not recharged adequately. This can have long-term implications for local water availability.
2. **Disturbance of Geological Layers**: Deep drilling can disrupt geological layers, potentially leading to the release of contaminants trapped in deeper strata. This can affect groundwater quality and the surrounding environment.

Cost and Time

Budget Constraints

The available budget for site investigation can significantly influence the choice of methods. More advanced and specialized investigation techniques may incur higher costs, which may not be feasible within the project budget. Therefore, cost-effective methods that provide sufficient data for decision-making may be prioritized, especially for preliminary investigations. It is essential to balance the need for comprehensive data with budgetary constraints to ensure that the investigation remains financially viable.

For the Utafiti Primary School project, the budget will need to account for drilling costs, equipment rental, and any necessary environmental assessments. Careful planning and budgeting will be essential to ensure that the project can proceed without financial overruns.

Environmental Impacts of Budget Constraints

1. **Compromised Environmental Standards**: Budget constraints may lead to the adoption of less rigorous environmental standards, increasing the risk of negative impacts on local ecosystems and groundwater quality.

2. **Reduced Monitoring and Mitigation**: Limited budgets may restrict the ability to implement comprehensive monitoring and mitigation strategies, potentially leading to long-term environmental degradation.

Project Timeline

The time available for site investigation can also impact the choice of methods. Rapidly changing project timelines may necessitate the use of quicker investigation methods, such as geophysical surveys or existing data reviews, rather than extensive drilling and laboratory testing. The urgency of the project may also affect the scheduling of fieldwork and laboratory analyses, requiring a flexible approach to ensure that critical deadlines are met without compromising data quality.

In the case of Utafiti Primary School, the timeline for drilling the borehole will need to align with the school's operational schedule and the community's water needs. Efficient planning and execution of the investigation will be crucial to meet these timelines.

Environmental Impacts of Project Timelines

1. **Rushed Assessments**: Tight timelines may lead to rushed assessments, increasing the likelihood of overlooking critical environmental factors and potential risks.

2. **Inadequate Community Engagement**: Limited time for community engagement can result in a lack of stakeholder input, which is essential for identifying local concerns and ensuring that the project aligns with community needs.

Hydrogeology

Groundwater Occurrence

Groundwater typically occurs in the pores and interstices of various rock formations, depending on the geological conditions and physiography of the area. Factors such as permeability, porosity, the degree and depth of weathering, and historical tectonic conditions play a significant role in groundwater occurrence. The recharge conditions, including precipitation patterns and evapotranspiration rates, are also critical in determining groundwater availability. Groundwater exists in two primary zones: the vadose zone (or zone of aeration) and the phreatic zone (or zone of saturation). The vadose zone contains three types of water: soil water, intermediate vadose water, and capillary water. Soil water is essential for plant growth, while the intermediate zone retains water that is not readily available for use. Capillary water is held in the smallest spaces between soil particles and percolates downwards to the zone of saturation when the holding capacity is exceeded.

There are two types of groundwater based on origin: connate water, which is trapped in sediments at the time of deposition, and meteoric water, which infiltrates the rock through subsurface processes. Groundwater can be found in one continuous body or in distinct layers, with the thickness of the groundwater-bearing zone influenced by local geology, weathering, porosity, and recharge dynamics.

Aquifer Characteristics

Aquifer Transmissivity: The transmissivity of an aquifer, often denoted by T, is a measure of how much water can be transmitted horizontally in a pumping well. It is directly proportional to average horizontal permeability (K_h) and aquifer thickness (D). Its approximate measurement is obtained from dividing the borehole yield, Q, by the drawdown, s_w. The main source of error is made by inaccurate estimates of aquifer thickness. T is thus estimated by the following equation:

$$T=Q/ s_w$$

The transmissivity of the proposed site can be estimated using borehole; **C-4865:** located 1.4km southeast from proposed site.

Thus T = 13.08*24/80.99 = **3.9m²/day.**

Borehole Specific Capacity: Specific capacity is a material physical property that characterizes the capacity of an aquifer to release groundwater from storage in response to a decline in hydraulic head. It is estimated using test pumping data and is crucial for understanding the aquifer's response to extraction.

Storage Coefficient/Specific Yield: The storage coefficient is the volume of water released from storage in a confined aquifer per unit decline in hydraulic head. This property is essential for assessing how much water can be sustainably extracted from the aquifer.

Hydraulic Conductivity: Hydraulic conductivity (K) describes the ease with which water can move through pore spaces or fractures in the aquifer. It is influenced by the intrinsic permeability of the material and the degree of saturation. The hydraulic conductivity can be estimated using test pumping data.

The Hydraulic Conductivity (K) is estimated as follows:

$$K = T/\text{Aquifer Thickness}$$

$$3.9/16 = \mathbf{0.24m/day}$$

Groundwater Flux: The groundwater flux of an aquifer is the rate of flow across a unit area ($m^3.m^{-2}.s^{-1}$) which is obtained from Darcy's Law of groundwater flow. The estimated groundwater flux, *v*, may be estimated using the test pumping data such as the drawdown, borehole diameter and discharge per given time.

$V = Q/A$ Q=discharge per second, A=drawdown x borehole circumference

The groundwater flux, , *v*, = 0.013m (14/8400). 16. 8100

$$\boldsymbol{v} = \mathbf{2.81m^3/day}$$

Previous Groundwater Development

There are several boreholes in the vicinity of the project site whose depths vary from 80 to 122m depth. They also have varying yields of between 4.63 and 16.3m^3 of water per hour. These boreholes have struck aquifers between 69.5 and 100m bgl The data for the boreholes drilled in the vicinity of the project area is given in Table 1.4 below from the Water Resources Management Authority (WRMA).[Map Sheet 148/1 – LIMURU 1:50,000]

Table 1.4: Data of Boreholes in the Vicinity of the Project Site

Bh No.	Owner	Distance from proposed site (Km)	Depth (m)	WSL (m)	WRL (m)	Q (m^3/hr)	PWL (m)
Ref: Utafiti Primary School.							
394	Kikuyu Estate	2.4NNE	80	76	13	12.3	-
3615	Kinyanjui Thuku	5.3N	85.3	79.2,85.3	71.9	4.36	-
3710	Sigona Golf	2.3ESE	86.9	69.5, 86.8	57.9	16.3	-
4865	M.O.W.D	1.4SE	122	95	85.09	13.08	80.99
13462	Kamuguga Water Project	2.2NE	120	100	81.39	6.48	82.2
Range			80-122	69.5-100	13-85.09	4.36-16.3	80.99-82.2

Environmental Impacts of Aquifer Characteristics

1. **Over-extraction Risks**: If the aquifer's transmissivity and specific capacity are not accurately assessed, there is a risk of over-extraction, leading to depletion of groundwater resources and long-term sustainability issues.

2. **Contamination Potential**: Poorly understood aquifer characteristics can increase the risk of contamination, particularly if extraction leads to changes in hydraulic gradients that allow pollutants to migrate into the aquifer.

3. **Ecosystem Disruption**: Changes in groundwater levels due to extraction can disrupt local ecosystems, particularly in areas dependent on consistent water availability, such as wetlands and riparian zones.

Water Quality

Overall, the groundwater quality in the Nairobi area is generally good; however, it is important to note that fluoride levels (Table 1.5) consistently exceed the World Health Organization (WHO) upper limit of 1.5 parts per million (ppm). This elevated fluoride concentration is primarily associated with the main aquifer system, which comprises the Athi lakebeds and surrounding sediments. High fluoride intake, particularly in growing infants, can lead to dental or skeletal fluorosis, which poses significant health risks.

Before any use of water from the borehole is initiated, it is essential to conduct a comprehensive chemical analysis to assess its quality. This analysis will help identify any contaminants and ensure that the water meets safety standards for consumption. Below are the recommended thresholds for drinking water quality.

Table 1.5 Recommended Drinking Water Action Levels

Quality Variable		Measuring Unit	WHO Recommended concentration	Comments
Colour		TCU	15	
Taste/Odour		-	Un-objectionable	
pH		pH Units	6.5 - 8.5	
Turbidity		NTU	5	
T. D. S		Mg/l	1,500	
Suspended material		-	Nil	
Hardness	($CaCO_3$)	Mg/l	500	
Arsenic/	(As).	mg/l.	0.05.	Toxic in excess,. Bronchial diseases.
Lead	(Pb)	Mg/l	0.05	Toxic to animals
Selenium	(Se)	"	0.01	Toxic in excess

Alumininum	(Al)	"	0.1	Soluble salts exhibit neurotoxicity.
Manganese	(Mn)	Mg/l	0.1	
Iron	(Fe)	"	0.3	
Mercury	(Hg)	"	0.001	
Magnesium	(Mg)	"	100	May cause diarrhoea in new users
Calcium	(Ca)	"	250	
Copper	(Cu)	"	0.1	
Ammonia organisms	(N)	Mg/l	0.5	Toxic, particluarly to aquatic
Fluoride	(F)	Mg/l	1.5	Cause dental & Skeletal Fluorosis
Chloride	(Cl)	"	250	
Sulphide	(H_2)	µg/l	Undetectable	
Nitrate	(NO_3)	"	10	Infant blue baby Syndrome
Sulphate	(SO_4)	Mg/l	400	Taste, Odour, cathartic effects
Sodium	(Na)	"	200	Chronic, long term toxic
Zinc	(Zn)	"	5	Toxic in excess
Total Coliforms		per 100ml	Nil	
Faecal Coliforms		per 100ml	Nil	
Total Phenols		Mg/l	0.002	

Conclusion

The assessment of groundwater conditions, topographical features, and specific data needs is crucial for effective site investigations. The case study of Utafiti Primary School underscores the significance of understanding these factors when drilling a borehole for water supply. By considering the environmental impacts associated with groundwater extraction and site accessibility, stakeholders can make informed decisions that balance resource needs with environmental stewardship. This comprehensive approach is vital for ensuring the sustainability of groundwater resources and protecting local ecosystems for future generations.

The selection of site investigation methods is influenced by various factors, including the geological and topographical characteristics of the site, the specific information required, and considerations of cost and time. By carefully evaluating these elements, engineers and geologists can choose the most appropriate and effective methods for site investigation, ensuring that the data collected meets project requirements while optimizing resources and minimizing risks. A well-planned site investigation lays the foundation for successful project execution and enhances the overall safety and efficiency of construction activities.

Concluding Remarks

In summary, site investigation is a fundamental process that establishes the groundwork for successful engineering and construction projects. A systematic approach to site investigation—encompassing preliminary assessments, reconnaissance, detailed investigations, and comprehensive reporting—ensures that thorough data is collected regarding subsurface conditions. Each stage serves a specific purpose, from initial assessments and validation of geological interpretations to detailed field and laboratory investigations, culminating in a thorough reporting of findings.

The choice of site investigation methods is shaped by a multitude of factors, including geological and topographical characteristics, specific information needs, and considerations of cost and time. Environmental factors also play a critical role in shaping the investigation approach. Understanding the potential for soil and groundwater contamination, assessing ecological sensitivities, and adhering to regulatory requirements are essential for responsible site investigation practices.

By carefully evaluating these factors, engineers and geologists can select the most appropriate and effective methods for site investigation. This not only ensures that the data collected meets project needs but also promotes sustainable practices and minimizes environmental impacts. Ultimately, a well-executed site investigation contributes to the safety, efficiency, and success of construction activities, fostering responsible management of natural resources and the built environment.

Recommendations

1. **Conduct Comprehensive Preliminary Assessments**: Prioritize thorough preliminary investigations to gather existing data and identify potential geological hazards early in the project lifecycle.

2. **Utilize a Multi-Method Approach**: Employ a combination of investigation methods, including geophysical surveys, drilling, and sampling, to obtain a holistic understanding of subsurface conditions.

3. **Incorporate Environmental Considerations**: Assess environmental impacts as part of the site investigation process. Implement measures to mitigate potential contamination and protect sensitive ecosystems.

4. **Engage Stakeholders Early**: Involve local stakeholders, including landowners and regulatory agencies, in the planning stages to facilitate access and ensure compliance with environmental regulations.

5. **Invest in Training and Technology**: Stay updated with advancements in site investigation technologies and methodologies. Invest in training for personnel to enhance their skills and knowledge in modern investigation techniques.

6. **Document and Report Findings Thoroughly**: Ensure that all findings are documented comprehensively and communicated clearly in reports. This transparency aids in decision-making and fosters trust among stakeholders.

7. **Review and Adapt Practices**: Regularly review site investigation practices and adapt them based on lessons learned from previous projects and emerging industry standards.

By implementing these recommendations, practitioners can enhance the effectiveness of site investigations, ensuring that they are not only efficient and cost-effective but also environmentally responsible and aligned with best practices in the field.

Fieldwork & Geophysics

Introduction

This method utilizes the existing natural electric field as a source for probing and detecting underground rocks, minerals, and groundwater. By leveraging resistivity contrasts, the instrument can identify subsurface features at various depths. The technique involves measuring the natural electric field at the surface between two selected points (M & N), allowing for the recording of different frequency electric field components. Variations in the collected data can lead to further interpretations, enabling the identification of abnormal geological features or bodies based on significant changes in the data.

The working principle relies on the electrical differences in the natural earth's magnetic field, utilizing specific frequencies ranging from 0 to 170 kHz. These frequencies provide information about specific depths. The natural electric field method measures the electrical component of the electromagnetic field of the earth, and the frequency selection method is commonly referred to as the natural potential frequency method. Instruments designed to operate on this principle are known as natural selected frequency electric field instruments, which can be configured to meet specific requirements.

The instrument (PQWT-TC700) used in conducting the above Hydrogeological survey can conduct the following measurement/specifications;

1) Single Frequency Measurements.
2) Three Frequency Measurements. 170 Hz (shallow), 67 Hz (middle level), 25h Hz (Deep level).
3) Band (many) Frequency Measurements-used for Profile Imaging-. 0 to 170 Hz.

Note:

a) The Equipment has a fixed probing Depth of 500 meter and specifications of measuring the above frequencies.
b) This Survey was undertaken using the Profile Imaging (Band Frequency Measurements).

Methodology

The methodology for measuring the natural electric field involves placing wired electrodes in a specific direction and location at 10-meter intervals (Fig. 32). The straight line under investigation should be perpendicular to tectonic lines. A calibrated tape measure is used to mark the points/sites accurately.

1. **Initial Setup**: The first data collection involves positioning one electrode (M) at the 0-meter mark and the other electrode (N) at the 10-meter mark on the tape.

2. **Data Collection**: After collecting data at the initial point, both electrodes (M & N) are moved 1 meter forward for the next data collection. This process is repeated systematically, with each subsequent data collection point marked at 1-meter intervals along the tape.

3. **Equidistant Connection**: The electrodes (M and N) should maintain an equidistant connection while collecting data, ensuring consistency in measurements.

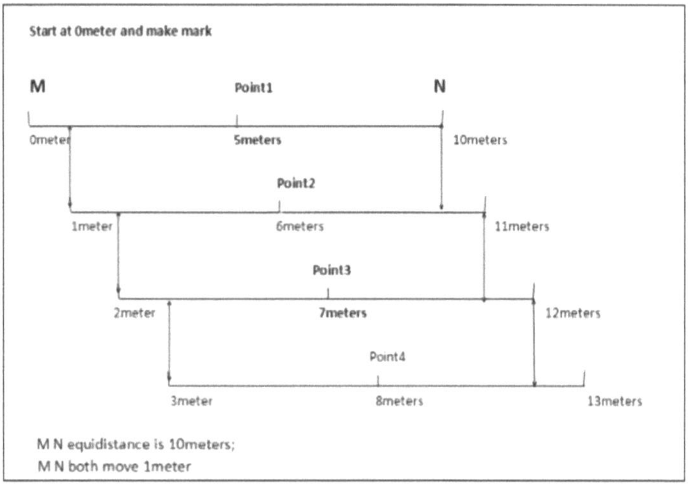

(Note: It is better to measure and tap at 10Meters of M N equidistance, and both M N move 1 meter, because the water detector was designed at 10meters of M N equidistance, and M N both move 1meters. changing the MN equidistance has effect on the depth of measurement)

Figure 32: Procedure for Exploration using PQWT instrument.

How to calculate the meters on the ground, according to point in the profile map? Example what meters on the ground at Point 7 in the profile map?

M N=10M and both M N move 2 meters, point 7 is at 17 meters on the ground. (point number multiply by 2 +3meters).

The systematic approach to data collection allows for a comprehensive analysis of the subsurface conditions, facilitating the identification of geological features and groundwater resources. The collected data will be analyzed to interpret the resistivity variations, which can provide insights into the geological and hydrological characteristics of the area under investigation.

Field Work

Fieldwork was conducted on July 29, 2022, during which observations were made regarding the general topography, drainage patterns, geological setup, and the proximity to neighboring boreholes. The profile imaging survey commenced with a horizontal profile across the school's land. Utilizing the PQWT-TC underground water detector, a comprehensive profile imaging survey of the site area was performed, as illustrated in the results figures below.

The selection of the drilling point was influenced by several factors, including accessibility, geophysical results, and proposed infrastructure. The area under investigation is situated in the Nguriunditu area of Kiambu County.

Results

The data collection process was successful, yielding profile curves and images that were instrumental in interpreting the underground water prospects of the area. The results indicate varying resistivity levels, which provide insights into the subsurface geological conditions and the potential for groundwater availability. These findings will guide the decision-making process for the drilling location, ensuring that the selected site is optimal for accessing groundwater resources.

Table 1.6: Fielswork data collection process and observations

PROFILE No.	SELECTED POINTS	GRID REFRENCES	UTM-37M
Profile 1 – selected point	Point 7 on the profile	**0237349,9865954**	**E 0237349** **N 9865954**

NOTE: Data collection for profiles of both sites had the data collected under the command of +2 Gain function and therefore improving on quality of signature of these specific lines. The Gain function was adopted due values noticed at the start of every data set collection of these lines.

i. ***Profile Imaging survey-LINE 1***

A profile survey was done across the client parcel of land. A total of 11 Point station was taken as displayed below by the profile curve (Figs. 33 and 34).

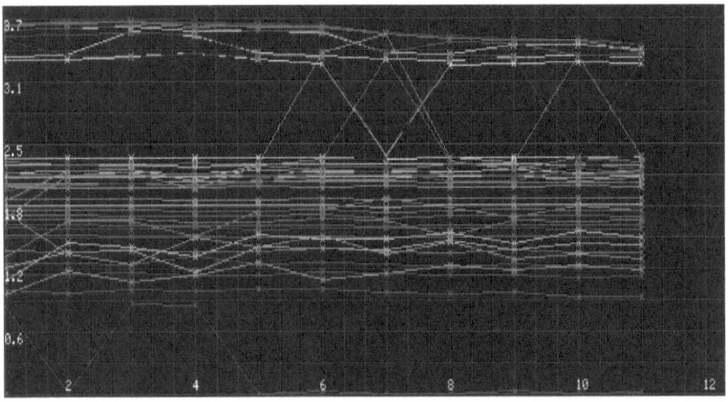

Figure 33: Raw Original curves produced on collection after measurement of the14 points.

1) Sub-Surface Profile of the project area

PROPOSED
BH SITE

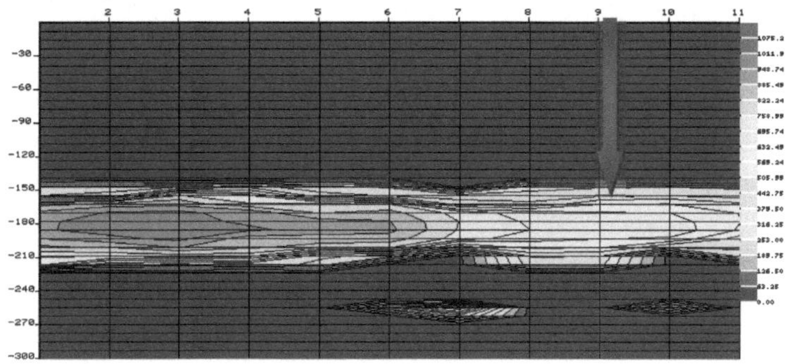

Figure 34: Original Map for the Surveyed Line

Following consideration of the geological set up at the site area, horizontal profile 1 (Figs. 33 and 24, above) was done at a constant separation of 2m. The profile was able to identify a section that has a well define zone of weathering/Fracture. As indicated in the Figure 35 the best drilling position was marked within profile 1 is at point/station 7 (on ground 17m). The site location shows fair potential sub-surface/underground zone, thus a good potential site for groundwater/borehole drilling.

Based on the geophysical data, first aquifer prospects on point 7 is expected at depth 60m to 90m. lower saturation parts of this first aquifer is expected at 120m to depth 150m, a column with high probability of encountering a thick clay formation. The deeper aquifer is expected from from depth 230m to 250m to deeper levels. Notably the famous clay is usually 80m to 120m thick, and should be penetrated during drilling to enable abstracting water from the deeper aquifer below it.

Groundwater prospects at this site are fair due to the low codded values observed and from the Profile, station 7 is the best point for profile 1. The station 7 has been recommended as the site for the proposed borehole to depth 250m

Conclusions and Recommendations

Based on the hydrogeological study conducted, several conclusions have been drawn regarding the project site. The geological and hydrogeological conditions indicate favorable prospects for accessing groundwater, making it suitable for borehole drilling. The aquifer systems in the area exhibit moderate to high productivity, primarily characterized by the highly fractured and weathered Kandizi Phonolytes and Limuru Trachytes. Borehole yields in the region are generally variable, reflecting differences in local hydrogeological conditions within the subsurface geological formations. The proposed borehole is estimated to yield at least 8,000 liters of water per hour, while the anticipated water requirement from the borehole is a maximum of 30,000 liters per day for domestic use.

While the proposed borehole project presents opportunities for improved water access, it is essential to consider the potential environmental impacts associated with groundwater extraction and borehole construction. One significant concern is groundwater depletion, as over-extraction can lead to a decline in the water table, affecting nearby wells, surface water bodies, and local ecosystems that depend on consistent groundwater levels. Additionally, there are contamination risks; improper construction and sealing of the borehole can lead to contamination from surface runoff or shallow groundwater, compromising water quality and posing health risks to users. Changes in groundwater levels can also disrupt local ecosystems, particularly in areas with sensitive habitats such as wetlands, leading to loss of biodiversity and alteration of natural habitats. Furthermore, the drilling process may disturb the soil and surrounding environment, potentially leading to erosion or instability, especially in areas with loose or unconsolidated materials. Lastly, the drilling operation may generate noise and vibrations that can disturb local wildlife and nearby residents, impacting their quality of life and the natural environment.

To ensure the successful implementation of the borehole project while minimizing environmental impacts, several recommendations are proposed. First, the borehole should be drilled at UTM coordinates 37M 0237349 E & 9865954 N (WGS 84), where a subsurface profile has been conducted and is known to the client, to a maximum depth of 250 meters. The borehole should have a minimum diameter of 8 inches and must be lined with appropriate steel casings and screens, following the design specifications recommended by the supervising hydrogeologist. To prevent contamination from shallow subsurface water, the borehole should be properly gravel packed and sealed, with special attention given to the guidelines provided in the accompanying addendum on borehole drilling, design, and construction. Additionally, the borehole should be equipped with a water master meter to monitor groundwater abstraction, as well as an airline for future monitoring of water levels. It is crucial that the drilling, construction, and test-pumping of the borehole be supervised by a qualified hydrogeologist to ensure compliance with best practices and safety standards.

Implementing a monitoring program to assess the impacts of groundwater extraction on local ecosystems and water quality is also recommended. This should include regular assessments of groundwater levels and quality, as well as monitoring of nearby surface water bodies. Engaging with local stakeholders and communities is essential to inform them about the project, address concerns, and ensure that the project aligns with community needs and environmental stewardship. Furthermore, the client should be aware that before drilling commences, a groundwater abstraction permit must be obtained from the Water Resources Management Authority (WRMA), Athi Catchment in Kiambu, through the Kiambu Sub-Regional Office. This step is crucial for ensuring legal compliance and sustainable water resource management.

Geo-materials in Kenya

Kenya is rich in various geo-materials that can be utilized for construction and other industrial purposes. Major un-weathered rock masses for dimension stones include granite intrusives, basalt, and limestone. Granite intrusives, found predominantly in the Greenstone-Granite Belt, are ideal for dimension stones and crushed aggregates. Basalt, commonly found in areas with volcanic activity, is used for aggregates in concrete and road construction. Limestone, found in various regions, is utilized for cement production and as a dimension stone.

In terms of major rock quarries, Vihiga County is known for its massive granite intrusives, with several quarries exploiting these resources for building stones and aggregates. Nairobi County hosts numerous quarries that extract basalt and limestone for construction purposes, while Kajiado County is recognized for its extensive deposits of volcanic rocks, producing aggregates for concrete and road construction.

Case Studies

Case Study 1: Building Stones and Crushed Aggregates

In Vihiga County, granite is extracted through a combination of blasting and mechanical methods. The quarrying process begins with the identification of suitable granite deposits, followed by the use of explosives to break the rock into manageable blocks. Once fragmented, heavy machinery is employed to lift and transport the blocks to processing facilities. The extracted granite blocks are then cut into dimension stones for use in construction or crushed into aggregates for concrete production.

The granite blocks undergo processing in local facilities where they are cut, polished, and finished to meet specific construction standards. This processing involves several steps, including cutting large blocks into smaller, more manageable pieces using diamond saws or wire saws, which provide precision and reduce waste. After cutting, the stones are polished to achieve a smooth finish, enhancing their aesthetic appeal for use in buildings and monuments. Additional finishing processes may include honing, flaming, or sandblasting to achieve desired textures and appearances.

However, the quarrying activities in Vihiga County also have environmental impacts that must be addressed. Habitat destruction is a significant concern, as the establishment of quarries can lead to the destruction of local habitats, affecting flora and fauna in the area. The removal of vegetation and soil disrupts ecosystems and can lead to a loss of biodiversity. Additionally, quarrying activities can expose soil to erosion, particularly if proper rehabilitation measures are not implemented. This can result in sedimentation in nearby water bodies, negatively affecting water quality and aquatic life.

Water pollution is another potential impact, as the use of blasting and heavy machinery can contaminate nearby water sources with sediments, chemicals, and other pollutants. This contamination can have detrimental effects on local water quality and the health of communities relying on these sources. Furthermore, dust generated during quarrying and processing can contribute to air pollution, affecting the health of workers and nearby residents. Prolonged exposure to dust can lead to respiratory issues and other health problems. The noise generated from blasting and the operation of heavy machinery can also disturb local wildlife and impact the quality of life for nearby communities.

To mitigate these environmental impacts, rehabilitation measures are implemented after extraction. This involves filling voids left by quarrying with soil and other materials to restore the landscape and prevent the formation of hazardous pits. Additionally, native plants are introduced to stabilize the soil, enhance biodiversity, and restore the natural habitat. This helps prevent erosion and promotes ecological recovery. The granite dimension stones produced in Vihiga County are utilized in various applications, including residential and commercial buildings, where their durable and aesthetically pleasing nature makes them a popular choice for facades, countertops, and flooring. Crushed granite aggregates are essential for concrete production, providing strength and durability to structures, and are also used in road construction, contributing to the stability and longevity of roadways.

Case Study 2: Artisanal Dimensioning of Pyroclastics

In various regions of Kenya, artisanal dimensioning of pyroclastics occurs in areas characterized by volcanic activity, particularly along the East African Rift Valley. The availability of pyroclastic materials, such as tuffs and agglomerates, provides local communities with valuable resources for construction, especially in rural areas where access to conventional building materials may be limited. Tuffs, which are consolidated volcanic ash, offer lightweight and thermal insulation properties, while agglomerates, composed of larger volcanic fragments, are also lightweight and easy to work with, making them suitable for construction.

Local artisans engage in the extraction of pyroclastic materials using hand tools such as chisels, hammers, and picks. This artisanal approach allows for a controlled and selective extraction process, minimizing disturbance to the surrounding environment. The extraction process typically involves identifying suitable deposits, carefully removing overburden, and selectively extracting the tuffs and agglomerates with minimal damage to the surrounding landscape, thereby preserving the integrity of the ecosystem.

Once extracted, the tuffs and agglomerates are often processed on-site. Artisans cut and shape the materials into usable blocks tailored to specific construction needs. This processing may involve manual techniques, where artisans use hand saws or chisels to create uniform blocks that can be easily transported and assembled. On-site processing reduces transportation costs and allows artisans to maintain a close connection to the materials, ensuring that the final products meet local construction standards and preferences.

Despite the lower environmental impact of artisanal mining compared to large-scale operations, there are still concerns. Land degradation can occur if not managed properly, as over-extraction can lead to the loss of vegetation and soil quality. Erosion and sedimentation can increase due to the removal of vegetation and soil during extraction, negatively impacting aquatic ecosystems and water quality. Unsustainable extraction practices can lead to resource depletion, making it difficult for future generations to access these materials. Additionally, the processing of pyroclastics can generate waste materials, which, if not managed properly, can contribute to environmental pollution and landscape degradation. Health risks are also a concern, as artisanal miners may be exposed to dust and noise generated during extraction and processing, leading to respiratory issues and other health problems.

To address these environmental impacts, local artisans are encouraged to adopt sustainable practices, including selective extraction methods that focus on minimizing land disturbance and preserving the surrounding ecosystem. By avoiding over-extraction and maintaining vegetation cover, artisans can reduce the risk of habitat destruction and soil erosion. Community training programs are also implemented to educate miners on sustainable practices and rehabilitation techniques. These programs aim to raise awareness about the importance of environmental stewardship and the long-term benefits of sustainable practices, while also equipping artisans with the knowledge and skills needed to restore mined areas after extraction, including replanting native vegetation and managing waste.

The pyroclastic blocks produced through artisanal dimensioning are primarily used in various applications, particularly in low-cost housing. The lightweight nature of tuffs and agglomerates makes them an affordable alternative to conventional building materials, which is especially valuable in rural areas where budget constraints are common. These materials are also utilized in community infrastructure projects, such as constructing schools, health clinics, and community centers, thereby supporting local development initiatives and enhancing the quality of life for residents.

In summary, both case studies highlight the importance of sustainable practices in the extraction and utilization of geo-materials in Kenya. The granite dimension stones from Vihiga County and the artisanal pyroclastic materials from various regions demonstrate the potential for local resources to contribute to construction and development. However, it is crucial to balance resource extraction with environmental stewardship to ensure that these activities do not compromise the health of ecosystems or the well-being of local communities. By implementing effective mitigation measures, engaging in community training, and adhering to sustainable practices, the negative impacts of quarrying and artisanal mining can be minimized, allowing for responsible resource management that benefits both current and future generations.

Ultimately, the successful integration of geo-materials into construction practices in Kenya hinges on a commitment to sustainability, community engagement, and environmental protection. As the demand for construction materials continues to grow, it is imperative that stakeholders work collaboratively to develop strategies that promote the responsible use of natural resources while safeguarding the environment and supporting local economies. The use of locally sourced materials not only supports the local economy but also fosters a sense of community ownership and pride in construction projects. By providing affordable building materials, artisanal dimensioning of pyroclastics contributes to improved living conditions and supports sustainable development in the regions where these materials are extracted.

Conclusion

The artisanal dimensioning of pyroclastics in Kenya exemplifies how local communities can harness natural resources for construction while promoting sustainable practices. By focusing on selective extraction, on-site processing, and community engagement, artisans can minimize environmental impacts and contribute to the development of affordable housing solutions. This case study highlights the importance of integrating sustainable practices into artisanal mining, ensuring that the benefits of geo-materials are realized without compromising the health of local ecosystems or communities.

In contrast, the quarrying of granite in Vihiga County demonstrates the potential for large-scale extraction to meet construction demands while also emphasizing the need for responsible practices, such as quarry rehabilitation and environmental management. Both case studies illustrate the critical balance between resource utilization and environmental stewardship in the construction materials sector. By fostering a culture of sustainability and collaboration among stakeholders, Kenya can effectively leverage its rich geo-material resources to support economic growth and community development while protecting its natural environment for future generations.

Summary of Key Environmental Impacts and Mitigation Strategies

Case Study 1: Building Stones and Crushed Aggregates

Environmental Impacts: Quarrying activities in Vihiga County have profound environmental consequences that extend beyond the immediate area of extraction. One of the most pressing issues is habitat destruction, which leads to the displacement of various species and a significant loss of biodiversity. The removal of vegetation not only disrupts local ecosystems but also eliminates critical habitats for wildlife. Additionally, the extraction process increases soil erosion, as the protective layer of vegetation is stripped away. This erosion can lead to sedimentation in nearby water bodies, which can smother aquatic habitats, disrupt the natural flow of rivers, and degrade water quality, ultimately affecting both human health and aquatic life.

Water pollution is another significant concern associated with quarrying. The use of explosives and heavy machinery can introduce harmful pollutants into local water sources, compromising water quality and posing risks to communities that rely on these sources for drinking water and agriculture. Furthermore, the dust generated during extraction and processing can severely degrade air quality, leading to respiratory issues among nearby residents and workers. Prolonged exposure to dust can result in chronic health problems, particularly in vulnerable populations such as children and the elderly. Noise pollution from blasting and heavy machinery operation can also disturb local wildlife, leading to changes in animal behavior and impacting the quality of life for nearby communities.

Mitigation Strategies: To address these environmental impacts, several effective mitigation strategies can be implemented. Quarry rehabilitation is a critical component, involving the development of comprehensive plans that include filling voids left by extraction and planting native vegetation to restore the landscape. This not only helps to reinstate the natural environment but also promotes biodiversity by providing habitats for local wildlife. Additionally, monitoring and controlling dust emissions and noise levels through the use of water sprays and noise barriers can significantly reduce the impact on air quality and community health.

Establishing buffer zones around quarries is another effective strategy to protect sensitive habitats and minimize environmental impacts. These zones act as a barrier, reducing the direct effects of quarrying on surrounding ecosystems and providing a space for wildlife to thrive. Regular environmental assessments are essential to ensure compliance with regulations and identify areas for improvement. These assessments can help in adapting practices to mitigate negative impacts effectively, ensuring that quarrying activities are conducted responsibly and sustainably.

Case Study 2: Artisanal Dimensioning of Pyroclastics

Environmental Impacts: In various regions of Kenya, the artisanal dimensioning of pyroclastics presents unique environmental challenges. One of the primary concerns is land degradation, which can occur if extraction is not managed properly. The extraction process can disturb the soil and vegetation, leading to long-term changes in the landscape and the loss of valuable ecosystems. Increased erosion is another significant impact, as the removal of vegetation can lead to sedimentation in nearby water bodies. This sedimentation can harm aquatic ecosystems by reducing light penetration and disrupting the habitats of fish and other aquatic organisms.

Resource depletion is a critical issue associated with unsustainable extraction practices. Over-extraction can lead to the depletion of local pyroclastic resources, limiting future availability and making it difficult for future generations to access these materials. Additionally, processing activities can generate waste materials that contribute to environmental pollution if not managed effectively. Improper disposal of waste can lead to contamination of soil and water sources, further exacerbating environmental degradation. Health risks are also a concern, as artisans may face exposure to dust and noise generated during extraction and processing, leading to respiratory issues and other health problems.

Mitigation Strategies: To mitigate these impacts, several strategies can be employed. Encouraging selective extraction methods that minimize land disturbance and preserve surrounding ecosystems is crucial. This approach helps maintain the integrity of the environment while allowing for resource use. Providing training programs to educate artisans on sustainable practices and rehabilitation techniques can empower local communities and lead to better resource management. These programs can cover topics such as responsible extraction methods, waste management, and the importance of ecosystem preservation.

Implementing effective waste management strategies is essential to reduce environmental pollution and promote recycling. Proper waste management practices can minimize the impact of mining activities on the surrounding environment and ensure that waste materials are handled responsibly. Additionally, promoting the replanting of native vegetation in mined areas can help restore ecosystems and enhance biodiversity. This practice not only aids in soil stabilization but also contributes to the overall health of the environment by providing habitats for local wildlife.

Case Study 3: Harvesting of Stream Sand and Gravel Materials

Location: Various riverbeds across Kenya

Materials: Stream sand and gravel

Activities: The extraction of stream sand and gravel involves harvesting materials from riverbeds using both manual and mechanical methods, including dredging and screening to separate fine sand from larger gravel. This process can significantly alter river dynamics and sediment transport, leading to potential ecological consequences. The extraction activities can disrupt the natural flow of rivers, affecting aquatic habitats and the organisms that depend on them. Once extracted, the materials are washed and sorted to meet construction specifications. However, this processing can generate additional waste, which requires careful management to prevent pollution and ensure that the surrounding environment is not adversely affected.

Mitigation Measures: To ensure sustainable extraction practices, regulations are in place that limit the depth of extraction and maintain buffer zones along riverbanks. These measures are crucial for preventing erosion and habitat destruction, protecting aquatic ecosystems, and maintaining the integrity of riverbanks. By limiting the depth of extraction, the natural sediment transport processes can continue, which is vital for the health of river ecosystems. Additionally, maintaining buffer zones helps to protect sensitive habitats and provides a space for wildlife to thrive, thereby minimizing the ecological footprint of sand and gravel harvesting.

Stream sand and gravel are primarily utilized in concrete production, road construction, and as fill material in various construction projects. Their availability is crucial for supporting infrastructure development in Kenya, making sustainable management of these resources essential. By implementing effective environmental management practices, stakeholders can ensure that the extraction of these materials does not compromise the health of river ecosystems or the communities that depend on them.

Final Thoughts

The case studies presented highlight the critical importance of sustainable practices in the extraction and use of natural resources for construction. By adopting responsible mining techniques and engaging local communities in environmental stewardship, the construction materials sector can contribute to economic development while preserving the integrity of ecosystems. The integration of sustainable practices not only benefits the environment but also enhances the resilience and well-being of communities that rely on these resources for their livelihoods.

As Kenya continues to develop its infrastructure and housing, prioritizing sustainable resource management is essential to ensure that future generations can benefit from the rich geological resources available in the country. This involves fostering a culture of sustainability in the extraction and processing of construction materials, where stakeholders—including government agencies, local communities, and industry players—collaborate to create a balanced approach that supports both economic growth and environmental conservation.

Moreover, the successful implementation of sustainable practices requires ongoing education and awareness-raising efforts. By equipping local communities with the knowledge and skills needed to engage in responsible resource management, stakeholders can empower them to take an active role in protecting their environment. This collaborative effort will be vital in addressing the challenges posed by resource extraction while ensuring that the benefits are equitably shared among all stakeholders involved.

In conclusion, the path toward sustainable resource management in Kenya's construction materials sector is not only a necessity for environmental protection but also a vital component of social and economic development. By prioritizing sustainability, Kenya can harness its natural resources responsibly, ensuring that they continue to support the nation’s growth while safeguarding the environment for future generations (Figure 35).

Figure 35: Typical Claystone Rock Sample (A), when used as Bricks for Roofing (R), Walling (W) and Fencing (F) in (B) or for Pottery needs as in (C).

Environmental Impacts of Geo-Materials Extraction

The extraction and use of geo-materials can have significant environmental impacts, which are crucial to understand for developing sustainable practices in the mining and construction industries. The following sections outline the key environmental impacts associated with geo-materials extraction.

1. Habitat Destruction

Loss of Biodiversity: Quarrying and mining activities often lead to the destruction of natural habitats, resulting in the displacement or loss of various plant and animal species. The removal of vegetation and alteration of land can disrupt local ecosystems, making it difficult for wildlife to thrive. This loss of biodiversity can have cascading effects on ecosystem services, such as pollination, water purification, and soil fertility.

Fragmentation of Ecosystems: The creation of quarries and extraction sites can fragment ecosystems, isolating wildlife populations and hindering their ability to migrate, find food, and reproduce. This fragmentation can lead to a decline in biodiversity and disrupt ecological balance, making ecosystems more vulnerable to external pressures such as climate change and invasive species.

2. Soil Erosion

Increased Erosion: The removal of vegetation and topsoil during extraction processes can lead to increased soil erosion. This erosion degrades land quality and can result in sedimentation in nearby water bodies, affecting aquatic ecosystems. The sediment can smother fish spawning grounds and disrupt the natural flow of rivers, leading to further ecological disruption.

Loss of Soil Fertility: Erosion can strip away nutrient-rich topsoil, significantly reducing the land's agricultural productivity. This loss of fertility can have long-term consequences for local agriculture and food security, as farmers may struggle to grow crops on degraded land.

3. Water Pollution

Contamination of Water Sources: Mining activities can introduce pollutants, such as heavy metals and chemicals, into nearby rivers and groundwater. This contamination affects water quality, posing risks to aquatic life and human health. Polluted water sources can lead to bioaccumulation of toxins in the food chain, impacting both wildlife and communities that rely on these water sources for drinking and irrigation.

Sedimentation:
Increased sedimentation from erosion can clog waterways, disrupt aquatic habitats, and reduce water quality. Sediment can smother fish spawning grounds and alter the natural flow of rivers, leading to further ecological disruption.

4. Air Pollution

Dust Generation: Quarrying and processing activities can generate significant amounts of dust, which can affect air quality and pose health risks to nearby communities. Prolonged exposure to dust can lead to respiratory issues, cardiovascular dise ases, and other health problems among workers and residents.

Emissions from Machinery: The use of heavy machinery in extraction processes contributes to air pollution through the emission of greenhouse gases and other pollutants. These emissions can exacerbate climate change and impact local air quality, leading to health problems and environmental degradation.

5. Noise Pollution

Disruption to Wildlife and Communities: The noise generated from blasting, machinery, and transportation can disturb local wildlife, affecting their behavior and habitat use. Additionally, noise pollution can negatively impact the quality of life for nearby residents, leading to stress, sleep disturbances, and other health issues.

6. Landscape Alteration

Visual Impact: The creation of quarries and extraction sites can significantly alter the landscape, leading to visual impacts that may affect tourism and local aesthetics. The alteration of natural scenery can diminish the appeal of an area for visitors and residents alike, potentially impacting local economies that rely on tourism.

Land Degradation: The physical alteration of land can lead to long-term degradation, making it difficult to restore the area to its original state. This degradation can result in loss of land value and reduced opportunities for future land use, limiting options for agriculture, recreation, and conservation.

7. Resource Depletion

Unsustainable Extraction: Over-extraction of geo-materials can lead to resource depletion, making it challenging to meet future demands. Unsustainable practices can result in economic challenges and increased competition for remaining resources, potentially leading to conflicts over access and use.

8. Climate Change

Carbon Emissions: The extraction and processing of geo-materials contribute to greenhouse gas emissions, particularly through the use of fossil fuels in machinery and transportation. These emissions play a role in climate change, impacting global weather patterns and ecosystems.

Land Use Changes: Changes in land use due to mining can affect the carbon sequestration capabilities of ecosystems, further contributing to climate change. The loss of vegetation reduces the land's ability to absorb carbon dioxide, exacerbating the greenhouse effect.

Mitigation Measures

To address these environmental impacts, several mitigation measures can be implemented:

- **Sustainable Quarrying Practices:** Implementing best practices for extraction, such as selective mining and minimizing land disturbance, can help reduce environmental impacts. Techniques like phased extraction and maintaining buffer zones can also be beneficial.
- **Rehabilitation and Restoration:** Developing plans for rehabilitating mined areas is essential. This includes replanting native vegetation, restoring habitats, and ensuring that the land is returned to a usable state post-extraction. Effective rehabilitation can help restore ecological functions and improve the landscape's aesthetic value.
- **Water Management:** Implementing measures to prevent water pollution, such as sedimentation ponds and proper waste disposal, can help protect local water sources. Regular monitoring of water quality is crucial for the early detection of contamination and for ensuring compliance with environmental standards.
- **Community Engagement:** Involving local communities in decision-making processes ensures that their concerns are addressed. Engaging stakeholders can lead to more sustainable practices and foster a sense of ownership over local resources. Community involvement can also enhance transparency and accountability in mining operations.
- **Regulatory Compliance:** Adhering to environmental regulations and standards is essential for minimizing impacts and promoting sustainable practices. Regular audits and assessments can help ensure compliance and identify areas for improvement. Establishing clear guidelines and enforcement mechanisms can help mitigate environmental risks associated with geo-material extraction.

Conclusion

The geo-materials found in Kenya, including dimension stones, aggregates, alluvial deposits, and other geological resources, play a crucial role in the construction industry. These materials are essential for infrastructure development, housing, and various construction projects. The environmental impacts associated with the extraction and use of these geo-materials—such as habitat destruction, soil erosion, water pollution, and air pollution—underscore the need for sustainable practices in quarrying and processing.

By recognizing the potential environmental impacts and implementing effective mitigation measures, stakeholders can work towards more sustainable practices that protect ecosystems and communities while meeting the demand for construction materials. Promoting responsible extraction practices, such as sustainable quarrying, rehabilitation of mined areas, and community engagement, will help ensure that Kenya can harness its rich geo-materials sustainably. This approach not only supports economic development but also protects the environment and preserves natural resources for future generations.

In summary, the integration of sustainable practices in the extraction and use of geo-materials is essential for balancing economic growth with environmental conservation. By prioritizing sustainability, the construction industry can contribute to a healthier planet and a more resilient society, ensuring that the benefits of geo-materials are realized without compromising the ecological integrity of the regions from which they are sourced.

Case Study: Geological Evaluation of Ainamoi Bauxite Deposit for Cement Manufacturing in Kericho County, Kenya

Introduction

The significance of minerals in the construction of structures continues to grow, particularly in the context of cement manufacturing, which is a critical component of the construction industry. This study focuses on the geological evaluation of the Ainamoi bauxite deposit, aiming to assess its viability as a natural mineral resource for cement production.

Study Area Location

The study area is located in the Poiywek area of Ainamoi constituency in Kericho County, Kenya. Kericho County is one of the fourteen counties in the Great Rift Valley Province and is renowned for its rich agricultural land and diverse mineral resources. The present study area is geographically defined by longitudes 35° 02' and 35° 40' East, extending from the Equator (0°) to the North down to latitude 0° 23' South (Figure 36).

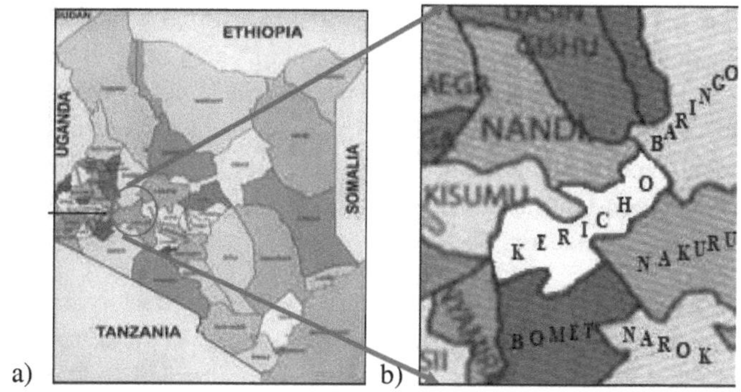

Figure 36: Location of Counties of Kenya (A) and Kericho County (B)

Kericho County is bordered by several other counties, including Uasin Gishu County to the north, Baringo County to the northeast, Nandi County to the northwest, Nakuru County to the east, and Bomet County to the south. Additionally, it is bordered to the southwest by Nyamira and Homa Bay Counties, while Kisumu County lies to the west. The total area of Kericho County is approximately 2,479 square kilometers, providing a diverse landscape that supports various economic activities, including agriculture and mining.

The strategic location of the Ainamoi bauxite deposit within Kericho County positions it favorably for potential exploitation, given its accessibility to transportation networks and proximity to markets for cement production. The geological characteristics of the region, combined with its favorable climate, further enhance the viability of the Ainamoi bauxite deposit as a natural resource for cement manufacturing.

Uses of Bauxite

Bauxite is primarily known as the principal ore of aluminum, but its applications extend beyond aluminum production. One of the most significant uses of bauxite is in the production of aluminum itself. Bauxite is refined to produce alumina (aluminum oxide) through the Bayer process, which is then smelted to produce aluminum metal. This aluminum is utilized across various industries, including transportation, where it is used in aircraft and automobiles, construction for windows, doors, and roofing, and packaging in products like cans and foils. The versatility of aluminum makes bauxite a critical resource for modern manufacturing.

In addition to aluminum production, bauxite is increasingly recognized as a valuable raw material in cement manufacturing. Its high aluminum content can replace other materials in cement formulations, contributing to the production of high-quality cement with desirable properties. The incorporation of bauxite in cement can enhance the strength and durability of the final product, making it an attractive option for construction projects that require robust materials.

Bauxite also plays a crucial role in the manufacture of refractory materials, which are essential for high-temperature applications. These materials are used in furnaces, kilns, and reactors, where they must withstand extreme heat and corrosive environments. Bauxite-based refractories are vital in industries such as steel, glass, and ceramics, where high-performance materials are necessary to ensure operational efficiency and safety.

Moreover, bauxite serves as a source of various chemicals in the chemical industry. For instance, aluminum sulfate, derived from bauxite, is used in water treatment, paper manufacturing, and as a mordant in dyeing processes. Additionally, bauxite can be processed to produce other aluminum compounds that find applications in various industrial sectors, further underscoring its versatility.

Due to its hardness, bauxite can also be utilized as an abrasive material in grinding and polishing applications. It is often used in the production of grinding wheels and other abrasive products, making it valuable in manufacturing and maintenance processes that require surface finishing.

While bauxite mining can lead to significant land disturbance, proper management can rehabilitate the land for agricultural or other uses. The residual bauxite residue can be used to improve soil quality and fertility, promoting sustainable land use practices. This aspect of bauxite mining highlights the potential for responsible resource management to benefit both the economy and the environment.

In some cases, bauxite can even be used as a feedstock for energy production. The combustion of bauxite in certain processes can generate heat, which can be harnessed for energy production, adding another dimension to its utility as a resource.

Despite the numerous economic benefits associated with bauxite extraction and utilization, it is essential to consider the potential environmental impacts of bauxite mining. One of the most significant concerns is deforestation and habitat destruction. Mining activities often require clearing large areas of land, leading to the loss of natural habitats and biodiversity. This loss can have long-term effects on local ecosystems and wildlife, disrupting the delicate balance that sustains various species.

Soil erosion is another critical environmental issue linked to bauxite mining. The removal of vegetation and topsoil during mining can result in increased soil erosion, which can lead to sedimentation in nearby water bodies. This sedimentation negatively affects water quality and aquatic life, as it can smother habitats and disrupt the natural flow of rivers and streams.

Water pollution is a significant risk associated with bauxite mining. The mining and processing of bauxite can generate waste materials that may contaminate local water sources. Chemicals used in the refining process, such as caustic soda, can leach into groundwater and surface water, posing risks to human health and the environment. Additionally, air pollution from dust and emissions generated by mining operations can contribute to respiratory issues and other health problems for nearby communities.

Waste management is another crucial aspect of bauxite mining. The production of bauxite generates significant amounts of waste, including red mud, a byproduct of alumina refining. Proper management of this waste is essential to prevent environmental contamination and ensure sustainable practices. Furthermore, mining activities can disrupt local communities, leading to displacement and changes in land use. It is vital to engage with local populations and consider their rights and needs in the planning and execution of mining projects to mitigate social impacts.

In summary, the Ainamoi bauxite deposit not only holds promise for cement manufacturing but also not only holds promise for cement manufacturing but also has broader implications for various industries. The diverse applications of bauxite highlight its importance as a natural resource, making it a valuable asset for economic development in Kericho County and beyond. However, it is crucial to balance the economic benefits of bauxite extraction with the potential environmental impacts.

The extraction of bauxite can lead to significant ecological consequences, including deforestation, soil erosion, water pollution, and air pollution. These environmental challenges necessitate a careful and responsible approach to mining practices. By implementing sustainable mining practices and engaging with local communities, Kenya can maximize the advantages of the Ainamoi bauxite deposit while minimizing adverse effects on the environment and society.

Sustainable practices may include conducting thorough Environmental Impact Assessments (EIAs) before mining operations begin, ensuring that potential environmental impacts are identified and addressed. Additionally, adopting best practices in mining operations can help reduce land disturbance and waste generation, while effective waste management strategies can mitigate the risks associated with byproducts like red mud.

Engaging local communities in the decision-making process is also essential. By involving them in discussions about mining activities, their rights and needs can be respected, and the benefits of mining can be shared more equitably. This collaborative approach can foster trust and support from local populations, which is vital for the long-term success of mining projects.

In conclusion, while the Ainamoi bauxite deposit presents significant opportunities for economic growth and industrial development, it is imperative to approach its extraction with a commitment to sustainability. By prioritizing environmental protection and community engagement, Kenya can harness the potential of its bauxite resources responsibly, ensuring that both the economy and the environment can thrive together. This balanced approach will not only benefit current generations but also safeguard the natural resources and ecosystems for future generations.

Recommendations for Sustainable Practices

To ensure that the exploitation of the Ainamoi bauxite deposit is conducted responsibly, several key recommendations should be considered. First and foremost, conducting comprehensive Environmental Impact Assessments (EIAs) before commencing mining operations is essential. These assessments should identify potential environmental impacts and develop effective mitigation strategies. Involving stakeholders, particularly local communities, in the EIA process is crucial to ensure that their concerns and insights are taken into account, fostering a sense of ownership and responsibility towards the mining activities.

Adopting sustainable mining practices is another critical recommendation. Mining operations should implement best practices that minimize land disturbance, reduce waste generation, and protect water resources. Techniques such as selective mining, which focuses on extracting only the necessary materials while preserving surrounding ecosystems, and progressive rehabilitation, where mined areas are restored concurrently with extraction, can significantly mitigate environmental impacts. These practices not only help maintain ecological balance but also enhance the long-term viability of the mining operations.

Rehabilitation and restoration of mined areas must be prioritized as part of the mining process. Developing and implementing comprehensive plans for rehabilitation is essential to ensure that the land can be restored to its natural state or repurposed for agricultural or other uses post-mining. This includes replanting native vegetation, restoring soil quality, and implementing measures to prevent erosion. Such efforts can help revitalize local ecosystems and promote biodiversity, contributing to the overall health of the environment.

Effective water management strategies are also vital to prevent contamination of local water sources. This involves monitoring water quality throughout the mining process and ensuring that wastewater is treated before being discharged into the environment. By implementing stringent water management practices, mining operations can protect vital water resources and safeguard the health of surrounding communities and ecosystems.

Community engagement is essential for the success of sustainable mining practices. Fostering strong relationships with local communities by involving them in decision-making processes related to mining activities can help build trust and address concerns. This engagement ensures that the benefits of mining are shared equitably with the community, enhancing social acceptance and reducing potential conflicts. By prioritizing community involvement, mining companies can create a more inclusive and responsible approach to resource extraction.

Establishing a robust monitoring system is crucial to track the environmental impacts of mining activities and ensure compliance with regulations. Regular audits and assessments can help identify areas for improvement and ensure that sustainable practices are being followed. This proactive approach allows for timely adjustments to operations, minimizing negative impacts and promoting continuous improvement in environmental performance.

Finally, investing in research and development is essential for advancing sustainable practices in bauxite extraction and processing. By exploring innovative technologies and methods that minimize environmental impacts, the industry can enhance its efficiency and reduce its ecological footprint. This includes investigating alternative uses for bauxite waste and developing more efficient refining processes that can contribute to a circular economy. Through ongoing research and innovation, the bauxite mining sector can evolve to meet the challenges of sustainability while continuing to provide valuable resources for economic development.

Environmental Impacts of Bauxite Mining

While the extraction and utilization of bauxite offer numerous economic benefits, it is essential to consider the potential environmental impacts associated with bauxite mining. One of the most significant impacts is deforestation and habitat destruction. Mining activities often require clearing large areas of land, which leads to the loss of natural habitats and biodiversity. This loss can have long-term effects on local ecosystems and wildlife, disrupting the delicate balance that sustains various species.

Another critical environmental concern is soil erosion. The removal of vegetation and topsoil during mining operations can result in increased soil erosion, which can lead to sedimentation in nearby water bodies. This sedimentation negatively affects water quality and aquatic life, as it can smother habitats and disrupt the natural flow of rivers and streams. Additionally, water pollution is a significant risk associated with bauxite mining. The mining and processing of bauxite can generate waste materials that may contaminate local water sources. Chemicals used in the refining process, such as caustic soda, can leach into groundwater and surface water, posing risks to human health and the environment.

Air pollution is another consequence of bauxite mining. Dust and emissions from mining operations can contribute to air pollution, affecting the health of nearby communities and ecosystems. Particulate matter and other pollutants can lead to respiratory issues and other health problems for individuals living in proximity to mining sites. Furthermore, waste management is a crucial aspect of bauxite mining. The production of bauxite generates significant amounts of waste, including red mud, a byproduct of alumina refining. Proper management of this waste is essential to prevent environmental contamination and ensure sustainable practices.

Social impacts also arise from mining activities, as they can disrupt local communities, leading to displacement and changes in land use. It is vital to engage with local populations and consider their rights and needs in the planning and execution of mining projects. This engagement can help mitigate the negative social consequences of mining and ensure that local communities benefit from the economic opportunities created by bauxite extraction.

Conclusion

In summary, the Ainamoi bauxite deposit not only holds promise for cement manufacturing but also has broader implications for various industries. The diverse applications of bauxite highlight its importance as a natural resource, making it a valuable asset for economic development in Kericho County and beyond. However, it is crucial to balance the economic benefits of bauxite extraction with the potential environmental impacts. By implementing sustainable mining practices and engaging with local communities, Kenya can maximize the advantages of the Ainamoi bauxite deposit while minimizing adverse effects on the environment and society.

Methodology

The evaluation process began with comprehensive field geological mapping, which serves as a foundational step for subsequent investigative methods. This process included the collection of historical data, geological mapping, field observations, and the gathering of representative field samples for further analysis.

Results

The evaluation of the Ainamoi bauxite deposit has yielded promising results, confirming its viability as a raw material for cement manufacturing. With an aluminum hydroxide content of 66.67%, the Ainamoi bauxite demonstrates a suitable chemical composition for cement production. The cement produced from this bauxite has shown commendable compressive strength values, achieving 17.42 kN at seven days, 20.46 kN at fourteen days, and 24.36 kN at twenty-one days of curing. These results align closely with the ordinary compressive strength standards required for effective cement performance, indicating that Ainamoi bauxite can be effectively utilized in cement formulations.

Moreover, the consistency of the cement at 35% of 400g, along with the initial and final setting times of 45 minutes and 600 minutes respectively, fall within acceptable ranges for cement applications. The measured expansion of 7.3mm further supports the suitability of Ainamoi bauxite for cement manufacturing, as it meets the necessary criteria for stability and performance.

Recommendations

Given these findings, it is recommended that the Ainamoi bauxite be considered for commercial exploitation in the cement manufacturing industry in Kenya. This local resource not only has the potential to enhance the sustainability of the construction sector by reducing reliance on imported materials but also contributes to the economic development of the region.

To fully realize the potential of the Ainamoi bauxite deposit, further assessments, including detailed geological and geophysical studies, are essential. Such investigations will provide a clearer understanding of the deposit's size and economic viability, ensuring that the extraction and utilization of this natural resource are conducted efficiently and sustainably.

Broader Implications

In addition to the immediate benefits of utilizing Ainamoi bauxite for cement manufacturing, the strategic development of this resource can have broader implications for the local economy and the environment. By tapping into local mineral resources, Kenya can reduce its dependence on imported raw materials, which often come with significant transportation costs and environmental impacts. This shift not only supports local industries but also creates job opportunities in mining, processing, and construction sectors, thereby contributing to community development and economic resilience.

Furthermore, the sustainable management of the Ainamoi bauxite deposit can serve as a model for responsible resource extraction in Kenya. By implementing best practices in mining and processing, stakeholders can ensure that environmental considerations are prioritized, minimizing the ecological footprint of cement production. This includes adopting technologies that reduce emissions and waste, as well as engaging local communities in decision-making processes to ensure that their needs and concerns are addressed.

Conclusion

The positive results from the Ainamoi bauxite evaluation also highlight the importance of ongoing research and innovation in the cement industry. As the demand for sustainable construction materials grows, there is an opportunity for Kenya to position itself as a leader in the development of eco-friendly cement alternatives. Continued investment in research and development can lead to the discovery of new applications for bauxite and other local minerals, further enhancing the country's industrial capabilities.

In conclusion, the Ainamoi bauxite deposit represents a significant opportunity for the cement manufacturing industry in Kenya. With its favorable chemical composition and performance characteristics, Ainamoi bauxite can play a crucial role in meeting the growing demand for cement while supporting local economic development. By prioritizing further geological assessments and sustainable practices, Kenya can harness the potential of this natural resource, paving the way for a more resilient and environmentally responsible construction sector. The journey towards fully realizing the benefits of the Ainamoi bauxite deposit is not only a step towards industrial growth but also a commitment to sustainable development that can benefit future generations.

Geological Environmental Impacts: *Case Study on Transportation, Erosion, and Diagenesis*

Implications of Transportation and Erosion on the Environment

The processes of transportation and erosion have significant environmental implications, affecting ecosystems, land use, and human activities. Understanding these impacts is crucial for effective land management and environmental conservation (Figure 37).

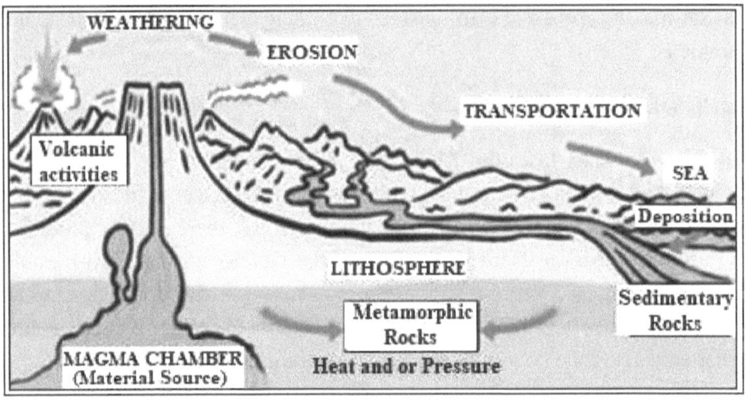

Figure 37: General Processes of Sedimentary Cycle

Ecosystem Disruption

Ecosystem disruption is a significant consequence of erosion and sediment transport, leading to habitat alteration and nutrient redistribution. Habitat alteration occurs when erosion and sediment deposition change the physical characteristics of environments, resulting in habitat loss for various species. For instance, the erosion of riverbanks can destroy critical nesting sites for birds and other wildlife, while sediment deposition can modify aquatic habitats, adversely affecting fish populations and other aquatic organisms. These changes can disrupt the delicate balance of ecosystems, leading to declines in biodiversity and the health of natural habitats.

Nutrient redistribution is another critical aspect of ecosystem disruption caused by erosion. The loss of topsoil, which is rich in essential nutrients for plant growth, can significantly reduce agricultural productivity and threaten food security. As nutrients are washed away, the fertility of the land diminishes, impacting crop yields and the livelihoods of farmers. Conversely, sediment deposition in other areas can enrich soils, potentially benefiting some ecosystems. However, this redistribution can create imbalances, favoring certain species over others and altering the natural dynamics of the ecosystem.

Water Quality Degradation

Water quality degradation is a direct consequence of increased sediment transport and erosion. One of the primary issues is the increased sediment load in water bodies, which can lead to higher turbidity levels in rivers, lakes, and oceans. This turbidity affects light penetration, hindering photosynthesis in aquatic plants and disrupting the food chain. Additionally, sediments can carry pollutants, such as heavy metals and agricultural chemicals, leading to further degradation of water quality and harming aquatic life. The health of aquatic ecosystems is closely tied to water quality, and any degradation can have cascading effects on biodiversity and ecosystem services.

Eutrophication is another significant concern associated with water quality degradation. Nutrients from eroded soils, particularly nitrogen and phosphorus, can enter water bodies and promote algal blooms. These blooms deplete oxygen levels in the water, creating dead zones where aquatic life cannot survive. The resulting hypoxic conditions can lead to fish kills and a decline in overall biodiversity, further impacting the health of aquatic ecosystems. Managing nutrient runoff and sediment transport is essential to prevent eutrophication and protect water quality.

Land Use and Human Activities

Erosion has profound implications for land use and human activities, particularly in agriculture. Agricultural practices can be significantly impacted by erosion, leading to reduced crop yields and increased costs for farmers. The loss of topsoil diminishes soil fertility, making it more challenging to grow healthy crops. To mitigate erosion and maintain soil health, practices such as contour farming, terracing, and cover cropping can be implemented. These methods help stabilize the soil, reduce runoff, and enhance agricultural productivity, ultimately supporting food security.

Infrastructure damage is another critical concern related to erosion. Erosion can undermine roads, bridges, and buildings, leading to costly repairs and increased maintenance. Understanding erosion patterns is essential for urban planning and infrastructure development, as it allows for the design of resilient structures that can withstand the impacts of erosion. By incorporating erosion management strategies into infrastructure planning, communities can reduce the risk of damage and ensure the longevity of their built environment.

Natural Hazards

Natural hazards, such as landslides and flooding, are exacerbated by increased erosion and sediment transport. Erosion can trigger landslides, particularly in steep terrain, posing significant risks to human safety and infrastructure. These events can lead to substantial economic losses, as they often require extensive recovery efforts and can disrupt local economies. Understanding the factors that contribute to landslides is crucial for developing effective risk management strategies and protecting vulnerable communities.

Flooding is another natural hazard influenced by changes in sediment transport. Alterations in sediment dynamics can modify river channels and floodplain characteristics, increasing the risk of flooding in certain areas. Effective management of sediment transport is crucial for flood risk reduction, as it helps maintain the natural flow of rivers and prevents the accumulation of sediment that can obstruct waterways. By implementing sediment management practices, communities can enhance their resilience to flooding and protect lives and property from the impacts of natural hazards.

Conclusion

The interplay between weathering, transportation, erosion, and diagenesis is fundamental to understanding geological processes and their environmental impacts. These processes shape landscapes, influence ecosystems, and affect human activities. By recognizing the significance of these interactions, we can develop better strategies for land management, conservation, and disaster risk reduction.

In summary, the transportation of weathered materials by agents such as running water, glaciers, wind, and gravity not only shapes the Earth's surface but also has profound implications for ecosystems, water quality, land use, and natural hazards. Understanding these dynamics is essential for sustainable environmental practices and effective resource management. The ongoing study of these processes will continue to inform our approach to mitigating their impacts and enhancing the resilience of both natural and human systems.

REFERENCES

Books

1. Adger, W. N., et al. (2005). Social-ecological resilience to coastal disasters. *Science*, 309(5737), 1036-1039. https://doi.org/10.1126/science.1112122

2. Aloimonos, J., et al. (2012). Visual perception for autonomous robots: Advances in robotic vision. In *Springer Handbook of Robotics* (pp. 1001-1024). Springer.

3. Berry, L. G., Mason, B., & Dietrich, R. V. (1983). *Mineralogy: Concepts, descriptions and determinations*. W. H. Freeman and Company.

4. Berkman, A., & Xu, X. (2017).Sensor fusion for autonomous robotic systems. *IEEE Sensors Journal*, 17(12), 3812-3820. https://doi.org/10.1109/JSEN.2017.2688330

5. Billings, M. P. (2016). *Structural geology*. Pearson Education Dorling.

6. Bloss, F. D. (1971). *Crystallography and crystal chemistry: An introduction*. Holt, Rinehart & Winston.

7. Brannon, C. H. (2016). *A look at erosion and weathering: The rock cycle*. Enslow Publishing.

8. Brian, J., & Noel, P. J. (2015). *Origin of carbonate sedimentary rocks*. American Geophysical Union.

9. British Standards Institute. (1980). *Methods of test for soils for civil engineering purposes (BS 1377)*. London.

10. Bullard, R. D. (2005). *The quest for environmental justice: Human rights and the politics of pollution*. Sierra Club Books.

11. Cadena, C., et al. (2016). Past, present, and future of simultaneous localization and mapping: Toward the robust-perception age. *IEEE Transactions on Robotics*, 32(6), 1309-1332. https://doi.org/10.1109/TRO.2016.2572015

12. Chebet, J., Ndegwa, F., & Rop, B. (2024). *Bauxite deposits for cement manufacturing in Poiywek, Ainamoi, Kenya: An industrial mineral perspective*. LAP LAMBERT Academic Publishing. ISBN 978-620-8-41623-2.

13. Craig, R. F., & Knappett, J. (2012). *Soil mechanics* (8th ed.). CRC Press.

14. Das, B. M. (2012). *Soil mechanics laboratory manual* (8th ed.). Oxford University Press.

15. Das, B. M. (2019). *Advanced soil mechanics* (5th ed.). CRC Press.

16. Dissanayake, M., et al. (2001). A solution to the simultaneous localization and mapping (SLAM) problem. *IEEE Transactions on Robotics and Automation*, 17(3), 229-241. https://doi.org/10.1109/70.93311

17. DuBois, C., & Walsh, J. (1970). *Minerals of Kenya*. Geological Survey of Kenya Bulletin No. 11.

18. Evett, J. (2013). *Soils and foundations* (8th ed.). Pearson.

19. Famine Early Warning Network. (n.d.). Retrieved from http://www.ids.ac.uk/fews.net

20. Food and Agriculture Organization (FAO). (2016). *Climate change and food security: Risks and responses*. Retrieved from FAO website.

21. Fossen, H. (2016). *Structural geology* (2nd ed.). Cambridge University Press.

22. Gill, R. (2012). *Igneous rocks and processes: A practical guide*. Wiley-Blackwell.

23. Gupta, S., & Sharma, N. (2019). Multi-robot systems in autonomous mining: Collaborative navigation and task allocation. *Robotics and Autonomous Systems*, 114, 113-127. https://doi.org/10.1016/j.robot.2018.06.005

24. Harlan, S. L., et al. (2009). In the shade of affluence: The role of community in climate change adaptation. *Environmental Science & Policy*, 12(5), 1-12. https://doi.org/10.1016/j.envsci.2009.06.001

25. Hazarika, H., & Ishibashi, I. (2015). *Soil mechanics fundamentals and applications* (2nd ed.). CRC Press.

26. Head, K. H. (2011). *Manual of soil laboratory testing: Vol. 2, Permeability, shear strength and compressibility test.* Whittles.

27. Hobbs, B. E., & Ord, A. (2014). *Structural geology: The mechanics of deforming metamorphic rocks.* Elsevier.

28. Htoo, K. M., Htoon, S., & Khin, T. (2004). Study on physical properties and chemical composition of some Myanmar gems. *The Journal of Myanmar Academy of Arts & Sciences*, 2(1), 175.

29. Hughes, T. P., Kerry, J. T., Alvarez-Noriega, M., et al. (2017). Global warming and recurrent mass bleaching of corals. *Nature*, 543(7645), 373-377.

30. Jacobson, M. Z., & Delucchi, M. A. (2011). Providing all global energy with wind, water, and solar power, Part I: Technologies, energy resources, quantities and areas of infrastructure, and materials. *Energy Policy*, 39(3), 1154-1169. https://doi.org/10.1016/j.enpol.2010.11.040

31. Jiang, R., & Xu, L. (2020). Development and deployment of autonomous underground mining robots. *Journal of Mining Science*, 56(3), 358-370. https://doi.org/10.1134/S106273911703033X

32. Keller, E. A., & Blodgett, R. H. (2006). *Natural hazards: Earth's processes as hazards, disasters, and catastrophes.* Prentice Hall.

33. Keter, C. M. (2025). *Hydrogeological survey report for Giftmart Supermarket in Kericho County, Kenya.* Kapkatet Area, Kapkatet Location, Bureti Division, Bureti Sub-County, Kericho County, Kenya.

34. Keter, C. M. (2025). *Hydrogeological survey report for Utafiti Primary School, located in the Nguriunditu area of Kiambu County, Kenya.*

35. Kiamba, E., Kasomo, R. M., Rop, B. K., & Mwendwa, J. M. (2019). Limestone production optimization by fragmentation control – Case study of Bissel Quarry. In *Proceedings of the 2nd Annual International Conference* (pp. 1-10). Machakos University, Machakos, Kenya.

36. Kipkiror, L. J., Rop, B. K., & Elias, U. (2020). Impact related geology of Silali Basin (Kenya) as an extra-terrestrial impact crater (ETIC). In *Proceedings of the Sustainable Research and Innovation Conference* (pp. 59-71). JKUAT Main Campus, Kenya.

37. Li, Z., & Wei, J. (2022). Energy management and path planning for autonomous mining robots. *International Journal of Mining Technology*, 28(2), 156-171. https://doi.org/10.1016/j.jmin.2022.05.005

38. Mulligan, T., & Jones, J. (2018). Robotics and automation in the mining industry: Overcoming the challenges of harsh environments. *Mining Technology Review*, 39(3), 75-81.

39. Mitchell, R. H. (2013). *Kimberlites: Mineralogy, geochemistry, and petrology*. Springer Science & Business Media.

40. National Environment Management Authority. (1999). *Environmental management and co-ordination act, 1999*. Government of Kenya Printer.

41. National Oceanic & Atmospheric Administration. (n.d.). Retrieved from http://www.photoolib.noaa.gov/

42. National Research Council. (2012). *Sustainable water and environmental management in the California Bay-Delta*. The National Academies Press.

43. Pellow, D. N. (2002). *Garbage wars: The struggle for environmental justice in Chicago*. MIT Press.

44. Patwardhan, A. M. (1999). *The dynamic earth system*. Prentice-Hall of India.

45. Powrie, W. (2013). *Soil mechanics: Concepts and applications* (3rd ed.). CRC Press.

46. Prothero, D. R., & Schwab, F. (2013). *Sedimentary geology* (3rd ed.). W. H. Freeman.

47. Rop, B. K. (2021). *Petroleum perspective of the NW-Kenya basins: Evidence and emerging oil exploration issues*. Saarbrücken, Germany: LAP LAMBERT Academic Publishing. ISBN 978-620-3-86294-2.

48. Rop, B. K. (2025). *Navigating the Climate Crisis: Integrating Environmental Sustainability and Social Justice*. (233 p.). Scholar's Press. ISBN 978-3-659-8217-7.

49. Rop, B. K. (2025). *Sedimentary Rocks: A Comparative Illustrative Approach*. (233 p.) LAP LAMBERT Academic Publishing. ISBN 978-620-8-44043-5.

50. Rop, K. B. (2025). Ainamoi's bauxite boom: Economic and employment opportunities. *International Journal of Geological and Geotechnical Engineering*, 11(1), 18-37. https://journalspub.com/journal/ijgge/ ISSN: 2581-5598 DOI: 10.37628/IJGGE

51. Rop, B. K., Fedha, I. M., & Namwiba, W. H. (2021). *Workshop and laboratory practices: A competency and evaluation based guide* (96 p.). Scholar's Press.

52. Rop, B. K., & Namwiba, W. H. (2023). *Laboratory manual in soil mechanics: Essential procedures for testing and reporting* (112 p.). Scholar's Press.

53. Ruddiman, W. F. (2013). *Earth's climate: Past and future*. W. H. Freeman and Company.

54. Sandra, E., & Robles, C. (2008). *Kimberlites associated with the Lucapa structure, Angola* (PhD dissertation). Universitat De Barcelona, Spain.

55. Schwager, M., et al. (2011). Cooperative path planning for multiple robots in dynamic environments. *Autonomous Robots*, 31(1), 1-12. https://doi.org/10.1007/s10514-011-9194-0

56. Sidiki, S. (2019). Bauxite Mining in the Boké Region (Western Guinea): Method Used and Impacts on Physical Environment. *European Journal of Sustainable Development Research*, 3(3).

57. Smith, G. N. (1990). *Elements of soil mechanics* (6th ed.). Blackwell Science.

58. Smith, I. (2014). *Elements of soil mechanics* (9th ed.). Wiley-Blackwell.

59. Sutton, B. H. C. (1993). *Solving problems in soil mechanics* (2nd ed.). Longman Pub Group.

60. Terzaghi, K., Peck, R. B., & Mesri, G. (1996). *Soil mechanics in engineering practice* (3rd ed.). Wiley-Interscience.

61. Theissen, K., & Swanson, J. (2014). *Metamorphic rocks: Rocks and minerals* (48 p.). Core Library.

62. Thrun, S., et al. (2004). A probabilistic approach to concurrent mapping and localization for mobile robots. *Autonomous Robots*, 10(3), 255-263. https://doi.org/10.1023/B:AURO.0000021045.27750.1f

63. United Nations. (1972). *UN Conference on Human Environment (Stockholm); UNEP established.*

64. United Nations. (1992). *UN Conference on Environment and Development (the Earth Summit) in Rio de Janeiro, Brazil.*

65. United Nations. (2000). *Launch of the International Strategy for Disaster Reduction (ISDR).*

66. United Nations. (2002). *World Summit on Sustainable Development, Johannesburg, South Africa.*

67. United Nations. (2005). *World Conference on Disaster Reduction, Kobe, Japan.*

68. United Nations. (2015). *Transforming our world: The 2030 Agenda for Sustainable Development.* Retrieved from UN website.

69. United Nations Environment Programme (UNEP). (2016). *Global Environment Outlook: Regional Assessments.* Retrieved from UNEP website.

70. United Nations Environment Programme (UNEP). (n.d.). Retrieved from http://www.unep.org

71. UNFCCC. (2015). *Paris Agreement.* Retrieved from UNFCCC website.

72. Vallati, M., et al. (2016). A review of path planning algorithms for autonomous robots in mining. *International Journal of Robotics and Automation*, 33(5), 531-547. https://doi.org/10.1109/ROBOT.2016.7852153

73. Wang, H., & Liu, Y. (2020). Autonomous robots in hazardous mining environments: A survey. *Mining Engineering Journal*, 72(6), 324-338.

74. World Bank. (n.d.). Retrieved from http://www.worldbank.org

75. World Health Organization. (n.d.). Retrieved from http://www.who.int/pub/en

Journals

1. California Geological Survey. (2018). *California's geologic hazards: A guide for local governments*. California Department of Conservation.
2. *Environmental Science & Policy.*
3. *Geology.*
4. *Journal of Environmental Management.*
5. *Journal of Geotechnical and Geo-environmental Engineering.*
6. *Journal of Rock Mechanics and Geotechnical Engineering* (ISSN: 1674-7755).
7. *Journal of Soil Mechanics and Foundations Division* (ISSN: 0044-7994).
8. *Journal of Sustainable Mining.*
9. *Robotics and Autonomous Systems.*
10. *Soil Mechanics and Foundation Engineering* (ISSN: 1573-9279).
11. *Soils and Foundations* (ISSN: 0038-0806).

Additional References

1. British Standards Institute. (1980). *Methods of test for soils for civil engineering purposes (BS 1377)*. London.
2. Craig, R. F., & Knappett, J. (2012). *Soil mechanics* (8th ed.). CRC Press.
3. Das, B. M. (2012). *Soil mechanics laboratory manual* (8th ed.). Oxford University Press.
4. Hazarika, H., & Ishibashi, I. (2015). *Soil mechanics fundamentals and applications* (2nd ed.). CRC Press.
5. Mitchell, R. H. (2013). *Kimberlites: Mineralogy, geochemistry, and petrology*. Springer Science & Business Media.
6. Ondieki J. O. (2020). Mapping Radioactive Minerals Using Remote Sensing: A Case study Mrima Hill Kwale County, Kenya

7. Otwoma, D., Patel, J. P., Bartilol, S. and Mustapha, A. O. (2013). "Estimation of annual effective dose and radiation hazards due to natural radionuclides in mount Homa, southwestern Kenya," *Radiation Protection Dosimetry*, vol. 155, no. 4, pp. 497–504.

8. Onuonga, I.O. (1996) Geochemistry and mineralization of Buru and Kuge volcanic carbonatite centres, western Kenya. PhD thesis, Univ. St. Andrews, Scotland, U.K.

9. Onuonga, I.O. (2000). Hot-spring and supergene lanthanide mineralizationat of the Buru Carbonatite Centre, Western Kenya *Mineralogical Magazine, August 2000, Vol. 64(4), pp. 663–673*

10. Pabello N. G. and Bolivar V. J. (2005). Young Brains on Lead: Adult Neurological Consequences? *Toxicological Sciences,* 86(2): 211-213.

11. Patel J. P. and Mangala M. J., 1994. Elemental Analysis of Carbonatite Samples from Mrima Hill, Kenya, by Energy Dispersive X-Ray Fluorescence. *Nuclear Geophysics*, **8**: 389-393.

12. Patrick L. (2006). Lead Toxicity Part II: The Role of Free Radical Damage and the Useof Antioxidants in the Pathology and Treatment of Lead Toxicity. *Alternative Medicine Review*, **11**(2): 124-127.

13. Rop, B. K., & Namwiba, W. H. (2018). *Fundamentals of applied geology: Competency and evaluation approach* (345 p.). LAP LAMBERT Academic Publishing. ISBN 978-613-9-57896-2.

14. Rop, B. K., & Namwiba, W. H. (2022). *Applied geology and evaluation perspective* (368 p.). Scholar's Press. ISBN 978-613-8-97461-1.

15. Rop, B. K., & Namwiba, W. H. (2023). *Fundamentals of soil mechanics: A competency and evaluation based class lecture manual* (223 p.). Scholar's Press. ISBN 978-620-5-52401-5.

16. Rop, B. K., & Namwiba, W. H. (2023). *Laboratory manual in soil mechanics: Essential procedures for testing and reporting* (112 p.). Scholar's Press. ISBN 978-620-55240-53.

17. Rop, B. K., Digne, E. R., & Namwiba, W. H. (2020). *Applied geology in mining and construction practice: A competency and geohazard mitigation platform* (213 p.). Scholar's Press. ISBN 978-613-8-93406-6.

18. Rop, B. K., Namwiba, W. H., & Keter, M. C. (2023). *Economic potential of Kericho Ainamoi bauxite: A geological evaluation approach report* (80 p.). LAP LAMBERT Academic Publishing.

19. Rop, B. K., Namwiba, W. H., & Maina, C. W. (2019). *Introduction to geology* (129 p.). LAP LAMBERT Academic Publishing. ISBN 978-613-9-82770-1.

20. Rop, B. K., Rwatangabo, D. E. R., & Namwiba, W. H. (2023). *Properties, occurrences and uses of minerals and rocks: A comparative illustrative approach* (257 p.). Scholar's Press. ISBN 978-620-5-52235-6.

21. Rop, B. K., Seroni, A., & Krop, I. (2020). *Economic impacts of ASM in Taita Taveta County, Kenya: Gemstones mining* (129 p.). LAP LAMBERT Academic Publishing. ISBN 978-620-2-51546-7.

15. Rop, B. K., Wangari, J. M., & Namwiba, W. H. (2022). *Fundamentals of laboratory practices in mineralogy and petrology: Mineralogy and petrology (A handbook for comparative studies)* (216 p.). Scholar's Press. ISBN 978-613-8-96862-7.

16. Rop, B. K., Mayik, D. A., & Gach, G. K. (2023). *Petroleum source rocks characterisation: LT-1 well Northern Kenya.* LAP LAMBERT Academic Publishing. ISBN 978-620-6-15177-7.

17. Rop, B. K., Mayik, D. A., & Gach, K. (2023). *Petroleum source rocks characterisation: LT-1 well Northern Kenya.* LAP LAMBERT Academic Publishing. ISBN 978-620-6-15177-7.

18. Rop, B. K., & Kipkiror, L. J. (2021). *Marakwet East Lagam escarpment landslides: Mitigation measures, Kenya.* Scholar's Press. ISBN 978-3-639-66207-8.

19. Rop, B. K., Ketter, M. C., & Sawe, S. J. (2021). *Review of oil and gas prospectivity: Kenya Rift basins* (104 p.). Scholar's Press. ISBN 978-613-8-94692-2.

20. Rop, B. K., Nyamai, C. M., & Namwiba, W. H. (2020). *Narrative glossary of fundamentals for applied geology: A construction perspective* (117 p.). Scholar's Press. ISBN 978-613-8-93606-0.

21. Rop, B. K., Anyona, S., Kipseba, E. K., Namwiba, W. H., & Kipkiror, L. J. (2020). Review of geothermal resources as manifestations of volcanism in Eastern Africa. In *Proceedings of the Sustainable Research and Innovation Conference* (pp. 79-88). JKUAT Main Campus, Kenya. ISSN 2079-6226.

22. Rop, B. K., Kipkiror, L. J., & Kipseba, E. K. (2020). Geoenviron tourism sites in Northern Kenya Rift Valley: The case of Lake Bogoria, Lake Baringo, Kerio Valley, and Silali. In *Proceedings of the Sustainable Research and Innovation Conference* (pp. 126-135). JKUAT Main Campus, Kenya. ISSN 2079-6226.

23. Rop, B. K., Mwendwa, J. M., Maina, C., & Namwiba, W. H. (2019). Applied geology in mining engineering. In *Proceedings of the Sustainable Research and Innovation (SRI) International Conference* (pp. 1-10). JKUAT College of Engineering and Technology (CoETEC), Juja, Kenya.

24. Rop, B. K., Mwendwa, J. M., Maina, C., & Namwiba, W. H. (2019). Importance of applied geology in mining perspective. In *Proceedings of the 2nd Annual International Conference* (pp. 1-10). Machakos University, Machakos, Kenya.

25. Kiamba, E., Kasomo, R. M., Rop, B. K., & Mwendwa, J. M. (2019). Limestone production optimization by fragmentation control – Case study of Bissel Quarry. In *Proceedings of the 2nd Annual International Conference* (pp. 1-10). Machakos University, Machakos, Kenya.

26. Sandra, E., & Robles, C. (2008). *Kimberlites associated with the Lucapa structure, Angola* (PhD dissertation). Universitat De Barcelona, Spain.

27. Schwager, M., et al. (2011). Cooperative path planning for multiple robots in dynamic environments. *Autonomous Robots*, 31(1), 1-12. https://doi.org/10.1007/s10514-011-9194-0

28. Sidiki, S. (2019). Bauxite Mining in the Boké Region (Western Guinea): Method Used and Impacts on Physical Environment. *European Journal of Sustainable Development Research*, 3(3).

29. Smith, G. N. (1990). *Elements of soil mechanics* (6th ed.). Blackwell Science.

30. Smith, I. (2014). *Elements of soil mechanics* (9th ed.). Wiley-Blackwell.

31. Sutton, B. H. C. (1993). *Solving problems in soil mechanics* (2nd ed.). Longman Pub Group.

32. Terzaghi, K., Peck, R. B., & Mesri, G. (1996). *Soil mechanics in engineering practice* (3rd ed.). Wiley-Interscience.

33. Theissen, K., & Swanson, J. (2014). *Metamorphic rocks: Rocks and minerals* (48 p.). Core Library.

34. Thrun, S., et al. (2004). A probabilistic approach to concurrent mapping and localization for mobile robots. *Autonomous Robots*, 10(3), 255-263. https://doi.org/10.1023/B:AURO.0000021045.27750.1f

35. United Nations. (1972). *UN Conference on Human Environment (Stockholm); UNEP established.*

36. United Nations. (1992). *UN Conference on Environment and Development (the Earth Summit) in Rio de Janeiro, Brazil.*

37. United Nations. (2000). *Launch of the International Strategy for Disaster Reduction (ISDR).*

38. United Nations. (2002). *World Summit on Sustainable Development, Johannesburg, South Africa.*

39. United Nations. (2005). *World Conference on Disaster Reduction, Kobe, Japan.*

40. United Nations. (2015). *Transforming our world: The 2030 Agenda for Sustainable Development*. Retrieved from UN website.

41. United Nations Environment Programme (UNEP). (2016). *Global Environment Outlook: Regional Assessments*. Retrieved from UNEP website.

42. United Nations Environment Programme (UNEP). (n.d.). Retrieved from http://www.unep.org

43. UNFCCC. (2015). *Paris Agreement*. Retrieved from UNFCCC website.

44. Vallati, M., et al. (2016). A review of path planning algorithms for autonomous robots in mining. *International Journal of Robotics and Automation*, 33(5), 531-547. https://doi.org/10.1109/ROBOT.2016.7852153

45. Wang, H., & Liu, Y. (2020). Autonomous robots in hazardous mining environments: A survey. *Mining Engineering Journal*, 72(6), 324-338.

46. World Bank. (n.d.). Retrieved from http://www.worldbank.org

47. World Health Organization. (n.d.). Retrieved from http://www.who.int/pub/en

Journals

1. California Geological Survey. (2018). *California's geologic hazards: A guide for local governments.* California Department of Conservation.

2. *Environmental Science & Policy.*

3. *Geology.*

4. *Journal of Environmental Management.*

5. *Journal of Geotechnical and Geo-environmental Engineering.*

6. *Journal of Rock Mechanics and Geotechnical Engineering* (ISSN: 1674-7755).

7. *Journal of Soil Mechanics and Foundations Division* (ISSN: 0044-7994).

8. *Journal of Sustainable Mining.*

9. *Robotics and Autonomous Systems.*

10. *Soil Mechanics and Foundation Engineering* (ISSN: 1573-9279).

11. *Soils and Foundations* (ISSN: 0038-0806).

ABOUT THE BOOK

Environmental Geology: ***Insights from Applied Geology on Environmental Impact*** is a meticulously crafted academic and professional resource that adeptly bridges theoretical geology with actionable environmental management strategies. In response to the escalating demand for interdisciplinary solutions to global environmental challenges, this volume synthesises cutting-edge research, compelling case studies, and industry best practices, equipping readers with the essential tools to navigate the complexities of geological and environmental systems.

Interdisciplinary Approach

This volume adopts a robust interdisciplinary approach, integrating principles from geology, hydrology, geochemistry, and environmental science. It underscores the pivotal role of geological processes in shaping ecosystems, influencing resource availability, and mediating human-environment interactions. Key theoretical frameworks include Systems Theory, which analyses the Earth as an interconnected system wherein geological, hydrological, and climatic processes interact, thereby providing a holistic understanding of environmental dynamics. Additionally, the book employs both quantitative and qualitative models for evaluating geological hazards, such as probabilistic seismic hazard analysis, to inform decision-making and enhance risk management strategies. It also aligns geological practices with global sustainability targets, particularly Sustainable Development Goals (SDGs) 6 (Clean Water), 13 (Climate Action), and 15 (Life on Land), thereby promoting responsible resource management.

Socio-Economic Environmental Aspects

A significant focus of this book is the socio-economic environmental aspects of environmental geology. It explores how geological resources underpin economic development and social well-being, while also addressing the potential conflicts that arise from resource extraction. The text examines the socio-economic implications of mining activities, including job creation, community development, and the equitable distribution of benefits. It also highlights the importance of stakeholder engagement and community participation in decision-making processes, ensuring that local populations have a voice in how geological resources are managed.

Furthermore, the book discusses the environmental justice implications of geological practices, particularly in marginalized communities that may bear the brunt of environmental degradation. By integrating socio-economic considerations into environmental geology, the book advocates for sustainable practices that not only protect the environment but also promote social equity and economic resilience.

Structure and Content

The book is structured into several chapters that contribute to its scholarly rigor. Chapters 1 to 4 establish the theoretical underpinnings of environmental geology, encompassing soil mechanics, hydrogeology, and rock weathering processes. These chapters introduce advanced analytical techniques, such as X-ray diffraction for mineral identification and Geographic Information Systems (GIS) for spatial analysis, thereby enhancing the reader's technical proficiency.

Chapters 5 to 7 critically examine Environmental Impact Assessment (EIA) frameworks, including ISO 14001 standards and the Equator Principles, to ensure compliance and promote sustainable practices. This section also explores the geochemical cycling of carbon and metals, linking geological processes to climate change and resource depletion, thereby highlighting the relevance of geology in contemporary environmental issues.

Chapters 8 to 10 present case studies on geohazard mitigation, such as landslide early warning systems in the Himalayas, illustrating the practical applications of theoretical concepts. They evaluate the efficacy of international environmental policies, including the Kyoto Protocol and the Minamata Convention on Mercury, to inform future policy development.

Industry Applications

Finally, Chapters 11 to 13 discuss industry-focused applications, including rock blasting constraints, focusing on fracture mechanics and computational modelling (e.g., finite element analysis) to optimise blasting while minimising ecological disruption. This section also analyses the geoeconomic implications of rare earth element (REE) extraction in Kenya, including life cycle assessments (LCAs) of mining operations, to address sustainability challenges in resource extraction. Additionally, it details ASTM/ISO standards for soil sampling, in-situ testing (e.g., Standard Penetration Test (SPT), Cone Penetration Test (CPT)), and geophysical surveys (e.g., Electrical Resistivity

Tomography (ERT), Ground Penetrating Radar (GPR)), providing practical guidance for field investigations.

Enhancing Learning

To enhance learning, the book incorporates various pedagogical tools, featuring over 20 global case studies, including the 2010 Haiti earthquake and Kenya's Mrima Hill rare earth deposits, which provide real-world context. End-of-chapter exercises are designed to reinforce analytical skills, such as calculating slope stability using Bishop's method, encouraging active engagement with the material. High-resolution diagrams, stratigraphic cross-sections, and 3D geological models facilitate comprehension of complex concepts and enhance the learning experience.

Professional Relevance

The professional relevance of this book is evident in its applications across various industries. In the realm of mining and resource extraction, it provides protocols for sustainable practices, including tailings management and post-closure land rehabilitation (e.g., phytoremediation of heavy metals), to mitigate environmental impacts. It also addresses the geopolitics of critical minerals, focusing on Kenya.

The book also addresses the geopolitics of critical minerals, focusing on Kenya's role in the global rare earth element (REE) supply chain, to inform strategic resource management. In urban planning and infrastructure, the book guides engineers in assessing liquefaction risks for foundation design in seismic zones, ensuring safety and resilience in construction. Furthermore, it offers frameworks for integrating green infrastructure (e.g., permeable pavements) into urban stormwater management, promoting sustainable urban development.

Regulatory Compliance and Risk Management

The book also provides step-by-step guidance on compliance with EU Directive 2014/52/EU and NEPA (US) regulations, ensuring adherence to legal standards in Environmental Impact Assessments (EIAs). It includes templates for scoping reports, baseline studies, and mitigation plans that streamline the EIA process and enhance efficiency. In terms of risk management, it introduces ISO 31000 risk management standards for geohazards, providing a structured approach to identifying and mitigating risks. Additionally, it offers tools for cost-benefit analysis of mitigation strategies (e.g., comparing retaining walls versus slope regrading) to support informed decision-making.

Technological Integration

Technological integration is another key aspect of the book. It demonstrates the application of InSAR (Interferometric Synthetic Aperture Radar) for monitoring ground subsidence, showcasing innovative technological applications in geology. The exploration of machine learning algorithms for predicting landslide susceptibility highlights the role of technology in enhancing risk assessment. Furthermore, protocols for drone-based LiDAR surveys and borehole televiewer logging provide practical guidance for modern geological investigations, improving data collection and analysis.

Target Audience

The target audience for this book is diverse. It serves as a core text for undergraduate and graduate courses in environmental geology, geotechnical engineering, and natural resource management, providing foundational knowledge and practical applications. Additionally, it acts as a reference for researchers seeking methodologies in geochemical analysis, hazard modelling, and sustainability science, thereby facilitating advanced research. For industry professionals, the book offers practical tools for site characterisation and slope stability analysis, enhancing engineering practices.

Environmental consultants will find frameworks for regulatory compliance and contaminated site remediation, supporting effective environmental management. Policymakers will benefit from evidence-based insights for drafting mineral extraction policies and climate adaptation strategies, informing public policy. The book is also accessible to a broader audience, including NGOs, journalists, and environmentally conscious citizens, as its clear explanations of complex geological processes foster a greater public understanding of environmental issues, promoting awareness and engagement.

Conclusion

In conclusion, **Environmental Geology:** ***Insights from Applied Geology on Environmental Impact*** transcends traditional academic boundaries by merging rigorous scientific inquiry with pragmatic solutions for industry challenges. Its dual focus on theory and practice positions it as an indispensable resource for fostering resilience in a rapidly changing world. By equipping readers with both the knowledge to understand Earth's systems and the skills to protect them, this book advances the global dialogue on sustainability and responsible resource management.

Through its comprehensive approach, the book not only addresses the pressing environmental challenges associated with geo-material extraction but also emphasises the importance of integrating sustainable practices across various sectors. By fostering a deeper understanding of the interconnectedness of geological processes and environmental impacts, the book encourages stakeholders to adopt a holistic perspective in their decision-making.

The key principles from this work include the interdisciplinary approach that integrates geology, hydrology, geochemistry, and environmental science, providing a comprehensive framework for understanding and addressing environmental issues related to geo-materials. The book offers practical applications and actionable insights that can be directly applied in industry settings, enhancing the relevance of geological research to real-world challenges. By aligning geological practices with the Sustainable Development Goals, it underscores the critical role of responsible resource management in achieving global sustainability targets.

Emphasising the importance of community engagement fosters a sense of ownership and accountability, leading to more effective and sustainable outcomes. The exploration of advanced technologies highlights the potential for innovation to improve environmental monitoring and risk assessment, while providing clear guidelines for compliance with environmental regulations ensures that industry practices align with legal and ethical standards, promoting accountability and transparency.

In a world increasingly beset by environmental pressures, this book serves as a vital resource for academics, industry professionals, and policymakers alike. Its comprehensive coverage of theoretical foundations, applied methodologies, and practical applications equips readers with the knowledge and tools necessary to navigate the complexities of environmental geology. By fostering a culture of sustainability and responsible resource management, this book contributes to the ongoing dialogue about how best to balance economic development with environmental protection.

As we move forward, the insights and strategies presented in this work will be essential for addressing the challenges of our time and ensuring a sustainable future for generations to come. This book not only enriches the academic landscape but also serves as a catalyst for change in the mining and construction industries, promoting practices that protect our planet while meeting the needs of society. Through its rigorous scholarship and practical relevance, **Environmental Geology** stands as a cornerstone for advancing the field of environmental geology and enhancing our collective efforts toward a sustainable future.

ENGAGING WITH ENVIRONMENTAL GEOLOGY: CRITICAL QUESTIONS AND ANSWERS

Environmental geology is a vital field that examines the interactions between geological processes and environmental systems. As we face unprecedented challenges such as climate change, resource depletion, and biodiversity loss, understanding the principles of environmental geology becomes increasingly important. This section aims to engage readers in critical thinking by posing essential questions that explore the foundational concepts, tools, applications, and future directions of environmental geology. By addressing these questions, we can deepen our understanding of how geological processes influence our environment and inform sustainable practices for the future.

Preface

1. **What are the key factors contributing to the imbalance between geological stability and environmental sustainability in the 21st century?**
   Key factors include rapid population growth, which increases demand for resources; industrialisation, which often leads to environmental degradation; and unsustainable resource consumption patterns that deplete natural resources faster than they can be replenished. Additionally, climate change exacerbates these issues by altering geological processes, leading to increased frequency and intensity of natural disasters.

2. **How does the integration of scientific rigor and ethical considerations enhance our understanding of environmental geology?**

   Integrating scientific rigor ensures that geological studies are based on empirical evidence and sound methodologies, while ethical considerations promote responsible stewardship of natural resources. This combination helps to ensure that geological practices do not harm the environment or marginalized communities, fostering a more equitable approach to resource management and environmental protection.

3. **In what ways can this book serve as a resource for fostering responsible interactions with our planet?**

   The book provides comprehensive frameworks, methodologies, and case studies that guide readers in making informed decisions about environmental management. By illustrating best practices and highlighting the consequences of poor decision-making, it encourages a proactive approach to sustainability, empowering readers to engage in responsible practices that protect both the environment and human health.

Context and Motivation

1. **How do contemporary environmental challenges, such as climate change and biodiversity loss, relate to geological processes?**

   Geological processes, such as erosion, sedimentation, and tectonic activity, play a crucial role in shaping ecosystems and influencing climate patterns. For instance, changes in landforms can affect water drainage and soil fertility, while geological events like volcanic eruptions can release greenhouse gases. Understanding these processes is essential for addressing environmental challenges, as they often have geological underpinnings that must be considered in management strategies.

2. **What gaps exist in traditional geological approaches when addressing modern environmental issues?**

   Traditional geological approaches often focus on static models and historical data, which may not adequately account for the dynamic nature of contemporary environmental issues. Additionally, these approaches may lack interdisciplinary perspectives, failing to integrate insights from fields such as ecology, sociology, and economics that are essential for understanding the complexities of environmental challenges.

3. **How can interdisciplinary solutions improve the effectiveness of environmental management strategies?**

   Interdisciplinary solutions bring together diverse expertise and perspectives, allowing for a more comprehensive understanding of environmental issues. By integrating knowledge from various fields, such as geology, biology, and social sciences, practitioners can develop more effective and holistic management strategies that address the root causes of environmental problems rather than just their symptoms.

Structure and Scope

1. **How does the structure of the book facilitate a progressive deepening of knowledge in environmental geology?**

   The book is organised into sections that build upon one another, starting with foundational concepts and gradually introducing more complex applications. This structure allows readers to develop a solid understanding of basic principles before tackling advanced topics, ensuring that they are well-prepared to engage with the material and apply it in real-world contexts.

2. **In what ways do the foundational principles outlined in the first section prepare readers for more advanced applications in later chapters?**
   Foundational principles provide essential context and terminology that are critical for understanding more complex concepts. By establishing a strong base in geological processes, environmental systems, and key methodologies, readers are better equipped to grasp the intricacies of advanced applications, such as environmental impact assessments and risk modelling.

3. **Why is it important to include both theoretical and practical elements in the study of environmental geology?**

   Including both theoretical and practical elements ensures that readers can connect abstract concepts to real-world applications. This dual approach enhances learning by allowing students to see the relevance of theoretical knowledge in practical scenarios, thereby preparing them for careers in environmental management and geology.

Section I: Foundations (Chapters 1–4)

1. **What are the key geological processes that influence environmental systems, and how do they interact?**

   Key geological processes include erosion, which shapes landscapes and affects soil quality; sedimentation, which influences water quality and habitat formation; and tectonic activity, which can lead to natural disasters like earthquakes and volcanic eruptions. These processes interact in complex ways, as changes in one can affect others, ultimately influencing ecosystem health and stability.

2. **How does an understanding of soil mechanics contribute to effective environmental stewardship?**

   Understanding soil mechanics is essential for assessing land stability, which is crucial for construction, agriculture, and land use planning. Knowledge of soil properties helps in managing erosion, preventing landslides, and ensuring sustainable agricultural practices, thereby promoting environmental stewardship and protecting natural resources.

3. **Why is it essential to comprehend geochemical cycles in the context of nutrient and contaminant transport?**

   Geochemical cycles govern the movement of nutrients and contaminants through ecosystems. Understanding these cycles is vital for managing soil fertility, water quality, and pollution. By comprehending how substances cycle through the environment, practitioners can develop strategies to enhance nutrient availability while minimizing the impact of contaminants. This knowledge is crucial for sustainable land use and environmental protection, as it allows for informed decision-making regarding agricultural practices, waste management, and ecosystem restoration.

Section II: Tools and Techniques (Chapters 5–7)

1. **What methodologies are most effective for conducting environmental impact assessments (EIAs), and why?**

   Effective methodologies for conducting EIAs include comprehensive stakeholder engagement, baseline environmental studies, impact prediction, and mitigation planning. These methodologies ensure that all potential environmental impacts are identified and assessed, allowing for informed decision-making. Engaging stakeholders helps to incorporate diverse perspectives and local knowledge, which can lead to more effective and accepted outcomes.

2. **How can risk modelling enhance our understanding of potential environmental hazards?**

   Risk modelling quantifies potential hazards and their impacts, providing a framework for assessing vulnerability and exposure. By simulating various scenarios, risk modelling helps identify high-risk areas and informs the development of mitigation strategies. This proactive approach enables communities and policymakers to prepare for and respond to environmental hazards more effectively.

3. **In what ways do advanced tools like geospatial analytics improve the accuracy of environmental assessments?**

   Geospatial analytics enhance the accuracy of environmental assessments by providing spatial context and visualisation of data. These tools allow for the analysis of geographic patterns and relationships, enabling practitioners to identify trends and correlations that may not be apparent through traditional methods. By integrating geospatial data with other environmental information, decision-makers can make more informed and effective management choices.

Section III: Applications and Case Studies (Chapters 8–12)

1. **How do case studies illustrate the real-world relevance of geological principles in environmental management?**

   Case studies provide concrete examples of how geological principles are applied in practice, showcasing successful interventions and highlighting lessons learned. They demonstrate the practical implications of theoretical concepts, allowing readers to see the effectiveness of various strategies in addressing environmental challenges. By examining real-world scenarios, readers can better understand the complexities of environmental management and the importance of context-specific solutions.

2. **What lessons can be learned from successful geohazard mitigation strategies implemented in various regions?**

   Successful geohazard mitigation strategies often involve early warning systems, community engagement, and adaptive management practices. These strategies highlight the importance of preparedness and resilience in the face of natural disasters. Lessons learned include the value of collaboration among stakeholders, the need for continuous monitoring and assessment, and the importance of integrating local knowledge and experiences into planning and response efforts.

3. **How does the geopolitics of mineral extraction impact local communities and global supply chains?**

   The geopolitics of mineral extraction can lead to significant economic benefits for some regions, but it may also result in social and environmental injustices. Local communities may face displacement, loss of livelihoods, and environmental degradation due to mining activities. Additionally, global supply chains can be affected by geopolitical tensions, leading to fluctuations in resource availability and prices. Understanding these dynamics is essential for developing equitable and sustainable resource management practices.

Section IV: Innovation and Future Directions (Chapter 13)

1. **What emerging technologies are shaping the future of ground investigation, and how can they mitigate environmental risks?**

   Emerging technologies such as artificial intelligence, remote sensing, and drone-based surveys are revolutionising ground investigation by enhancing data collection and analysis. These technologies improve the efficiency and accuracy of assessments, allowing for more effective identification of environmental risks. By leveraging these innovations, practitioners can develop targeted strategies for risk mitigation and environmental management.

2. **How can interdisciplinary collaboration drive innovation in environmental geology?**
   Interdisciplinary collaboration fosters the exchange of ideas and resources among experts from various fields, leading to innovative solutions for complex environmental challenges. By integrating knowledge from geology, ecology, social sciences, and engineering, practitioners can develop more comprehensive and effective management strategies that address the multifaceted nature of environmental issues.

3. **In what ways can standardised protocols in ground investigation enhance global resilience to environmental challenges?**

   Standardised protocols ensure consistency and reliability in ground investigations, facilitating better data sharing and collaboration across regions. By establishing common practices, practitioners can enhance the comparability of data and findings, leading to more effective responses to environmental challenges. This standardisation also promotes best practices and encourages the adoption of innovative approaches in environmental management.

Unique Contributions

1. **How does the book's interdisciplinary integration contribute to a more comprehensive understanding of environmental challenges?**

   The book's interdisciplinary integration allows for a nuanced understanding of environmental challenges by combining insights from various fields, including geology, ecology, sociology, and economics. This holistic approach enables readers to appreciate the complexity of issues and develop innovative solutions that consider multiple perspectives. Ultimately, this fosters more effective environmental management practices that are responsive to the interconnected nature of ecological and human systems.

2. **Why is it important to balance global principles with local case studies in the context of resource management?**

   Balancing global principles with local case studies ensures that solutions are contextually relevant and culturally sensitive. This approach allows for the adaptation of best practices to local conditions, enhancing the effectiveness of resource management strategies. By respecting the unique needs and circumstances of local communities, practitioners can promote sustainable development that aligns with both environmental health and social equity.

3. **How can the insights gained from this book inform future research and policy development in environmental geology?**

   The insights gained from this book can guide future research by identifying key areas where further investigation is needed, such as the impacts of climate change on geological processes or the effectiveness of various environmental management strategies. Additionally, these insights can inform policy development by providing evidence-based recommendations that promote sustainable practices and responsible resource management. By bridging the gap between research and policy, the book aims to contribute to more informed decision-making that prioritises both environmental health and social equity,

Conclusion

1. **What role does public engagement play in advancing the goals of environmental geology?**

   Public engagement is crucial for advancing the goals of environmental geology, as it fosters awareness and understanding of geological processes and their impact on the environment. By involving communities in discussions about environmental issues, practitioners can encourage active participation in decision-making and promote stewardship of natural resources. Educational initiatives, outreach programs, and community-based projects can empower individuals to take action, leading to more sustainable practices at the local level.

2. **How can the principles outlined in this book be applied to address emerging environmental challenges?**

   The principles outlined in this book can be applied to address emerging environmental challenges by providing a framework for understanding the interactions between geological processes and environmental systems. For instance, as new challenges arise from climate change, such as increased flooding or soil degradation, the methodologies and case studies presented can guide practitioners in developing adaptive strategies that mitigate risks and enhance resilience. By applying these principles, stakeholders can better navigate the complexities of modern environmental issues.

3. **What is the future outlook for environmental geology in the context of global sustainability?**

   The future outlook for environmental geology is promising, particularly as the field continues to evolve in response to global sustainability challenges. As awareness of the interconnectedness of geological and environmental systems grows, there is an increasing emphasis on integrating geological knowledge into sustainable development practices. This shift presents opportunities for innovation in resource management, disaster risk reduction, and environmental restoration. By fostering collaboration among scientists, policymakers, and communities, environmental geology can play a pivotal role in shaping a sustainable future for our planet.

Call to Action

In conclusion, this book serves as a vital resource for anyone interested in understanding the intricate relationship between geology and the environment. It is designed not only to impart knowledge but also to inspire critical engagement with the material presented. Readers are encouraged to delve deeply into the concepts discussed, applying the knowledge gained to real-world scenarios and contributing to the ongoing dialogue surrounding environmental sustainability.

As we navigate the complexities of the 21st century, characterized by rapid environmental change, resource depletion, and social inequities, the insights and frameworks provided in this book are more relevant than ever. The integration of geological principles with environmental management practices equips individuals and communities with the tools necessary to address pressing challenges such as climate change, habitat destruction, and pollution.

1. **Engagement with the Material**: Readers are urged to approach the content with a critical mindset, questioning assumptions and exploring the implications of geological processes on environmental systems. This engagement can take many forms, including participation in discussions, conducting independent research, or collaborating with peers to explore local geological issues. By fostering a culture of inquiry, readers can deepen their understanding and contribute to the collective knowledge base.

2. **Application of Knowledge**: The frameworks and methodologies outlined in this book are intended to be applied in various contexts, from academic research to community-based projects. Readers are encouraged to implement these principles in their own work, whether it involves conducting environmental impact assessments, developing sustainable land-use plans, or advocating for policies that promote environmental justice. Practical application not only reinforces learning but also demonstrates the real-world relevance of geological insights.

3. **Contributing to the Dialogue**: The challenges we face today require collaborative solutions that draw on diverse perspectives and expertise. Readers are invited to engage in interdisciplinary dialogues that bridge the gap between geology, ecology, sociology, and policy-making. By participating in forums, workshops, and community meetings, individuals can share their insights, learn from others, and contribute to the development of innovative strategies for sustainable resource management.

4. **Empowerment for Informed Decision-Making**: The knowledge gained from this book empowers individuals and communities to make informed decisions that promote a healthier, more sustainable planet. Understanding the geological underpinnings of environmental issues enables stakeholders to advocate for evidence-based policies and practices that prioritize ecological integrity and social equity. This empowerment is crucial for fostering resilience in the face of environmental challenges, as it encourages proactive rather than reactive approaches to sustainability.

5. **Commitment to Future Generations**: Finally, the call to action extends beyond individual and community engagement; it encompasses a commitment to future generations. By prioritizing sustainable practices and advocating for responsible stewardship of natural resources, readers can help ensure that future generations inherit a planet that is not only habitable but thriving. This long-term perspective is essential for addressing the interconnected challenges of environmental degradation, climate change, and social justice.

In summary, this book is not merely a collection of information; it is a call to action for all who seek to understand and address the intricate relationship between applied geology and the environment. Together, we can rise to the challenges of our time and work towards a healthier, more equitable world for generations to come.

GEOLOGICAL PERSPECTIVES ON ENVIRONMENTAL CHALLENGES: AN IN-DEPTH QUESTIONS AND ANSWERS

Introduction

In an era marked by rapid environmental change and increasing human impact on the planet, the field of environmental geology has emerged as a critical discipline for understanding and addressing the complex interactions between geological processes and human activities. This comprehensive Q&A guide aims to explore the fundamental concepts, challenges, and solutions within environmental geology, providing insights into how geological knowledge can inform sustainable practices and effective environmental management.

As we navigate issues such as climate change, resource depletion, and geological hazards, it becomes increasingly important to recognize the role of geology in shaping our environment and influencing our decisions. From the assessment of natural resources to the evaluation of geological hazards, the principles of environmental geology offer valuable tools for mitigating risks and promoting sustainability.

This guide is structured to address key questions across various topics, including the fundamentals of applied geology, the significance of soil and water resources, the implications of geological hazards, and thc intersection of geology with environmental policy. Each section is designed to foster a deeper understanding of the geological factors that underpin environmental challenges and to highlight the importance of integrating geological insights into decision-making processes.

We invite you to explore these questions and answers as a means of enhancing your knowledge of environmental geology and its vital contributions to addressing the pressing challenges of our time.

Chapter 1: Introduction to Environmental Geology

1. What is environmental geology, and how does it differ from traditional geology?

Environmental geology is the study of the interactions between geological processes and human activities, focusing on how geological factors influence environmental issues. Unlike traditional geology, which primarily examines the Earth's materials, structures, and processes in a more theoretical context, environmental geology emphasizes the application of geological knowledge to solve practical environmental problems. This includes assessing risks, managing natural resources, and understanding the impacts of human activities on the geological environment.

2. Why is a geological understanding crucial for addressing environmental issues?

A geological understanding is crucial for addressing environmental issues because it helps identify the natural processes that shape the environment, such as erosion, sedimentation, and tectonic activity. This knowledge is essential for assessing risks associated with geological hazards, such as landslides and earthquakes, and for informing land-use planning and resource management. By understanding geological processes, we can develop effective strategies to mitigate environmental impacts, promote sustainability, and enhance resilience to natural disasters.

3. How do geological processes influence human activities and environmental management?

Geological processes influence human activities in various ways, including land stability, water availability, and resource distribution. For example, the presence of certain rock types can affect groundwater quality and availability, while soil composition can determine agricultural productivity. Understanding these processes allows for better planning and management of natural resources, infrastructure development, and disaster preparedness. This knowledge is vital for creating policies that balance human needs with environmental protection.

Chapter 2: Fundamentals of Applied Geology

1. What are the key concepts in applied geology, and how do they relate to environmental geology?

Key concepts in applied geology include geological mapping, resource assessment, hazard evaluation, and environmental remediation. These concepts relate to environmental geology by providing the tools and methodologies needed to analyze geological conditions, assess environmental impacts, and inform sustainable management practices. For instance, geological mapping helps identify areas at risk of flooding or landslides, while resource assessment informs decisions about sustainable extraction of minerals and water.

2. In what ways does geology play a role in environmental management practices?

Geology plays a critical role in environmental management by providing insights into soil and water resources, identifying geological hazards, and guiding land-use planning. For example, understanding the geological context of an area can help determine suitable locations for waste disposal, construction, and agriculture. Additionally, geological assessments can inform strategies for pollution control, habitat restoration, and sustainable land development, ensuring that human activities do not compromise environmental integrity.

3. How can applied geology inform decision-making in land use and resource management?

Applied geology provides essential data on soil types, groundwater availability, and geological hazards, which are critical for informed decision-making. This information helps policymakers and planners make choices that minimize environmental impacts and promote sustainable land use and resource management. For example, knowledge of local geology can guide the placement of infrastructure to avoid areas prone to flooding or landslides, thereby reducing risks and enhancing community safety.

Chapter 3: Geological Hazards and Risks

1. What are the different types of geological hazards, and how do they impact communities?

Geological hazards include earthquakes, landslides, volcanic eruptions, tsunamis, and sinkholes. These hazards can cause significant damage to infrastructure, disrupt communities, and pose risks to human life. For instance, earthquakes can lead to building collapses and loss of life, while landslides can block roads and damage homes. Understanding these hazards is essential for developing effective preparedness and response strategies, as well as for designing resilient infrastructure.

2. How can we assess the risks associated with geological hazards?

Risk assessment involves identifying potential hazards, analyzing their likelihood and potential impacts, and evaluating vulnerabilities within communities. This process often includes geological surveys, historical data analysis, and modeling to predict future events and their consequences. By assessing risks, communities can prioritize mitigation efforts, allocate resources effectively, and develop emergency response plans to enhance resilience against geological hazards.

3. What strategies can be implemented to mitigate the effects of geological hazards on human populations?

Strategies for mitigation include land-use planning to avoid high-risk areas, implementing early warning systems, conducting public education campaigns, and investing in infrastructure that can withstand geological events. For example, building codes can be enforced to ensure structures are earthquake-resistant, and community drills can prepare residents for emergency situations. Additionally, restoring natural landscapes, such as wetlands, can help absorb floodwaters and reduce the impact of hazards.

Chapter 4: Soil and Water Resources (continued)

1. Why are soil and water considered critical components of environmental geology?

Soil and water are essential for sustaining ecosystems, agriculture, and human health. They are also influenced by geological processes, making their management crucial for environmental sustainability. Understanding their interactions helps in addressing issues like soil erosion, water quality, and resource depletion. For instance, soil health directly affects agricultural productivity, while water quality impacts both human health and ecosystem integrity. Geological factors, such as rock type and topography, influence the distribution and movement of both soil and water, making it essential to consider these elements in environmental geology.

2. What are the best practices for the management and conservation of soil and water resources?

Best practices for the management and conservation of soil and water resources include implementing sustainable agricultural techniques, such as crop rotation and cover cropping, to enhance soil health and reduce erosion. Water conservation measures, such as rainwater harvesting and efficient irrigation systems, can help manage water resources effectively. Additionally, restoring degraded land through reforestation and wetland restoration can improve soil and water quality. Regular monitoring and assessment of soil and water quality are also important for effective management, allowing for timely interventions when issues arise.

3. How do geological factors influence the availability and quality of soil and water resources?

Geological factors such as rock type, soil composition, and topography affect water drainage, retention, and quality. For example, sandy soils may drain quickly and require more frequent irrigation, while clay soils retain water but can become compacted and less permeable. Geological formations can also influence groundwater recharge and contamination; for instance, porous rock layers can facilitate groundwater movement, while impermeable layers can trap pollutants. Understanding these relationships is vital for effective resource management and environmental protection.

Chapter 5: Environmental Impact Assessment (EIA)

1. What are the fundamental principles and methodologies of Environmental Impact Assessment (EIA)?

Fundamental principles of EIA include transparency, public participation, and consideration of alternatives. Methodologies often involve scoping to identify key issues, conducting baseline studies to understand existing conditions, predicting potential impacts, and proposing mitigation measures. The goal is to assess potential environmental impacts before project implementation, ensuring that decision-makers have the necessary information to make informed choices that minimize harm to the environment.

2. What are the key steps involved in conducting an effective EIA?

Key steps in conducting an effective EIA include:

- **Screening:** Determining whether a project requires an EIA based on its potential environmental impacts.
- **Scoping:** Identifying the key issues and impacts to be assessed.
- **Baseline Studies:** Collecting data on existing environmental conditions.

- **Impact Assessment:** Evaluating potential environmental impacts and their significance.
- **Mitigation Measures:** Proposing strategies to avoid, minimize, or compensate for adverse impacts.
- **Public Participation:** Engaging stakeholders and the public in the assessment process.
- **Reporting:** Documenting the findings in an EIA report, which is submitted for review and decision-making.

3. How can EIA contribute to sustainable development and environmental protection?

EIA contributes to sustainable development by ensuring that environmental considerations are integrated into project planning and decision-making. By identifying potential impacts and proposing mitigation measures, EIA helps prevent environmental degradation and promotes the responsible use of natural resources. Additionally, public participation in the EIA process fosters community engagement and accountability, leading to more sustainable and equitable outcomes.

Chapter 6: Geology and Climate Change

1. How does climate change impact geological processes and systems?

Climate change affects geological processes by altering weather patterns, increasing the frequency and intensity of extreme weather events, and influencing erosion and sedimentation rates. For example, rising temperatures can lead to thawing permafrost, which destabilizes land and increases the risk of landslides. Additionally, changes in precipitation patterns can affect groundwater recharge and surface water availability, impacting ecosystems and human activities.

2. What geological contributions can be made to mitigate climate change effects?

Geology can contribute to climate change mitigation through carbon sequestration, where geological formations are used to store carbon dioxide emissions from industrial processes. Additionally, understanding geological processes can inform the development of renewable energy resources, such as geothermal energy. Sustainable land management practices, informed by geological knowledge, can also enhance carbon storage in soils and vegetation, further contributing to climate change mitigation efforts.

3. In what ways can geological research inform climate adaptation strategies?

Geological research can inform climate adaptation strategies by providing insights into how geological processes interact with climate change impacts. For example, understanding the geological context of an area can help identify vulnerabilities to flooding, erosion, and landslides, allowing for targeted adaptation measures. Additionally, geological studies can inform the design of resilient infrastructure and land-use planning that considers future climate scenarios, enhancing community resilience.

Chapter 7: Geology and Natural Resources

1. What is the relationship between natural resources and their geological context?

Natural resources, such as minerals, fossil fuels, and groundwater, are often directly linked to geological formations and processes. The distribution, quality, and accessibility of these resources are influenced by the geological history of an area, including tectonic activity, sedimentation, and weathering processes. Understanding the geological context helps in identifying potential resource deposits and assessing their economic viability while considering environmental impacts.

2. How can we achieve sustainable management of natural resources?
Sustainable management of natural resources involves balancing economic development with environmental protection. This can be achieved through practices such as responsible extraction methods, recycling and reuse of materials, and implementing conservation strategies. Engaging local communities in resource management decisions and promoting policies that prioritize sustainability are also essential for ensuring that natural resources are available for future generations.

3. What challenges do we face in balancing resource extraction with environmental protection?

Challenges include the potential for environmental degradation, such as habitat destruction, pollution, and depletion of resources. Conflicts may arise between economic interests and environmental conservation, particularly in regions rich in natural resources. Additionally, regulatory frameworks may be inadequate to address the complexities of resource management, and there may be a lack of public awareness or engagement in sustainable practices. Addressing these challenges requires collaboration among stakeholders, including governments, industries, and local communities.

Chapter 8: Geohazards and Risk Management

1. What strategies are effective for managing geohazards?

Effective strategies for managing geohazards include risk assessment and mapping, land-use planning to avoid high-risk areas, and implementing early warning systems. Community education and preparedness programs are also crucial for enhancing resilience. Additionally, engineering solutions, such as building retaining walls or improving drainage systems, can help mitigate the impacts of geohazards. Collaboration among government agencies, scientists, and local communities is essential for developing comprehensive risk management plans.

2. Can you provide examples of successful case studies in geohazard risk management?

One example is the implementation of early warning systems for earthquakes in Japan, which has significantly reduced casualties and damage during seismic events. Another case is the use of controlled drainage systems in the Netherlands to manage flood risks in low-lying areas. In California, community-based programs that educate residents about wildfire risks and preparedness have proven effective in reducing losses during fire seasons. These case studies highlight the importance of proactive measures and community involvement in geohazard risk management.

3. How do community engagement and education play a role in geohazard preparedness?

Community engagement and education are vital for building awareness of geohazards and promoting preparedness. Informed communities are better equipped to respond to emergencies, understand risks, and participate in mitigation efforts. Educational programs can provide residents with knowledge about local hazards, emergency response plans, and safety measures. Engaging communities in planning processes fosters a sense of ownership and responsibility, leading to more effective risk management strategies.

Chapter 9: Geology and Environmental Policy

1. How does geology intersect with environmental policy-making?

Geology intersects with environmental policy-making by providing the scientific basis for understanding natural resources, hazards, and environmental impacts. Policymakers rely on geological data to inform regulations related to land use, resource extraction, and environmental protection. For example, geological assessments can guide decisions on site selection for waste disposal or infrastructure development, ensuring that policies are grounded in scientific understanding.

2. What are the key regulatory frameworks that govern geological practices and their implications?

Key regulatory frameworks include environmental protection laws, land-use planning regulations, and resource management policies. These frameworks establish guidelines for sustainable practices, risk assessment, and public participation in decision-making. For instance, the National Environmental Policy Act (NEPA) in the United States requires federal agencies to assess environmental impacts before undertaking projects. The implications of these regulations can shape how geological resources are managed and how communities respond to environmental challenges.

3. How can geologists influence environmental policy to promote sustainable practices?

Geologists can influence environmental policy by providing expert knowledge and data that inform decision-making. By engaging with policymakers, participating in public forums, and collaborating with environmental organizations, geologists can advocate for science-based policies that prioritize sustainability. Additionally, conducting research that highlights the importance of geological considerations in environmental management can help raise awareness and drive policy changes.

Chapter 10: The Future of Geology and Sustainability

1. What emerging trends in geology are shaping the future of environmental sustainability?

Emerging trends include the integration of technology in geological research, such as remote sensing, GIS, and machine learning, which enhance data collection and analysis. Additionally, there is a growing focus on interdisciplinary approaches that combine geology with fields such as ecology, economics, and social sciences to address complex environmental challenges. The rise of sustainable mining practices, including the use of less invasive extraction methods and the recycling of materials, is also significant. Furthermore, the increasing emphasis on climate change adaptation and mitigation strategies, such as carbon capture and storage, highlights the role of geology in developing solutions for a sustainable future. These trends reflect a shift towards more holistic and integrated approaches to managing geological resources and environmental impacts.

2. How can geology contribute to sustainable development goals?

Geology contributes to sustainable development goals (SDGs) by providing essential resources for clean water, energy, and materials while ensuring that their extraction and use are environmentally responsible. For example, geological studies can inform sustainable land-use planning, help manage natural resources efficiently, and support disaster risk reduction efforts. By integrating geological knowledge into development projects, we can promote economic growth while safeguarding ecosystems and communities, ultimately contributing to the achievement of SDGs related to poverty alleviation, environmental protection, and climate action.

3. What role do geoscientists play in addressing global sustainability challenges?

Geoscientists play a critical role in addressing global sustainability challenges by conducting research that informs policy and practice. They provide insights into the geological processes that underpin natural resource availability, assess environmental impacts, and develop strategies for sustainable management. Geoscientists also engage with communities and stakeholders to raise awareness of geological issues and promote best practices. Their expertise is essential for understanding the interactions between human activities and the Earth system, enabling informed decision-making for a sustainable future.

Chapter 11: Rock Blasting Constraints

1. What are the primary techniques used in rock blasting, and what challenges do they present?

Primary techniques used in rock blasting include controlled blasting, smooth blasting, and pre-splitting. Controlled blasting involves carefully calculating the amount and placement of explosives to minimize vibration and fly rock. Smooth blasting aims to create a clean rock face with minimal damage to surrounding areas. Pre-splitting is used to create a fracture line that reduces overbreak. Challenges associated with these techniques include managing noise and vibration, ensuring safety, and minimizing environmental impacts, such as air quality degradation and habitat disruption.

2.What environmental considerations must be taken into account during rock blasting operations?

Environmental considerations during rock blasting operations include assessing the potential impacts on air quality, noise levels, water resources, and local ecosystems. It is essential to monitor dust emissions and implement measures to control them, such as using water sprays. Additionally, the effects of vibration on nearby structures and wildlife must be evaluated, and appropriate mitigation measures should be employed. Compliance with environmental regulations and community engagement are also critical to address concerns and minimize negative impacts.

3. How can the impacts of rock blasting on the environment be minimized?
Impacts of rock blasting can be minimized through careful planning and implementation of best practices. This includes conducting thorough environmental assessments before blasting, using advanced blasting techniques that reduce noise and vibration, and scheduling blasts during times that minimize disturbance to local communities. Additionally, employing dust control measures, such as water sprays and dust suppressants, can help mitigate air quality impacts. Engaging with local stakeholders and providing information about blasting schedules and safety measures can also foster community trust and cooperation.

Chapter 12: Strategic Metals, Minerals, and Rare Earth Elements (REEs) in the Mining Sector in Kenya

1. What are the strategic resources found in Kenya, and why are they important?

Kenya is rich in various strategic resources, including rare earth elements (REEs), titanium, and other minerals such as gold and fluorspar. These resources are important for several reasons: REEs are critical for high-tech applications, including electronics and renewable energy technologies; titanium is used in aerospace and medical applications; and other minerals contribute to economic development and job creation. The sustainable management of these resources is essential for maximizing their benefits while minimizing environmental impacts.

2. What are the environmental and economic implications of mining rare earth elements (REEs)?

The mining of REEs can have significant environmental implications, including habitat destruction, water pollution, and soil degradation. The extraction process often involves the use of toxic chemicals, which can pose risks to local ecosystems and communities. Economically, while mining can generate revenue and create jobs, it can also lead to boom-and-bust cycles that affect local economies. Sustainable practices, such as responsible mining techniques and community engagement, are crucial to balancing economic benefits with environmental protection.

3. How can Kenya balance the extraction of strategic minerals with environmental sustainability?

Kenya can balance the extraction of strategic minerals with environmental sustainability by implementing comprehensive regulatory frameworks that enforce environmental protection standards. This includes conducting thorough environmental impact assessments (EIAs) before mining projects are approved, ensuring that potential impacts on ecosystems and communities are carefully evaluated. Additionally, adopting sustainable mining practices, such as minimizing land disturbance, using less harmful extraction methods, and rehabilitating mined areas, can help mitigate environmental impacts. Engaging local communities in decision-making processes and ensuring that they benefit from mining activities can also promote social equity and support sustainable development.

Chapter 13: Fundamentals of Ground Investigation

1. Why is ground investigation important in the context of environmental geology?

Ground investigation is crucial in environmental geology because it provides essential data about subsurface conditions, including soil composition, groundwater levels, and geological structures. This information is vital for assessing site suitability for construction, identifying potential contamination, and understanding the risks associated with geological hazards. Ground investigations help inform land-use planning, resource management, and environmental protection efforts, ensuring that decisions are based on accurate and comprehensive geological data.

2. What are the key objectives of conducting a ground investigation?
The key objectives of conducting a ground investigation include:

- **Characterizing the subsurface conditions:** Understanding the geological and hydrological characteristics of the site.
- **Assessing contamination:** Identifying any existing contamination and evaluating its extent and potential impacts on human health and the environment.
- **Evaluating geotechnical properties:** Determining the physical and mechanical properties of soil and rock to inform construction and engineering decisions.
- **Informing risk assessments:** Providing data necessary for assessing risks related to geological hazards, such as landslides or ground instability.

3. What methodologies are most effective for carrying out ground investigations?

Effective methodologies for ground investigations include:

- **Drilling and sampling:** Collecting soil and rock samples from various depths to analyze their properties in the laboratory.
- **Geophysical surveys:** Using techniques such as seismic, resistivity, and ground-penetrating radar to map subsurface conditions without extensive excavation.

- **Monitoring groundwater:** Installing piezometers to measure groundwater levels and quality over time.
- **Site characterization:** Combining data from various methods to create a comprehensive understanding of the site's geological and hydrological conditions.

Conclusion Summary

The exploration of environmental geology reveals the intricate relationships between geological processes and human activities, highlighting the critical role that geology plays in addressing contemporary environmental challenges. As we face issues such as climate change, resource depletion, and geological hazards, a comprehensive understanding of geological principles becomes essential for sustainable development and effective environmental management.

Throughout the chapters, we have seen how environmental geology differs from traditional geology by emphasizing practical applications that inform land-use planning, resource management, and disaster preparedness. The importance of geological assessments in mitigating risks associated with natural hazards, such as earthquakes and landslides, underscores the need for proactive strategies that enhance community resilience.

The management of soil and water resources, as well as the sustainable extraction of strategic minerals and rare earth elements, illustrates the delicate balance between economic development and environmental protection. Best practices in resource management, informed by geological knowledge, are vital for ensuring that natural resources are utilized responsibly and equitably.

Furthermore, the integration of environmental impact assessments (EIA) into project planning emphasizes the necessity of transparency, public participation, and scientific rigor in decision-making processes. As we look to the future, emerging trends in geology, including technological advancements and interdisciplinary approaches, offer promising pathways for addressing global sustainability challenges.

In conclusion, geoscientists and policymakers must collaborate to leverage geological insights for sustainable practices that protect ecosystems, support communities, and promote resilience in the face of environmental change. By prioritizing the integration of geological knowledge into environmental policy and management, we can work towards a more sustainable and equitable future for all.

FSC
www.fsc.org
MIX
Papier | Fördert gute Waldnutzung
FSC® C083411

Kolibri 360 GmbH
Ferdinand-Jühlke-Straße 7
99095 Erfurt, Deutschland
produktsicherheit@kolibri360.de

Druck:
CPI Druckdienstleistungen GmbH
im Auftrag der
Kolibri 360 GmbH
Ferdinand-Jühlke-Str. 7
99095 Erfurt